FOUNDATIONS
of
CONSTRUCTIVE ANALYSIS

Errett Bishop

Department of Mathematics
University of California, San Diego

With a New Foreword by
Michael Beeson

ISHI PRESS
INTERNATIONAL

Foundations of Constructive Analysis

by Errett Bishop

Department of Mathematics

University of California, San Diego

Published in 1967 in by McGraw-Hill

Copyright © 1967 by Errett Bishop

**This Printing in July, 2012
by Ishi Press in New York and Tokyo**

with a new introduction by Michael Beeson

Copyright © 2012 by Michael Beeson

ISBN 4-87187-714-0
978-4-87187-714-5

Ishi Press International
1664 Davidson Avenue, Suite 1B
Bronx NY 10453-7877
USA

1-917-507-7226

Printed in the United States of America

Foreword

Michael Beeson

1 Significance of this book

This book made an important contribution to a debate about the meaning of mathematics that started about a century before its 1967 publication, and is still going on today. That debate is about the meaning of "non-constructive" proofs, in which one proves that something exists by assuming it does not exist, and then deriving a contradiction, without showing a way to construct the thing in question. As an example of the kind of proof in question, we might consider the "fundamental theorem of algebra", according to which every non-constant polynomial equation $f(z) = 0$ has a solution (among the complex numbers). To prove this theorem, do we have to show how to calculate a solution? Or is it enough to derive a contradiction from the assumption that $f(z)$ is never zero?

My favorite quotation from Errett Bishop is this:

Meaningful distinctions need to be preserved.[5]

He is talking about the distinction between constructive and non-constructive proof. This quotation encapsulates his approach to the matter. His predecessors had taken an all-or-nothing approach, either maintaining that *only* constructive proofs are correct, and non-constructive proofs are illusory or just wrong, or else maintaining that non-constructive proofs are valid, and while computational information might be interesting in special cases, it is of no philosophical significance. There was, naturally, very little productive interplay between these camps, only exchanges of polemics.

Bishop changed that situation with this book. He simply applied a technique that has been used many times in mathematics: he worked in the common ground, so that both "classical mathematics" (that is, with proofs of existence by contradiction allowed) and previously existing varieties of constructive mathematics (which made claims contradicting classical mathematics) could be viewed as generalizing the body of mathematics that Bishop developed in this book. The surprising thing was that this common body of mathematics turned out to be quite large! Bishop showed that it encompassed the main tools of mathematical analysis. That surprised everyone, constructivists and classical mathematicians alike.

Previously, both sides believed one had to make a choice: Either

(1) reject non-constructive proofs, and with it reject much of modern mathematics, but keep your philosophical purity; or

(2) deny that there is any philosophical problem with proving things exist without constructing them.

Since very few were willing to reject most of modern mathematics, choice (1) had been made by almost nobody; in practice the mathematical community was proving existence theorems by any means possible, and not worrying whether the proof provided a construction of whatever was proved to "exist." But not everyone wholeheartedly believed that the distinction between constructive and nonconstructive proof was meaningless; it just seemed that the price of attaching any importance to that distinction was unaffordably high. Bishop showed the mathematical community a way to acknowledge the importance of that distinction, without putting the main body of mathematics at risk.

Bishop was not neutral on the issue: he made it clear that he believed that if a human proves something exists, he or she should show how to construct it. But the *mathematics* he wrote allows one to take a low-key, non-confrontational attitude towards the issue, and simply provide constructive proofs that can be accepted as valid by *everyone*.

To understand the significance of Bishop's book, one needs

some understanding of the past history of constructivity. A proper treatment of that history is far too long for a foreword, where a succinct summary is required. In his review [21], Abraham Robinson gave such a summary in *just one paragraph*. We quote the first half of this masterpiece of brevity:

> In the second half of the nineteenth century, Leopold Kronecker made a determined attempt to turn mathematics away from its trend of ever increasing abstraction. His approach was based on the principle that in order to be meaningful, an existential assertion has to be buttressed by the actual construction of the object in question. Thus, a procedure that leads us to infer the existence of a mathematical object from purely formal-deductive considerations, e.g. by the use of the principle of the excluded middle, is regarded as inadequate or even misleading. Kronecker lent substance to his point of view by actually realizing the constructive approach in his lectures. Later, Brouwer, whose approach was based on the same attitude, went beyond Kronecker by developing a theory of the continuum, which may be called conditionally constructive, since it accepts the idea of a sequence of free choices as the basis of the theory of real numbers (just as even the most restrictive point of view accepts the unlimited counting process as the basis of arithmetic). Brouwer's school of thought—intuitionism—has remained the most vigorous of the several constructivist trends that have developed since Kronecker.

What we shall add to Robinson's paragraph is a few quotations to document the point that *everyone else* (other than Bishop) believed that if one accepted a constructivist philosophy, then one would have to give up much of classical mathematics. One of the chief advocates of that viewpoint was David Hilbert, one of the most famous mathematicians of the twentieth century. Here is Hilbert in 1927, speaking to the Hamburg Mathematical Seminar [23] (page 426).

For, compared with the immense expanse of modern mathematics, what would the wretched remnants mean, the few isolated results, incomplete and unrelated, that the intuitionists have obtained?

That was forty years before Bishop; but, in the next thirty years, despite the efforts of Heyting, Markov, and Kolmogorov, nothing changed in the prevalent viewpoint. When Fraenkel and Bar-Hillel wrote their famous book *Foundations of Set Theory* [14], they devoted a whole chapter (whose primary author was Fraenkel) to the *Intuitionistic conceptions of mathematics*; on page 263 we find

> In view of the mutilated shape which mathematics assumes according to intuitionistic principles it is not astonishing that a small minority only of mathematicians has been ready to accept the intuitionistic attitude; this situation will probably not change very much when the adherents ... will succeed in formulating their principles and inferences in a less dogmatic and more comprehensible and consistent form than done up to now.

They continue with the following remarks that seem, in a way, to foreshadow Bishop:

> Even after the partial failure of the Hilbert school regarding consistency proofs one must not forget that the very existence of mathematics and the wide range of its applications during many centuries seem to show that modern analysis is not just nonsensical or meaningless ... Hence we may trust that, sooner or later, methods will be found by which those methodical doubts which have not been answered so far can be dissolved. In short, the prevailing conviction is that one should not cut off the leg to heal the toe.

It was to be sooner rather than later: Bishop's book was only nine years in the future. Bishop's preface was quite explicit about his purpose:

to present the constructive point of view, to show
that the constructive program can succeed, and to
lay a foundation for further work. These immediate ends tend to an ultimate goal—to hasten the inevitable day when constructive mathematics will be
the accepted norm.

2   How this book was received

Bishop's book was reviewed by four quite different people.[1] The
first reviewer was Abraham Robinson [21], the founder of nonstandard analysis, who could be expected to be unsympathetic
to the constructivist viewpoint, since it is so far from the viewpoint of nonstandard analysis. Incidentally, Bishop also reviewed Robinson's book, and was definitely not sympathetic;
Keisler said that choosing Bishop to review Robinson was like
choosing a teetotaler to taste wines. Something similar probably applies to the choice of Robinson to review Bishop. Robinson's review begins with the summary of the work of Kronecker,
Brouwer, and others quoted above; then he continues with the
following tribute, all the more meaningful coming from Robinson:

> The present author's point of view is essentially Kro-
> necker's. He rejects the formalized versions of intu-
> itionism produced by Heyting . . . as well as Brouwer's
> theory of the continuum. On this basis, he provides a
> constructive development of some of the most impor-
> tant areas of classical and modern analysis and uses
> his great knowledge and power as an analyst to cope
> constructively with topics as advanced as the dual-
> ity theory of locally compact abelian groups and the
> theory of Banach algebras.

But Robinson ended his review with some negative remarks,
which are so unspecific and unsubstantiated that they must be

---

[1]Surprisingly, there was no review in *Zentralblatt*. The book's publication was listed in *Zentralblatt* **183**, p. 15, without a review.

taken to show Robinson's fundamental philosophical differences
with Bishop rather than be taken at face value:

> The sections of the book that attempt to describe
> the philosophical and historical background of this
> remarkable endeavor are more vigorous than accu-
> rate and tend to belittle or ignore the efforts of others
> who have worked in the same general direction.

The second reviewer was Gabriel Stolzenberg [22], who was
sympathetic to Bishop's viewpoint and had worked with him
during the book's preparation. His review explained Bishop's
viewpoint in detail, with attention to the points that often caused
confusion among classically-trained mathematicians. He also
placed Bishop's work in historical context with a discussion of
Brouwer and Weyl. Stolzenberg's review is too long to quote
extensively in a foreword, but the reader with the time to study
this book would do well to read Stolzenberg's review as well.
Stolzenberg's high opinion of the book is clear: He says that
Bishop has "demonstrate[d] to the classical mathematician what
the intuitionists ... did not: that to replace the classical system
by the constructive one does not in any way mutilate the great
classical theories of mathematics. Not at all. If anything, it
strengthens them, and shows them ... to be far grander than we
had known."

The third reviewer was John Myhill [19], a logician who later
developed formal set theories suitable for formalizing Bishop's
mathematics. Myhill was even more lavish in his praise than
Stolzenberg: "[T]he reviewer believes this book to be the most
important work on constructive mathematics ever written." Myhill reviewed both the book and Bishop's essay [4], and since
Stolzenberg had, in Myhill's opinion, done a good job of review-
ing the book, his review focused on the logical aspects of Bishop's
work, which he explained as well as possible using tools available
at that time. Logicians studying this book will certainly want
to read Bishop's essay and Myhill's review as well.

The fourth reviewer (not chronologically) was B. van Rootselaar, a Dutchman familiar with the intuitionistic tradition of

Brouwer. He notes [24] that Bishop "takes continuity in the sense of uniform continuity", and says that "this policy accounts to a large extent for the smoothness of the development." In Brouwer's intuitionism, continuity and uniform continuity are guaranteed by the "bar theorem" and the "fan theorem". Van Rootselaar took the view, which I have heard expressed orally in Amsterdam, Utrecht, and Nijmegen, that Bishop was in part merely rediscovering what intuitionists had long known:

> A comparison with intuitionistic mathematics modulo the bar theorem shows not too great differences, although some material has been developed in a more general setting in intuitionism.

However, van Rootselaar concluded his review with words of praise:

> It contains a substantial piece of constructive analysis and will take a worthy place among other textbooks on analysis written from a classical point of view. It certainly deserves to attract the attention of students and working mathematicians and it is likely that it will do so because of its elegance of exposition and because it advances well beyond the elementary facts of analysis.

Van Rootselaar levied a more substantial criticism as well:

> [Bishop] stresses the equal hypothesis interpretations of classical theorems, which is misleading since the hypotheses are for the greater part only verbally equal.

To explain this criticism: any classical theorem is likely to have more than one constructive formulation, all classically equivalent. An "equal hypothesis" version has the same hypotheses, but constructively weaker conclusions; an "equal conclusion" version has constructively stronger hypotheses. Of course there can be hybrid versions as well. Equal conclusion versions are usually more useful. Van Rootselaar's criticism was exactly the reason why Bishop and Cheng, and later Bishop and Bridges, revised the treatment of measure theory.

Finally, van Rootselaar agreed with Robinson about the more philosophical parts of the book: "The first chapter, the appendices and the notes reveal a dogmatic attitude of the author." In other words, neither Robinson nor van Rootselaar agreed with Bishop on philosophical matters. They would, of course, have disagreed even more with each other!

Bishop had a considerable reputation as a mathematician already in 1967–enough to command respect and attention when he began to espouse a minority view of the meaning of mathematics. I was in the audience when Bishop lectured at Stanford in 1969. He was received as a celebrity and spoke in a large lecture hall to a standing-room only audience, including all the senior professors of mathematics.

Stanford was not the only place where Bishop's lectures attracted large audiences and caused discussions. For example, he was invited to address the International Congress of Mathematicians in Moscow, 1966; he gave the Hedrik Lectures at the Mathematical Association of America's summer meeting in 1969; he gave the Colloquium Lectures at the 1969 summer meeting of the American Mathematical Society; in addition to these lectures he gave numerous one-hour invited addresses at national and regional meetings, according to Stefan Warschawski [25]. After the publication of his book (according to Nerode [20]), he made a tour of the eastern (U.S.) universities, and told Nerode that he was trying to communicate his viewpoint directly to the mathematical community, rather than through the logicians. After the trip was over, he told Nerode that the trip may have been counterproductive; he felt that his mathematical audiences were not taking the work seriously. He was surprised to get a more sympathetic hearing from the logicians.

Bishop also told Nerode about

> tribulations in the reviewing process when he submitted the book for publication. He mentioned that one of the referee's reports said explicitly that it was a disservice to mathematics to contemplate publica-

tion of this book. He could not understand, and was
hurt by, such a lack of appreciation of his ideas. One
of the reasons for the lecture tour was to be sure that
his ideas got a hearing.

Bishop also told Nerode that his students had experienced similar difficulties in developing their careers, and that he had ceased
to take students because of these problems. On the other hand,
Bishop was pleased at the warm reception he got from Myhill,
Friedman, Constable, and other logicians.

When Bishop lectured at Stanford in 1969, I was a graduate
student, and my thesis in some sense grew out of Bishop's lecture: it dealt with a logical problem that was related to Bishop's
work. I subsequently wrote a book [3] collecting the various approaches to axiomatizing Bishop's work. The timeline for my
own book illustrates that the mathematical community has quite
a bit of inertia: Bishop published his book in 1967, and mine,
which was based on the work of others whose work was directly
stimulated by Bishop's book, appeared 18 years later. We will
see, when we look to the literature for the effect of Bishop's
book, that this time lag was not unusual.

3   What happened afterwards

The first to react were those who had given Bishop's lectures
a warm reception: the logicians. John Myhill and then Harvey
Friedman developed intuitionistic set theories, adequate to formalize Bishop's book. They kept the framework of classical set
theory, changing only the logic and making minor changes to the
axioms. Solomon Feferman developed theories of "explicit mathematics" with the same purpose, but closer to Bishop's ideas.
Martin-Löf had already been working on his theories, and probably was not influenced by Bishop, although his theories could
be viewed as suitable for formalizing Bishop's work. (See [3] for
descriptions of all these theories and references to the original
publications.) In the summer of 1968, one year after the publication of Bishop's book, there was a conference on *Intuitionism and*

Proof Theory in Buffalo, New York. Bishop himself presented a paper [4], in which he indicated an approach to providing a logical foundation for constructive mathematics. All four of the above-mentioned logicians were present at the conference, and very likely all four were in the audience at Bishop's talk. Probably the Buffalo conference played an important role in attracting their attention to the problem of formulating theories adequate to formalize Bishop's work.

But not all logicians reacted, even within a few years. Six years after Bishop's book, in 1973, a second edition of Fraenkel and Bar-Hillels' book was published. The chapter on intuitionistic conceptions of mathematics is extensively revised; there are references to numerous works that appeared after the first edition, mostly by logicians. But Bishop's book is not mentioned or referenced! That seems surprising, especially in view of the widespread discussion of his ideas in the years immediately following 1967. I offer this partial explanation: Fraenkel and Bar-Hillel's second edition reflects only the work of logicians, and, by 1973, the papers of Friedman, Myhill, and Feferman had not yet appeared.

There were also mathematicians who took up Bishop's research program more directly, rather than studying its logical foundations. This group included Douglas Bridges, Bill Julian, Y. K. Chan, Ray Mines, Fred Richman, and Wim Ruitenberg. They produced several books and many papers extending Bishop's work. Bridges's book, *Constructive Functional Analysis*, appeared in 1979, but was mostly written in New Zealand in 1976, according to the preface. Then Bridges lectured in Oxford in 1981, and invited Richman to help him write *Varieties of Constructive Mathematics* [8], which, except for one chapter on algebra, is a work about the different possible foundational positions whose common core is Bishop's constructive mathematics.

After the initial foray into algebra just mentioned (proving the existence and uniqueness of splitting fields), Richman plunged headlong into the difficulties of constructive algebra, and, with the help of Mines and Ruitenberg, wrote *A Course in Constructive Algebra* [18], published in 1988 and dedicated to

Errett Bishop.

Bishop's book contains a chapter on measure theory, but the treatment was already judged inadequate by Bishop himself, as discussed above in connection with van Rootselaar's review. A better constructive measure theory was developed by Bishop and Cheng [7]. Partly in order to include this new theory, Bishop was engaged in revising his book for a second edition. He took on Bridges as a co-author, but then died before the second edition was finished. Bridges completed the book and it was published under joint authorship in 1985 [6].

The bibliography of this second edition has 93 entries; the first edition (reprinted here) has 17. Only about ten of the 93 appeared before 1967; so after twenty years, Bishop's book had resulted in three or four other books and several dozen research papers. He had certainly demonstrated that the constructive program could succeed, and the further work for which he had laid the foundation was in progress. But his followers could be counted on the fingers of two hands. The "inevitable day when constructive mathematics will be the accepted norm" seemed as far away as ever, and it still seems so today. The revolution didn't happen.

But perhaps his influence has simply been more subtle. Remember that Bishop said "meaningful distinctions need to be preserved." Bishop clearly made the mathematical community more aware than it had been of the meaningful distinction between constructing something, and proving that something exists without constructing it. He did not convince more than a handful of people that the latter is meaningless, any more than did Brouwer. But he did show by example how to make that distinction in branches of mathematics where it had not been thought before to be a meaningful distinction.

It is difficult to assess the extent to which the mathematics community in 2012 is aware of that distinction. It is my opinion that most mathematicians are aware of whether a given existence proof provides an algorithm or not, and find that a meaningful distinction worth preserving, but only in individual proofs. They are not aware of the connection between logic and algorithms, and they are not aware of the subtleties of representing abstract objects such as metric spaces and operators in a numerically meaningful way, and how these ideas can be used to preserve that meaningful distinction "globally." For example, consider number theory as presented, say, in the two volume textbook [10, 11]; it is very careful to distinguish between "effective" proofs and non-effective proofs. (Number theorists have used the word "effective" to mean "constructive" since long before Bishop.)

Logical work on constructive mathematics is alive and well, witness for example Kohlenbach's book [17]; he has applied his logical methodology to prove theorems in approximation theory, whose proofs (freed from logical trappings) have appeared in mathematics journals with non-logician co-authors who brought the problems to his attention. In spite of isolated successes like these, the widespread change of philosophical attitude that Bishop was trying to provoke just did not take place; most mathematicians are unaware of the meaningful distinction that Bishop wanted to preserve.

In attempting to assess the impact of Bishop's book, there is a complicating factor: the computer. During those decades, the widespread availability and use of computers by mathematicians has influenced attitudes at least as much as Bishop did. Certainly number theory these days almost always involves extensive computer-assisted numerical experimentation. There is, for example, *A Course in Computational Algebraic Number Theory* [9], with no apparent influence of Bishop.

It is even difficult to tell whether there really is *increased* interest in "effectivity", since number theorists were always interested in that distinction. This claim can be demonstrated by the fact that two different mathematicians were awarded a Fields medal for proofs of the same theorem: the first proof non-constructive, the second one (partially) constructive. Klaus Roth was awarded the Fields medal in 1958, and Alan Baker was awarded the Fields medal in 1970 for the work in [1].[2]

---

[2]The theorem in question is this: given an algebraic number α and $\epsilon > 0$, there are only finitely many rational numbers p/q with $|\alpha - p/q| < q^{-(2+\epsilon)}$. Baker showed how to effectively bound the number (but not the size) of

4 Relations to computer proofs

Bishop began his work when computers were, as depicted in the film *Dr. Strangelove*, large and expensive, and possessed only by very large organizations. Nevertheless, in Appendix B of his book, pp. 356–357, he wrote:

> It is clear that many of the results in this book could be programmed for a computer, by some such procedure as that indicated above. In particular, it is likely that most of the results of Chaps. 2, 4, 5, 9, 10, and 11 could be presented as computer programs. As an example, a complete separable metric space X can be described by a sequence of real numbers, and therefore by a sequence of integers, simply by listing the distances between each pair of elements of a given countable dense set. . . . As written, this book is person-oriented rather than computer-oriented. It would be of great interest to have a computer-oriented version.

What did Bishop have in mind by a "computer-oriented version"? Presumably he meant a computer program that could check the correctness of constructive proofs, and extract the underlying algorithms from them, so that those algorithms could be executed, without programming them in the usual sense, and in such a way that their correctness would be guaranteed by the proofs from which they were extracted.

At about the same time, and as far as I know, completely independently, de Bruijn was developing the first computer system for representing and proving mathematical theorems in a systematic way. That system was called AUTOMATH. It was the grandfather of all modern proof-checking systems. The second generation of such systems was pioneered at Cornell by Robert Constable, and Constable explicitly acknowledges [20] the influence of Bishop on the design of the system, which was ultimately

---

solutions; to bound the size of p and q, and thus be able to compute the exact number of solutions, could earn another Fields medal.

known as NuPrl [12]. NuPrl was explicitly designed to "execute constructive proofs", i.e. to extract programs from proofs. Constable writes,

> Shortly after we had executed our first constructive proof, I wrote to Bishop informing him of what I took to be an historic event. I told him how much his writings and his encouragement had meant to us on the long road to this accomplishment. I was crushed to receive my letter back unopened, marked "recipient deceased."

In [26], NuPrl is said to be based on Martin-Löf's systems, which does not contradict Constable's statements in [20], as Bishop's influence on the design was at the "requirements" level: the system had to be adequate to formalize Bishop's work, and hence finite type theory would not do; a more elaborate system had to be developed, and Martin-Löf's theories met that requirement, in the judgment of Constable.

In turn, NuPrl and Bishop both influenced the design of later proof-checkers. By 2006 there were at least seventeen computer systems designed to check proofs; see [26] for descriptions and comparisons of seventeen proof checkers. But of those seventeen, only a few work with constructive logic and/or permit the extraction of algorithms from proofs. Besides NuPrl, we have Coq, Minlog, and Alfa/Agda.

The proof-checker that most nearly corresponds to a "computer oriented version" of constructive mathematics is Coq. The native logic of Coq is intuitionistic logic. Coq was developed in France, but in Nijmegen, there has been for some years an intensive project to develop the "Coq Repository Nijmegen" (CoRN). The design of CoRN has been heavily influenced by Bishop, and it has been used to formalize parts of real analysis in Bishop's style [13]. In 2008, Georges Gonthier used Coq to prove the four-color theorem [15]. Previous proofs made use of computer programs to carry out certain searches, but those programs were never proved correct, leaving some room for lingering doubt about the correctness of the overall proof. But Gonthier's proof in Coq formalized the entire proof, including the proof of correctness of the algorithms. One can extract a coloring algorithm from Gonthier's proof, but the word "extract" is misleading, or even wrong, since Gonthier had the algorithm in mind and works hard to make his algorithm as efficient as possible, given the Coq framework. Nevertheless, the application of Coq to a problem of such historical significance was certainly a milestone. If Bishop could have seen Coq and CoRN, and the work formalized with these tools, he would have been interested and pleased.

Minlog allows one to work in intuitionistic or in classical logic, and amazingly even allows one to extract algorithms from some classical proofs; e.g. an algorithm for the greatest common divisor of m and n can be extracted from the classical proof that the least positive linear combination of m and n is their greatest common divisor. The designer of Minlog is Helmut Schwichtenberg, who has been familiar with the work on formal systems for constructive mathematics ever since the time when Bishop's book was published, and was certainly influenced by Bishop and by the logical and mathematical work that followed. Alfa/Agda was created in 1990 and uses a logic based on Martin-Löf's theories, so one cannot point to a direct influence of Bishop.

5 Life of Errett Bishop

Errett Albert Bishop was born July 24, 1928, in Newton, Kansas. His father, Albert T. Bishop, graduated from the United States Military Academy at West Point. Wikipedia says that Albert ended his career as professor of mathematics at Wichita State University in Kansas. It also says that Errett grew up in Newton, Kansas, and that Albert died when Errett was five. Newton is farther from Wichita than anyone commuted in those days, which is consistent with the MacTutor web page on Bishop, which says (without a reference) that Albert was forced by illness to retire early. His father influenced Errett's eventual career by the textbooks he left behind, which is how Errett discovered mathematics. Bishop entered the University of Chicago at the age of 16, and in three years obtained both his B. S. and his

M .S. degrees in mathematics. Then, according to Wikipedia, he began studying for his doctorate; but he performed two years of service in the US Army, 1950–52, doing mathematical research at the National Bureau of Standards. He completed his Ph. D. in 1954 under Paul Halmos; his thesis was titled *Spectral Theory for Operations on Banach Spaces*. Halmos tells in [16] that Errett wrote *two* Ph. D. theses. After his first one was finished, it was discovered that the same results had been obtained earlier by a Russian. Halmos told Bishop that it was not a problem, because Errett's work was independent. But Errett said no; he would write another thesis, and he did. This story was also related to me by a fellow student of Bishop's, who said that Errett (even in his student days) was "a strong person; he knew what he wanted, and was respected by everyone."

Errett was 39 years old when this book was published in 1967. He died of cancer on April 14, 1983, at the age of only 55. According to Wikipedia, he "became interested in foundational issues" in the 1964-65 academic year, which he spent in Berkeley at the Miller Institute for Basic Research; but in the acknowledgements of this book, Bishop says his work on the book was supported by NSF grants for the summers of 1964, 1965, and 1966, as well as by the Miller Foundation, so it seems he was working on it for at least three years. It is natural to wonder whether there was some event or experience that turned his interest to constructivity. Metakides and Nerode asked him about that [20], and "he said that the fact of the matter was that he thought by nature constructively from the beginning of his mathematical life, and that the book was a natural outgrowth of his mathematical temperament." He also told them that "he had been influenced by Weyl's book, and had looked briefly at Brouwer, but had avoided detailed study of Brouwer for fear of being led away from his own natural lines of development." This reminds me of Feynman, who, when working on an unsolved problem, did not read the work of other physicists because "they didn't get the answer."[3]

---

[3]Related to some students over lunch in a Caltech cafeteria in about 1966. I was one of those students.

From 1964 on, constructivity seems to have occupied his attention, as he was engaged right up until his death in the preparations for a second edition of his book, as described above. In the middle of his work on his book (1965), he moved to San Diego, and remained professor at the University of California, San Diego, for the rest of his life. The decision to leave Berkeley preceded the decision to go to UCSD; faculty at UCSD heard of this decision and recruited Bishop, calling him while he was visiting Yale to urge him to not make a decision until he had seen UCSD [25]. Bishop was full professor at Berkeley–why did he decide to leave? I was told by someone who heard it directly from Bishop that it was because of the student unrest. The Free Speech Movement took place from October 1, 1964 through January of 1965, so the timing is consistent. If Errett wanted to avoid a campus climate that included unruly political activity, he was right to leave when he did, as Berkeley continued to be tumultuous until after the Vietnam war. He probably had other reasons as well; it is a major decision to make based on a few demonstrations.

He had three Ph. D. students at Berkeley, who wrote theses about uniform approximation and interpolation (in the time preceding his interest in constructive mathematics). At UCSD, he had eight more Ph. D. students (according to [25]); the Mathematics Genealogy Project lists six of these, five of whom wrote theses with the word "constructive" in the title, providing further evidence that constructivity continued to hold his attention. We have already mentioned a reason why Bishop stopped supervising such theses–it was not because he lost interest.

Errett's colleague Stefan Warschawski described him [25] in these words:

> Errett was a remarkable personality. Particularly outstanding traits were his independence and originality, apparent in everything he did, in his research, his teaching, in every aspect of daily life. He had strong principles by which he lived and a strong feeling for fairness in the treatment of other people— his colleagues, his students. He treated people with

kindness and consideration. He was a very private
person and did not talk much about himself.

Warschawski also tells us that Errett had a stamp collection and
a collection of Indian and Mexican artifacts.

Errett was survived by his wife Jane and two sons, Edward
and Thomas, and a daughter Rosemary. Edward earned a Ph. D.
in mathematics from the University of California at San Diego
in 1991.

References

[1] Alan Baker. Contributions to the theory of Diophantine
    equations. I. on the representation of integers by binary
    forms. *Philosophical Transactions of the Royal Society of
    London*, 263(1139):173–191, July 1968.

[2] Alan Baker. Contributions to the theory of diophan-
    tine equations. II. the diophantine equation $y^2 = x^3 + k$.
    Philosophical Transactions of the Royal Society of London,
    263(1139):193–208, 1968.

[3] Michael Beeson. *Foundations of Constructive Mathematics*.
    Springer-Verlag, Berlin Heidelberg, 1985.

[4] Errett Bishop. Mathematics as a numerical language. In
    A. Kino, John Myhill, and Richard E. Vesley, editors,
    *Intuitionism and Proof Theory: Proceedings of the Sum-
    mer Conference at Buffalo, New York, 1968*, pages 53–71.
    North-Holland, Amsterdam, 1970.

[5] Errett Bishop. Schizophrenia in contemporary mathemat-
    ics. In Murray Rosenblatt, editor, *Errett Bishop: Reflec-
    tions on Him and His Research*, volume 39 of *Conte*, pages
    1–32. American Mathematical Society, 1973.

[6] Errett Bishop and Douglas Bridges. *Constructive Anal-
    ysis*, volume 279 of *Gundlehren der Mathematische Wis-
    senschaften*. Springer-Verlag, Berlin Heidelberg, 1985.

[7] Errett Bishop and Henry Cheng. Constructive measure theory. *Memoirs of the American Mathematical Society*, 116, 1972.

[8] Douglas Bridges and Fred Richman. *Varieties of Constructive Mathematics*. Cambridge University Press, Cambridge, U. K., 1987.

[9] Henri Cohen. *A Course in Computational Algebraic Number Theory*. Graduate Texts in Mathematics. Springer-Verlag, 1993.

[10] Henri Cohen. *Number Theory, Volume 1: Tools and Diophantine Equations*, volume 239 of *Graduate Texts in Mathematics*. Springer, New York, 2007.

[11] Henri Cohen. *Number Theory, Volume 2: Analytic and Modern Tools*, volume 240 of *Graduate Texts in Mathematics*. Springer, New York, 2007.

[12] Robert Constable. *Implementing Mathematics with the Nuprl Proof Development System*. Prentice-Hall, Englewood Cliffs, New Jersey, 1986.

[13] Luís Cruz-Filipe. A constructive formalization of the fundamental theorem of calculus. In H. Geuvers and F. Wiedijk, editors, *Types for Proofs and Programs*, volume 2646 of *LNCS*, pages 108–126. Springer–Verlag, 2003.

[14] Abraham A. Fraenkel and Yehoshua Bar-Hillel. *Foundations of Set Theory*. North-Holland, Amsterdam, first edition, 1958.

[15] Georges Gonthier. Formal proof – the four-color theorem. *Notices of the American Mathematical Society*, 55(11):1382–1393, 2008.

[16] Paul Halmos. *I want to be a mathematician*. Springer-Verlag, New York, 1985.

[17] Ulrich Kohlenbach. *Applied Proof Theory: Proof Interpretations and their Use in Mathematics*. Springer-Verlag, 2008.

[18] Ray Mines. *A Course in Constructive Algebra*. Springer-Verlag, Heidelberg, 1988.

[19] John Myhill. Review of Foundations of Constructive Analysis by Errett Bishop. *Journal of Symbolic Logic*, 37(4):744–747, 1972.

[20] Anil Nerode, George Metakides, and Robert Constable. Remembrances of Errett Bishop. In Murray Rosenblatt, editor, *Errett Bishop: Reflections on Him and His Research*, volume 39 of *Contemporary Mathematics*, pages 79–84. American Mathematical Society, 1985.

[21] Abraham Robinson. Review of Foundations of Constructive Analysis by Errett Bishop. *The American Mathematical Monthly*, 75(8):920–921, October 1968.

[22] Gabriel Stolzenberg. Review of Foundations of Constructive Analysis by Errett Bishop. *Bulletin of the American Mathematical Society*, 76(2):301–323, 1970.

[23] Jean van Heijenoort. *From Frege to Gödel, a source book in mathematical logic, 1879–1931*. Harvard University Press, Cambridge, Massachusetts, 1967.

[24] B. van Rootselaar. Review of Foundations of Constructive Analysis by Errett Bishop. *Mathematical Reviews*, 36(4930; MR0221878), 1968.

[25] Stefan Warschawski. Errett Bishop: In memoriam. In Murray Rosenblatt, editor, *Errett Bishop: Reflections on Him and His Research*, volume 39 of *Contemporary Mathematics*, pages 33–39. American Mathematical Society, 1985.

[26] Freek Wiedijk, editor. *The Seventeen Provers of the World*, volume 3600 of *Lecture Notes in Artificial Intelligence*. Springer, Berlin Heidelberg, 2006.

Acknowledgments It is a pleasure to compliment my wife and children for their patience and forebearance while I wrote this book. I owe much to J. L. Kelley, without whose encouragement I might have abandoned mathematics long ago. Gabriel Stolzenberg has read the manuscript and given me good advice. Joan Lagioe has done a fine job of typing the manuscript.

Support for much of the writing of this book and much of the research that went into it was supplied by the National Science Foundation, in the form of grants for the summers of 1964, 1965, and 1966, and by the Miller Foundation of the University of California, in the form of a fellowship for the academic year 1964–1965.

Errett Bishop

PREFACE

Most mathematicians would find it hard to believe that there could
be any serious controversy about the foundations of mathematics, any
controversy whose outcome could significantly affect their own mathematical activity. Their attitude well represents the actual state of
affairs: during a half-century of splendid mathematical progress there
has been no deviation from the norm. The voices of dissent, never
much heeded, have long been silent.

Perhaps the times are not conducive to introspection. Mathematics
flourishes as never before. Its scope is immense, its quality high.
Mathematicians flourish as never before. Their profession is respec-
table, their salaries good. Mathematical methods are more fashionable than ever before. Witness the surge of interest in
mathematical logic, mathematical biology, mathematical economics,
mathematical psychology—in mathematical investigations of every
sort. The extent to which many of these investigations are premature
or unrealistic indicates the deep attraction mathematical exactitude
holds for the contemporary mind.

And yet there is dissatisfaction in the mathematical community.

The pure mathematician is isolated from the world, which has little need of his brilliant creations. He suffers from an alienation which is seemingly inevitable: he has followed the gleam and it has led him out of this world.

If every mathematician occasionally, perhaps only for an instant, feels an urge to move closer to reality, it is not because he believes mathematics is lacking in meaning. He does not believe that mathematics consists in drawing brilliant conclusions from arbitrary axioms, of juggling concepts devoid of pragmatic content, of playing a meaningless game. On the other hand, many mathematical statements have a rather peculiar pragmatic content. Consider the theorem that either every even integer greater than 2 is the sum of two primes, or else there exists an even integer greater than 2 that is not the sum of two primes. The pragmatic content of this theorem is not that if we go to the integers and observe we shall see certain things happening. Rather the pragmatic content of such a theorem, if it exists, lies in the circumstance that we are going to use it to help derive other theorems, themselves of peculiar pragmatic content, which in turn will be the basis for further developments.

It appears then that there are certain mathematical statements that are merely evocative, which make assertions without empirical validity. There are also mathematical statements of immediate empirical validity, which say that certain performable operations will produce certain observable results, for instance, the theorem that every positive integer is the sum of four squares. Mathematics is a mixture of the real and the ideal, sometimes one, sometimes the other, often so presented that it is hard to tell which is which. The realistic component of mathematics—the desire for pragmatic interpretation—supplies the control which determines the course of development and keeps mathematics from lapsing into meaningless formalism. The idealistic component permits simplifications and opens possibilities which would otherwise be closed. The methods of proof and the objects of investigation have been idealized to form a game, but the actual conduct of the game is ultimately motivated by pragmatic considerations.

For 50 years now there have been no significant changes in the rules of this game. Mathematicians unanimously agree on how mathematics should be played. Accepted standards of performance suffice to regulate the course of mathematical activity, and there is no prospect that these standards will be changed in any significant respect by a revision of the idealistic code. In fact no efforts are being made to impose such a revision.

There have been, however, attempts to constructivise mathematics, to purge it completely of its idealistic content. The most sustained attempt was made by L. E. J. Brouwer, beginning in 1907. The movement he founded has long been dead, killed partly by compromises of Brouwer's disciples with the viewpoint of idealism, partly by extraneous peculiarities of Brouwer's system which made it vague and even ridiculous to practicing mathematicians, but chiefly by the failure of Brouwer and his followers to convince the mathematical public that abandonment of the idealistic viewpoint would not sterilize or cripple the development of mathematics. Brouwer and other constructivists were much more successful in their criticisms of classical mathematics than in their efforts to replace it with something better. Many mathematicians familiar with Brouwer's objections to classical mathematics concede their validity but remain unconvinced that there is any satisfactory alternative.

This book is a piece of constructivist propaganda, designed to show that there does exist a satisfactory alternative. To this end we develop a large portion of abstract analysis within a constructive framework.

This development is carried through with an absolute minimum of philosophical prejudice concerning the nature of constructive mathematics. There are no dogmas to which we must conform. Our program is simple: to give numerical meaning to as much as possible of classical abstract analysis. Our motivation is the well-known scandal, exposed by Brouwer (and others) in great detail, that classical mathematics is deficient in numerical meaning.

Some familiarity with Brouwer's critique is essential. Chapter 1 is primarily devoted to an examination of the defects of classical mathematics, following Brouwer, and a presentation of the thesis that all mathematics should have numerical meaning. Chapter 3 presents constructive versions of the fundamental concepts of sets and functions, and examines some of the obstacles to the constructivization of general topology.

The remaining chapters are primarily technical, and constitute a course in abstract analysis from the constructive point of view. Very little formal preparation is required of the reader, although a certain level of mathematical sophistication is probably indispensible. Every effort has been made to follow the classical development as closely as possible; digressions have been relegated to notes at the ends of the various chapters.

The task of making analysis constructive is guided by three basic principles. First, to make every concept affirmative. (Even the concept of inequality is affirmative.) Second, to avoid definitions that are not relevant. (The concept of a pointwise continuous function is not relevant. A continuous function is one that is uniformly continuous on compact intervals.) Third, to avoid pseudogenerality. (Separability hypotheses are freely employed. The justification for this is discussed in Appendix A.)

Chapters 1, 3, 4, 6, and 9 are the core of the book. Of the remaining chapters, Chapter 2 (Calculus and the Real Numbers), Chapter 5 (Complex Analysis), and Chapter 7 (Integration) are of less interest because they represent somewhat routine constructivizations of the corresponding classical theories. The constructivizations involved in Chapter 8 (Limit Operations in Measure Theory), Chapter 10 (Locally Compact Abelian Groups), and Chapter 11 (Banach Algebras), on the other hand, are by no means routine, but these chapters are perhaps of less interest because they are concerned with material of a somewhat more special kind.

As already remarked, Chapters 1 and 3 are primarily philosophical. This leaves Chapters 4 (Metric spaces), 6 (Measure theory), and 9 (Banach spaces). These chapters contain the core of constructive abstract analysis.

In a first reading one could well omit Chapters 5, 8, 10, and 11.

The book has a threefold purpose: first, to present the constructive point of view; second, to show that the constructive program can succeed; third, to lay a foundation for further work. These immediate ends tend to an ultimate goal—to hasten the inevitable day when constructive mathematics will be the accepted norm.

We are not contending that idealistic mathematics is worthless from the constructive point of view. This would be as silly as contending that unrigorous mathematics is worthless from the classical point of view. Every theorem proved with idealistic methods presents a challenge: to find a constructive version, and to give it a constructive proof.

Errett Bishop

CONTENTS

CHAPTER 3 SET THEORY   62

CHAPTER 4 METRIC SPACES   75

CHAPTER 5 COMPLEX ANALYSIS   112

CHAPTER 6 MEASURE   153

CHAPTER 7 INTEGRATION   182

CHAPTER **8** LIMIT OPERATIONS IN MEASURE THEORY 214

CHAPTER **9** NORMED LINEAR SPACES 243

CHAPTER **10** LOCALLY COMPACT ABELIAN GROUPS 298

CHAPTER **11** COMMUTATIVE BANACH ALGEBRAS 335

CHAPTER 1

A CONSTRUCTIVIST MANIFESTO

1. THE DESCRIPTIVE BASIS OF MATHEMATICS

Mathematics is that portion of our intellectual activity which transcends our biology and our environment. The principles of biology as we know them may apply to life forms on other worlds, yet there is no necessity for this to be so. The principles of physics should be more universal, yet it is easy to imagine another universe governed by different physical laws. Mathematics, a creation of mind, is less arbitrary than biology or physics, creations of nature; the creatures we imagine inhabiting another world in another universe, with another biology and another physics, will develop a mathematics which in essence is the same as ours. In believing this we may be falling into a trap: Mathematics being a creation of our mind, it is, of course, difficult to imagine how mathematics could be otherwise without actually making it so, but perhaps we should not presume to predict the course of the mathematical activities of all possible types of intelligence. On the other hand, the pragmatic content of our belief in the transcendence of mathematics has nothing to do with alien forms of life. Rather it serves to give a direction to mathematical investigation, resulting from the insistence that mathematics be born of an inner necessity.

The primary concern of mathematics is number, and this means the positive integers. We feel about number the way Kant felt about space. The positive integers and their arithmetic are presupposed by the very nature of our intelligence and, we are tempted to believe, by the very nature of intelligence in general. The development of the theory of the positive integers from the primitive concept of the unit, the concept of adjoining a unit, and the process of mathematical induction carries complete conviction. In the words of Kronecker, the positive integers were created by God. Kronecker would have expressed it even better if he had said that the positive integers were created by God for the benefit of man (and other finite beings). Mathematics belongs to man, not to God. We are not interested in properties of the positive integers that have no descriptive meaning for finite man. When a man proves a positive integer to exist, he should show how to find it. If God has mathematics of his own that needs to be done, let him do it himself.

Almost equal in importance to number are the constructions by which we ascend from number to the higher levels of mathematical existence. These constructions involve the discovery of relationships among mathematical entities already constructed, in the process of which new mathematical entities are created. The relations which form the point of departure are the order and arithmetical relations of the positive integers. From these we construct various rules for pairing integers with one another, for separating out certain integers from the rest, and for associating one integer to another. Rules of this sort give rise to the notions of sets and functions.

A set is not an entity which has an ideal existence. A set exists only when it has been defined. To define a set we prescribe, at least implicitly, what we (the constructing intelligence) must do in order to construct an element of the set, and what we must do to show that two elements of the set are equal. A similar remark applies to the definition of a function: in order to define a function from a set A to a set B, we prescribe a finite routine which leads from an element of A to an element of B, and show that equal elements of A give rise to equal elements of B.

Building on the positive integers, weaving a web of ever more sets and more functions, we get the basic structures of mathematics: the rational number system, the real number system, the euclidean spaces, the complex number system, the algebraic number fields, Hilbert space, the classical groups, and so forth. Within the framework of these structures most mathematics is done. Everything attaches itself to number, and every mathematical statement ultimately expresses the fact that if we perform certain computations within the set of positive integers, we shall get certain results.

Mathematics takes another leap, from the entity which is constructed in fact to the entity whose construction is hypothetical. To some extent hypothetical entities are present from the start: whenever we assert that every positive integer has a certain property, in essence we are considering a positive integer whose construction is hypothetical. But now we become bolder and consider a hypothetical set, endowed with hypothetical operations subject to certain axioms. In this way we introduce such structures as topological spaces, groups, and manifolds. The motivation for doing this comes from the study of concretely constructed examples, and the justification comes from the possibility of applying the theory of the hypothetical structure to the study of more than one specific example. Recently it has become fashionable to take another leap and study, as it were, a hypothetical hypothetical structure—a hypothetical structure qua hypothetical structure. Again the motivations and justifications attach themselves to particular examples, and the examples attach themselves to numbers in the ultimate analysis. Thus even the most abstract mathematical statement has a computational basis.

The transcendence of mathematics demands that it should not be confined to computations that I can perform, or you can perform, or 100 men working 100 years with 100 digital computers can perform. Any computation that can be performed by a finite intelligence—any computation that has a finite number of steps—is permissible. This does not mean that no value is to be placed on the efficiency of a computation. An applied mathematician will prize a computation for its efficiency above all else, whereas in formal mathematics much attention is paid to elegance and little to efficiency. Mathematics should and must concern itself with efficiency, perhaps to the detriment of elegance, but these matters will come to the fore only when realism has begun to prevail. Until then our first concern will be to put as much mathematics as possible on a realistic basis without close attention to questions of efficiency.

2. THE IDEALISTIC COMPONENT OF MATHEMATICS

Geometry was highly idealistic from the time of Euclid and the ancients until the time of Descartes, unfolding from axioms taken either to be self-evident or to reflect properties of the real world. Descartes reduced geometry to the theory of the real numbers, and in the nineteenth century Dedekind, Weierstrass, and others, by the arithmetization of the real number system, brought space into the concrete realm of objects constructed by pure thought.

Unfortunately the promise held out to mathematics by the arithmetization of space was not fulfilled, largely due to the intervention, around the turn of the century, of the formalist program. The successful formalization of mathematics helped keep mathematics on a wrong course. The fact that space has been arithmetized loses much of its significance if space, number, and everything else are fitted into a matrix of idealism where even the positive integers have an ambiguous computational existence. Mathematics becomes the game of sets, which is a fine game as far as it goes, with rules that are admirably precise. The game becomes its own justification, and the fact that it represents a highly idealized version of mathematical existence is universally ignored.

Of course, idealistic tendencies have been present if not dominant in mathematics since the Greeks, but it took the full flowering of formalism to kill the insight into the nature of mathematics which its arithmetization could have given.

To see how some of the most basic results of classical analysis lack computational meaning, take the assertion that every bounded non-void set A of real numbers has a least upper bound. (The real number b is the *least upper bound* of A if $a \leq b$ for all a in A and if there exist elements of A that are arbitrarily close to b.) To avoid unnecessary complications, we actually consider the somewhat less general assertion that every bounded sequence $\{x_k\}$ of rational numbers has a least upper bound b (in the set of real numbers). If this assertion were constructively valid, we could compute b, in the sense of computing a rational number approximating b to within any desired accuracy; in fact we could program a digital computer to compute the approximations for us. For instance, the computer could be programmed to produce, one by one, a sequence $\{(b_k,m_k)\}$ of ordered pairs, where each b_k is a rational number and each m_k is a positive integer, such that (i) $x_j \leq b_k + k^{-1}$ for all positive integers j and k, and (ii) $x_{m_k} \geq b_k - k^{-1}$ for all positive integers k. Unless there exists a general method M that produces such a computer program corresponding to each bounded constructively given sequence $\{x_k\}$ of rational numbers, we are not justified, by constructive standards, in asserting that each of the sequences $\{x_k\}$ has a least upper bound. To see the scope such a method M would have, consider a constructively given sequence $\{n_k\}$ of integers, each of which is either 0 or 1. Using the method M, we compute a rational number b_3 and a positive integer $N \equiv m_3$ such that (i)

$n_j \le b_3 + \frac{1}{3}$ for all positive integers j, and (ii) $n_N \ge b_3 - \frac{1}{3}$. Either $n_N = 0$ or $n_N = 1$. If $n_N = 0$, then (i) and (ii) imply that

$$n_j \le b_3 + \frac{1}{3} \le n_N + \frac{2}{3} = \frac{2}{3}$$

for all j. Since each n_j is either 0 or 1, it follows that $n_j = 0$, for all j. Thus, for each of the sequences $\{n_k\}$ being considered, the method M either (i) produces a proof that the n_k are all equal to 0, or (ii) produces a positive integer N such that $n_N = 1$. Of course, such a method M does not exist, and nobody expects that one will ever be found. Such a method would solve most of the famous unsolved problems of mathematics—Fermat's last theorem, the four-color problem, and the Riemann hypothesis, in particular, since each of these problems can be reduced to finding, for a certain sequence $\{n_k\}$ of the type being considered, either a proof that $n_k = 0$ for all k or a proof that $n_k = 1$ for some k.

As another instance, consider the intuitively appealing theorem that every continuous function f on the closed interval [0,1], for which $f(0) < 0$ and $f(1) > 0$, vanishes at some point x_0. This theorem can be derived from the least-upper-bound principle: take x_0 to be the least upper bound of the set of all x for which $f(x) < 0$. The fact that we make use of the least-upper-bound principle does not mean our theorem is unconstructive; it only means the given proof is unconstructive. A closer examination demonstrates that our theorem itself is unconstructive. This demonstration, which we now give, uses facts from the constructive theory of continuous functions, with which the reader is probably not familiar. Nevertheless, it should provide some insight. Let $\{n_k\}$ be any constructively given sequence of integers, each of which is either -1, 0, or 1. Define the real number a by

$a \equiv \sum_{k=1}^{\infty} 3^{-k} n_k$. There exists a unique constructively given continuous function f on [0,1] such that $f(0) = -1$, $f(1) = 1$, $f(\frac{1}{3}) = f(\frac{2}{3}) = a$, and f is linear on each of the intervals $[0,\frac{1}{3}]$, $[\frac{1}{3},\frac{2}{3}]$, and $[\frac{2}{3},1]$. If our theorem is valid, there exists a point x_0 with $f(x_0) = 0$. By computing a sufficiently close rational approximation to x_0, we show that either (a) $x_0 < \frac{2}{3}$, or (b) $x_0 > \frac{1}{3}$. In the first case, $a \ge 0$, and therefore the first nonzero term of the sequence $\{n_k\}$, if one exists, equals 1. Similarly, in the second case, the first nonzero term, if one exists, equals -1. Thus our theorem gives a method, for each of the sequences $\{n_k\}$ being considered, of either (i) proving that any term that equals 1 is preceded by a term that equals -1, or (ii) proving that any term that equals -1 is preceded by a term that equals 1. Nobody believes that such a method will ever be found.

Brouwer fought the advance of formalism and undertook the disengagement of mathematics from logic. He wanted to strengthen mathematics by associating to every theorem and every proof a pragmatically meaningful interpretation. His program failed to gain support. He was an indifferent expositor and an inflexible advocate, contending against the great prestige of Hilbert and the undeniable fact that idealistic mathematics produced the most general results with the least effort. More important, Brouwer's system itself had traces of idealism and, worse, of metaphysical speculation. There was a preoccupation with the philosophical aspects of constructivism at the expense of concrete mathematical activity. A calculus of negation was developed which became a crutch to avoid the necessity of getting precise constructive results. It is not surprising that some of Brouwer's precepts were then formalized, giving rise to so-called intuitionistic number theory, and that the formal system so obtained turned out not to be of any constructive value. In fairness to Brouwer it should be said that he did not associate himself with these efforts to formalize reality; it is the fault of the logicians that many mathematicians who think they know something of the constructive point of view have in mind a dinky formal system or, just as bad, confuse constructivism with recursive function theory.

Brouwer became involved in metaphysical speculation by his desire to improve the theory of the continuum. A bugaboo of both Brouwer and the logicians has been compulsive speculation about the nature of the continuum. In the case of the logicians this leads to contortions in which various formal systems, all detached from reality, are interpreted within one another in the hope that the nature of the continuum will somehow emerge. In Brouwer's case there seems to have been a nagging suspicion that unless he personally intervened to prevent it the continuum would turn out to be discrete. He therefore introduced the method of free-choice sequences for constructing the continuum, as a consequence of which the continuum cannot be discrete because it is not well enough defined. This makes mathematics so bizarre it becomes unpalatable to mathematicians, and foredooms the whole of Brouwer's program. This is a pity, because Brouwer had a remarkable insight into the defects of classical mathematics, and he made a heroic attempt to set things right.

3. THE CONSTRUCTIVIZATION OF MATHEMATICS

A set is defined by describing exactly what must be done in order to construct an element of the set and what must be done in order to show that two elements are equal. There is no guarantee that the description will be understood; it may be that an author thinks he has described a set with sufficient clarity but a reader does not understand. As an illustration consider the set of all sequences $\{n_k\}$ of integers. To construct such a sequence we must give a rule which associates an integer n_k to each positive integer k in such a way that for each value of k the associated integer n_k can be determined in a finite number of steps by an entirely routine process. Now this definition could perhaps be interpreted to admit sequences $\{n_k\}$ in which n_k is constructed by a search, the proof that the search actually produces a value of n_k after a finite number of steps being given in some formal system. Of course, we do not have this interpretation in mind, but it is impossible to consider every possible interpretation of our definition and say whether that is what we have in mind. There is always ambiguity, but it becomes less and less as the reader continues to read and discovers more and more of the author's intent, modifying his interpretations if necessary to fit the intentions of the author as they continue to unfold. At any stage of the exposition the reader should be content if he can give a reasonable interpretation to account for everything the author has said. The expositor himself can never fully know all the possible ramifications of his definitions, and he is subject to the same necessity of modifying his interpretations, and sometimes his definitions as well, to conform to the dictates of experience.

The constructive interpretations of the mathematical connectives and quantifiers have been established by Brouwer.

To prove the statement (P and Q) we must prove the statement P and prove the statement Q, just as in classical mathematics. To prove the statement (P or Q) we must either prove the statement P or prove the statement Q, whereas in classical mathematics it is possible to prove (P or Q) without proving either the statement P or the statement Q.

The connective "implies" is more complicated. To prove (P implies Q) we must show that P necessarily entails Q, or that Q is true whenever P is true. The validity of the computational facts implicit in the statement P must insure the validity of the computational facts implicit in the statement Q, but the way this actually happens can only be seen by looking at the proof of the statement (P implies Q). Statements formed with this connective, for example, statements of the type ((P implies Q) implies R), have a less immediate meaning than the statements from which they are formed, although in actual practice this does not seem to lead to difficulties in interpretation.

The negation (not P) of a statement P is the statement (P implies $0 = 1$). Classical mathematics makes no distinction between the content of the statements P and not (not P), whereas constructively the latter is a weaker statement. Brouwer's system makes essential use of negation in defining, for instance, inequality and set complementation. Thus two elements of a set A are unequal according to Brouwer if the assumption of their equality somehow allows us to compute that $0 = 1$. It is natural to want to replace this negativistic definition by something more affirmative, phrased as much as possible in terms of specific computations leading to specific results. Brouwer himself does just this for the real number system, introducing an affirmative and stronger relation of inequality in addition to the negativistic relation already defined. Experience shows that it is not necessary to define inequality in terms of negation. For those cases in which an inequality relation is needed, it is better to introduce it affirmatively. The same remarks apply to set complementation.

Van Dantzig and others have gone as far as to propose that negation could be entirely avoided in constructive mathematics. Experience bears this out. In many cases where we seem to be using negation—for instance, in the assertion that either a given integer is even or it is not— we are really asserting that one of two finitely distinguishable alternatives actually obtains. Without intending to establish a dogma, we may continue to employ the language of negation but reserve it for situations of this sort, at least until experience changes our minds, and for counter-examples and purposes of motivation. This will have the advantage of making mathematics more immediate and in certain situations forcing us to sharpen our results.

Proofs by contradiction are constructively justified in finite situations. When we have proved that one of finitely many alternatives holds at a certain stage in the proof of a theorem, to finish the proof of the theorem it is enough to show that the theorem is a consequence of each of the alternatives. Should one of the alternatives lead to a contradiction, that is, imply $0 = 1$, we may either say that the alternative in question is ruled out and pass on to the consideration of the other alternatives, or we may be more meticulous and prove that the theorem is a consequence of the equality $0 = 1$.

A universal statement, to the effect that every element of a certain set A has a certain property P, has the same meaning in constructive as in classical mathematics. To prove such a statement we must show by some general argument that if x is any element of A, then x has property P.

Constructive existence is much more restrictive than the ideal existence of classical mathematics. The only way to show that an object exists is to give a finite routine for finding it, whereas in classical mathematics other methods can be used. In fact the following principle is valid in classical mathematics: *Either all elements of A have property P or there exists an element of A with property not P.* This principle, which we shall call the *principle of omniscience,* lies at the root of most of the unconstructivities of classical mathematics. This is already true of the principle of omniscience in its simplest form: if $\{n_k\}$ is a sequence of integers, then either $n_k = 0$ for some k or $n_k \neq 0$ for all k. We shall call this the *limited principle of omniscience.* Theorem after theorem of classical mathematics depends in an essential way on the limited principle of omniscience, and is therefore not constructively valid. Some instances of this are the theorem that a continuous real-valued function on a closed bounded interval attains its maximum, the fixed-point theorem for a continuous map of a closed cell into itself, the ergodic theorem, and the Hahn-Banach theorem. Nevertheless these theorems are not lost to constructive mathematics. Each of these theorems P has a constructive substitute Q, which is a constructively valid theorem Q implying P in the classical system by a more or less simple argument based on the limited principle of omniscience. For instance, the statement that every continuous function from a closed cell in euclidean space into itself admits a fixed point finds a constructive substitute in the statement that such a function admits a point which is arbitrarily near to its image.

The extent to which good constructive substitutes exist for the theorems of classical mathematics can be regarded as a demonstration that classical mathematics has a substantial underpinning of constructive truth.

When a classical mathematician claims he is a constructivist, he probably means he avoids the axiom of choice. This axiom is unique in its ability to trouble the conscience of the classical mathematician, but in fact it is not a real source of the unconstructivities of classical mathematics. A choice function exists in constructive mathematics, because a choice is *implied by the very meaning of existence.* Applications of the axiom of choice in classical mathematics either are irrelevant or are combined with a sweeping appeal to the principle of omniscience. The axiom of choice is used to extract elements from equivalence classes where they should never have been put in the first place. For instance, a real number should not be defined as an equivalence class of Cauchy sequences of rational numbers; there is no need to drag in the equivalence classes. The proof that the real numbers can be well ordered is an instance of a proof in which a sweeping use of the principle of omniscience is combined with an appeal to the axiom of choice. Such proofs offer little hope of constructivization. It is not likely that the theorem "the real numbers can be well ordered" will be given a constructive version consonant with the intuitive interpretation of the classical result.

Almost every conceivable type of resistance has been offered to a straightforward realistic treatment of mathematics, even by constructivists. Brouwer, who has done more for constructive mathematics than anyone else, thought it necessary to introduce a revolutionary, semimystical theory of the continuum. Weyl, a great mathematician who in practice suppressed his constructivist convictions, expressed the opinion that idealistic mathematics finds its justification in its applications to physics. Hilbert, who insisted on constructivity in metamathematics but believed the price of a constructive mathematics was too great, was willing to settle for consistency. Brouwer's disciples joined forces with the logicians in attempts to formalize constructive mathematics. Others seek constructive truth in the framework of recursive function theory. Still others look for a short cut to reality, a point of vantage which will suddenly reveal classical mathematics in a constructive light. None of these substitutes for a straightforward realistic approach has worked. It is no exaggeration to say that a straightforward realistic approach to mathematics has yet to be tried. It is time to make the attempt.

CALCULUS AND THE REAL NUMBERS

Section 1 establishes some conventions regarding sets and functions. The next three sections are devoted to constructing the real numbers, as certain Cauchy sequences of rational numbers, and investigating their order and arithmetic. The rest of the chapter concerns the basic ideas of the calculus of one variable. Topics covered include continuity, the convergence of sequences and series of continuous functions, differentiation, integration, Taylor's theorem, and basic properties of the exponential and trigonometric functions and their inverses. Most of the material is a routine constructivization of the corresponding part of classical mathematics. For this reason it affords a good introduction to the constructive approach.

11

The reader is assumed to be familiar with the order and the arithmetic of the integers and the rational numbers. For us a rational number will be an expression of the form p/q, where p and q are integers with $q \neq 0$. The rational numbers p/q and p_1/q_1 are *equal* if $pq_1 = p_1q$. The integer n is identified with the rational number $n/1$.

There are geometric magnitudes that are not represented by rational numbers, that can only be described by a sequence of rational approximations. Certain such approximating sequences are called real numbers. In this chapter we construct the real numbers and study their basic properties. Then we develop the fundamental ideas of calculus.

1. SETS AND FUNCTIONS

Before constructing the real numbers, we introduce some notions which are basic to much of mathematics.

A *sequence* is a rule which associates to each positive integer n a mathematical object a_n. The object a_n is called the nth *term* of the sequence. The rule can be given explicitly or it can be left to inference by writing, for instance, the elements of the sequence in order

$$\{a_1, a_2, a_3, \ . \ . \ .\}$$

until the rule of formation is clear. Different notations for the sequence whose nth term is a_n are $n \rightarrow a_n$, $\{a_1, a_2, a_3, \ . \ . \ .\}$, $\{a_n\}_{n=1}^{\infty}$, and $\{a_n\}$. Thus the sequence whose nth term is n^2 can be written $n \rightarrow n^2$, or $\{1, 9, 16, \ . \ . \ .\}$, or $\{n^2\}_{n=1}^{\infty}$, or simply $\{n^2\}$.

The *cartesian product*, or simply *product*,

$$X \equiv X_1 \times \cdot \ \cdot \ \cdot \ \times X_n$$

of sets $X_1, X_2, \ . \ . \ . \ , X_n$ is defined to be the set of all ordered n-tuples $(x_1, \ . \ . \ . \ , x_n)$ with $x_1 \in X_1, \ . \ . \ . \ , x_n \in X_n$. Elements $(x_1, \ . \ . \ . \ , x_n)$ and $(y_1, \ . \ . \ . \ , y_n)$ of X are *equal* if the *coordinates* x_i and y_i are equal for each i.

The cartesian product $Z \times Z$ of the set Z of integers with itself can be arranged in a sequence, as follows. We order the elements (m,n) of $Z \times Z$, first according to the value of $|m| + |n|$, then according to the value of m, and finally according to the value of n. This produces the sequence

$$(1.1) \quad \{(0,0), (-1,0), (0,-1), (0,1), (-2,0), (-1,-1), \ . \ . \ .\}$$

in which each element of $Z \times Z$ occurs exactly once. In the sequence (1.1) we omit every term (m,n) with $n = 0$ and replace each term (m,n) with $n \neq 0$ by m/n, producing the sequence

$$(1.2) \qquad \left\{ \frac{0}{-1}, \quad \frac{0}{1}, \quad \frac{-1}{-1}, \cdots \right\}$$

in which every expression p/q, with p and q integers and $q \neq 0$, occurs exactly once. Keeping only the term $0/1$ of (1.2), and those terms for which $q > 0$, $p \neq 0$, and p is relatively prime to q, we obtain a sequence

$$(1.3) \qquad \left\{ \frac{0}{1}, \quad \frac{-1}{1}, \quad \frac{1}{1}, \cdots \right\}$$

which has the property that for any given rational number r there exists exactly one term a_n with $a_n = r$.

The totality of all mathematical objects constructed in accord with certain requirements is called a *set*. The requirements of the construction, which vary with the set under consideration, determine the set. Thus the integers are a set, the rational numbers are a set, and the collection of all sequences each of whose terms is an integer is a set. Each set A will be endowed with a relation $=$ of equality. This relation is a matter of convention, except that it must be an *equivalence relation*. This means that the following properties must hold for all elements a, b, and c of A.

$$(1.4)$$

        (i) $a = a$

      (ii) $b = a$ if $a = b$

    (iii) $a = c$ if $a = b$ and $b = c$

The relation $pq_1 = p_1 q$ of equality for rational numbers p/q and p_1/q_1 is an equivalence relation. In this instance there is a finite, mechanical process for deciding whether two given rational numbers are equal. Such a process may not always exist. There are instances in which the problem of whether two given elements of a given set are equal is a nontrivial mathematical problem that we are unable to solve. Such an instance, in the theory of the real numbers, will be given later.

We use the standard notation $a \in A$ to say that a is an element or member of the set A, or that the construction defining a satisfies the requirements a construction must satisfy in order to define an element of the set A.

The dependence of one quantity on another is expressed in the basic notion of an operation. An *operation* from a set A to a set B is a rule f which assigns an element $f(a)$ of B to each element a of A. The rule must afford an explicit, finite, mechanical reduction of the procedure for constructing $f(a)$ to the procedure for constructing a. The set A is called the *domain* of f. The most important case occurs when

$$f(a_1) = f(a_2)$$

whenever a_1 and a_2 are equal elements of A. Such an operation f is called a *function*. A function is sometimes called a *map* or a *mapping*. A function whose domain is the set Z^+ of positive integers is, as above, called a *sequence*.

A function f is sometimes written in the form $a \rightarrow f(a)$, so that $n \rightarrow n^2$ is the function f from Z^+ to Z^+ defined by $f(n) \equiv n^2$.

The notation $f: A \rightarrow B$ indicates that f is a function from A to B. If $f_1: A \rightarrow B$ and $f_2: A \rightarrow B$, then $f_1 = f_2$ means that $f_1(a) = f_2(a)$ for all elements a of A.

Functions $f: A \rightarrow B$ and $g: B \rightarrow C$ can be composed, giving a function $g \circ f: A \rightarrow C$ defined by

$$(g \circ f)(a) \equiv g(f(a))$$

Composition is associative,

$$h \circ (g \circ f) = (h \circ g) \circ f$$

whenever the compositions are well defined. If

$$f: A \rightarrow B \qquad g: B \rightarrow A$$
$$g(f(a)) = a$$

and

for all a in A, then g is called a *left inverse* of f, and f is called a *right inverse* of g. When g is both a right and a left inverse of f, it is called an *inverse* of f. A function $f: A \rightarrow B$ which has an inverse is called a *one-one correspondence*, or a *bijection*, and the sets A and B are said to be in *one-one correspondence* or to be *equipollent*.

A set which is equipollent with the integers is said to be *countably infinite*. As an example, let f be the sequence (1.3). Define the function g from the set Q of rational numbers to Z^+ by $g(r) = n$, where n is the unique positive integer for which $f(n) = r$. Then g is an inverse of f. The set Q is therefore countably infinite. A similar proof shows that $Z \times Z$ is countably infinite.

A function $f: A \to B$ which has a right inverse is said to map A *onto* B. When f satisfies the weaker condition that there exists an operation g from B to A such that $f(g(b)) = b$ for all b in B, it is called *surjective* and said to be a *surjection*. A set A is *countable* if there exists a surjection $f: Z^+ \to A$. Intuitively, this means that the elements of A can be arranged in a sequence with possible duplications.

For each positive integer n let Z_n be the set $\{0, 1, \ldots, n - 1\}$. A set A which is equipollent with Z_n is said to have n elements. Such sets are called *finite*. Every finite set is countable. Contrary to customary usage, every finite or countable set has at least one element.

If there is a surjection from Z_n to a set A, we say that A has *at most n* elements. Such sets A are called *subfinite*.

It is not true that every countable set is either countably infinite or subfinite. As an example let A consist of all positive integers n such that both n and $n + 2$ are prime. Then A is countable, but it is neither countably infinite nor subfinite. This does not mean that sometime in the future A will not have become countably infinite or subfinite. It is possible that tomorrow someone will show that A is subfinite.

This set A has the property that if it is subfinite, then it is finite. Not all sets have this property.

2. THE REAL NUMBER SYSTEM

The following definition is basic to everything that follows.

Definition 1 A sequence $\{x_n\}$ of rational numbers is *regular* if

$$(2.1) \qquad |x_m - x_n| \leq m^{-1} + n^{-1} \qquad (m, n \in Z^+)$$

A *real number* is a regular sequence of rational numbers. Two real numbers $x \equiv \{x_n\}$ and $y \equiv \{y_n\}$ are *equal* if

$$(2.2) \qquad |x_n - y_n| \leq 2n^{-1} \qquad (n \in Z^+)$$

The set of real numbers is denoted by **R**.

Proposition 1 *Equality of real numbers is an equivalence relation.*

Proof Parts (i) and (ii) of (1.4) are obvious. Part (iii) is a consequence of the following lemma.

Lemma 1 *The real numbers $x \equiv \{x_n\}$ and $y \equiv \{y_n\}$ are equal if and only if for each positive integer j there exists a positive integer N_j such that*

$$(2.3) \qquad\qquad |x_n - y_n| \le j^{-1} \qquad (n \ge N_j)$$

Proof If $x = y$, then (2.3) holds with $N_j \equiv 2j$

Assume conversely that for each j in Z^+ there exists N_j satisfying (2.3). Consider a positive integer n. If m and j are any positive integers with $m \ge \max \{j, N_j\}$, then

$$|x_n - y_n| \le |x_n - x_m| + |x_m - y_m| + |y_m - y_n|$$
$$\le (n^{-1} + m^{-1}) + j^{-1} + (n^{-1} + m^{-1}) < 2n^{-1} + 3j^{-1}$$

Since this holds for all j in Z^+, (2.2) is valid.

Notice that the proof of Lemma 1 singles out a specific N_j satisfying (2.3). This is true of all our existence proofs, implicitly if not explicitly.

The rational number x_n is called the *nth rational approximation* to the real number $x \equiv \{x_n\}$. Note that the operation from $\mathbf{R}$ to Q which takes the real number x into its nth rational approximation is not a function.

For later use we wish to associate to each real number $x \equiv \{x_n\}$ an integer K_x such that

$$|x_n| < K_x \qquad (n \in Z^+)$$

This is done by letting K_x be the least integer which is greater than $|x_1| + 2$. We call K_x the *canonical bound* for x.

The development of the arithmetic of the real numbers offers no surprises: we operate with real numbers by operating with their rational approximations.

Definition 2 Let $x \equiv \{x_n\}$ and $y \equiv \{y_n\}$ be real numbers with respective canonical bounds K_x and K_y. Write

$$k \equiv \max \{K_x, K_y\}$$

Let α be any rational number. We define

$$(2.4) \qquad\qquad
\begin{aligned}
&(a) \quad x + y \equiv \{x_{2n} + y_{2n}\}_{n=1}^{\infty} \\
&(b) \quad xy \equiv \{x_{2kn}y_{2kn}\}_{n=1}^{\infty} \\
&(c) \quad \max \{x, y\} \equiv \{\max \{x_n, y_n\}\}_{n=1}^{\infty} \\
&(d) \quad -x \equiv \{-x_n\}_{n=1}^{\infty} \\
&(e) \quad \alpha^* \equiv \{\alpha, \alpha, \alpha, \ldots\}
\end{aligned}$$

Proposition 2 *The sequences $x + y$, xy, $\max\{x,y\}$, $-x$, and α^* of Definition 2 are real numbers.*

Proof

(a) Write $z_n \equiv x_{2n} + y_{2n}$. Then $x + y \equiv \{z_n\}$. For all positive integers m and n,

$$
\begin{aligned}
|z_m - z_n| &\leq |x_{2m} - x_{2n}| + |y_{2m} - y_{2n}| \\
&\leq (2n)^{-1} + (2m)^{-1} + (2n)^{-1} + (2m)^{-1} \leq n^{-1} + m^{-1}
\end{aligned}
$$

Thus $x + y$ is a real number.

(b) Write $z_n \equiv x_{2kn}y_{2kn}$. Then $xy \equiv \{z_n\}$. For all positive integers m and n,

$$
\begin{aligned}
|z_m - z_n| &= |x_{2km}(y_{2km} - y_{2kn}) + y_{2kn}(x_{2km} - x_{2kn})| \\
&\leq k|y_{2km} - y_{2kn}| + k|x_{2km} - x_{2kn}| \\
&\leq k((2km)^{-1} + (2kn)^{-1} + (2km)^{-1} + (2kn)^{-1}) = n^{-1} + m^{-1}
\end{aligned}
$$

Thus xy is a real number.

(c) Write $z_n \equiv \max\{x_n,y_n\}$. Then $\max\{x,y\} \equiv \{z_n\}$.
Consider positive integers m and n. For simplicity assume that

$$
x_m = \max\{x_m,x_n,y_m,y_n\}
$$

Then
$$
\begin{aligned}
|z_m - z_n| &= |x_m - \max\{x_n,y_n\}| \\
&= x_m - \max\{x_n,y_n\} < x_m - x_n < n^{-1} + m^{-1}
\end{aligned}
$$

Thus $\max\{x,y\}$ is a real number.

(d) For all positive integers m and n,

$$
|-x_m - (-x_n)| = |x_m - x_n| \leq m^{-1} + n^{-1}
$$

Thus $-x$ is a real number.

(e) This is obvious.

There is no trouble in proving that $(x,y) \to x + y$, $(x,y) \to xy$, and $(x,y) \to \max\{x,y\}$ are functions from $\mathbf{R} \times \mathbf{R}$ to $\mathbf{R}$, that $x \to -x$ is a function from $\mathbf{R}$ to $\mathbf{R}$, and that $\alpha \to \alpha^*$ is a function from Q to $\mathbf{R}$.

The operation

$$
(2.5) \qquad\qquad x \to |x| \equiv \max\{x,-x\}
$$

is therefore a function from $\mathbf{R}$ to $\mathbf{R}$, and the operation

$$(2.6) \qquad (x,y) \rightarrow \min \{x,y\} \equiv -\max \{-x,-y\}$$

is a function from $\mathbf{R} \times \mathbf{R}$ to $\mathbf{R}$.

The next proposition states that the real numbers obey the same rules of arithmetic as the rational numbers.

Proposition 3 *For arbitrary real numbers* x, y, *and* z *and rational numbers* α *and* β,

$$(2.7) \quad
\begin{aligned}
&(a) \quad x + y = y + x,\ xy = yx \\
&(b) \quad (x + y) + z = x + (y + z),\ x(yz) = (xy)z \\
&(c) \quad x(y + z) = xy + xz \\
&(d) \quad 0^* + x = x,\ 1^* \cdot x = x \\
&(e) \quad x - x = 0^* \\
&(f) \quad |xy| = |x|\,|y| \\
&(g) \quad (\alpha + \beta)^* = \alpha^* + \beta^*,\ (\alpha\beta)^* = \alpha^*\beta^*,\ \text{and}\ (-\alpha)^* = -\alpha^*
\end{aligned}$$

We omit the simple proofs of these results.

There are three basic relations defined on the set of real numbers. The first of these, the equality relation, has already been defined. The remaining relations, which pertain to order, are best introduced in terms of certain subsets $\mathbf{R}^+$ and $\mathbf{R}^{0+}$ of $\mathbf{R}$.

Definition 3 A real number $x \equiv \{x_n\}$ is said to be *positive*, or $x \in \mathbf{R}^+$, if

$$(2.8) \qquad\qquad x_n > n^{-1}$$

for some n in Z^+. A real number $x \equiv \{x_n\}$ is said to be *nonnegative*, or $x \in \mathbf{R}^{0+}$, if

$$(2.9) \qquad\qquad x_n \geq -n^{-1} \qquad (n \in Z^+)$$

The following criteria are often useful.

Lemma 2 *A real number* $x \equiv \{x_n\}$ *is positive if and only if there exists a positive integer* N *such that*

$$(2.10) \qquad\qquad x_m \geq N^{-1} \qquad (m \geq N)$$

A real number $x \equiv \{x_n\}$ is nonnegative if and only if for each n in Z^+ there exists N_n in Z^+ such that

$$(2.11) \qquad\qquad x_m \geq -n^{-1} \qquad (m \geq N_n)$$

Proof Assume that $x \in \mathbf{R}^+$. Then $x_n > n^{-1}$ for some n in Z^+. Choose N in Z^+ with

$$2N^{-1} \leq x_n - n^{-1}$$

Then
$$\begin{aligned}
x_m &\geq x_n - |x_m - x_n| \geq x_n - m^{-1} - n^{-1}\\
&\geq x_n - n^{-1} - N^{-1} > N^{-1}
\end{aligned}$$

whenever $m \geq N$. Therefore (2.10) is valid.

Conversely, if (2.10) is valid, then (2.8) holds with $n = N + 1$. Therefore $x \in \mathbf{R}^+$.

Assume next that $x \in \mathbf{R}^{0+}$. Then for each positive integer n,

$$x_m \geq -m^{-1} \geq -n^{-1} \qquad (m \geq n)$$

Therefore (2.11) is valid.

Assume finally that (2.11) holds. Then if k, m, and n are positive integers with $m \geq N_n$, we have

$$x_k \geq x_m - |x_m - x_k| \geq -n^{-1} - k^{-1} - m^{-1}$$

Since m and n are arbitrary, this gives $x_k \geq -k^{-1}$. Therefore $x \in \mathbf{R}^{0+}$.

As a corollary of Lemma 2, we see that if x and y are equal real numbers, then x is positive if and only if y is positive, and x is nonnegative if and only if y is nonnegative.

It is not strictly correct to say that a real number $\{x_n\}$ *is* an element of $\mathbf{R}^+$. An element of $\mathbf{R}^+$ consists of a real number $\{x_n\}$ *and* a positive integer n, such that $x_n > n^{-1}$, because an element of $\mathbf{R}^+$ is not presented until both $\{x_n\}$ and n are given. One and the same real number $\{x_n\}$ can be associated with two distinct (but equal) elements of $\mathbf{R}^+$. Nevertheless we shall continue to refer loosely to a positive real number $\{x_n\}$. On those occasions when we need to refer to an n for which $x_n > n^{-1}$, we shall take the position that it was there implicitly all along.

The proof of the following proposition is now easy, and will be left to the reader. For convenience $\mathbf{R}^*$ represents either $\mathbf{R}^+$ or $\mathbf{R}^{0+}$.

Proposition 4 *Let x and y be real numbers. Then*

$$(2.12) \quad (a) \quad x + y \in \mathbf{R}^* \text{ and } xy \in \mathbf{R}^* \quad \textit{whenever } x \in \mathbf{R}^* \textit{ and } y \in \mathbf{R}^*$$

$$(b) \quad x + y \in \mathbf{R}^+ \quad \textit{whenever } x \in \mathbf{R}^+ \textit{ and } y \in \mathbf{R}^{0+}$$

$$(c) \quad |x| \in \mathbf{R}^{0+}$$

$$(d) \quad \max\{x,y\} \in \mathbf{R}^* \quad \textit{whenever } x \in \mathbf{R}^*$$

$$(e) \quad \min\{x,y\} \in \mathbf{R}^* \quad \textit{whenever } x \in \mathbf{R}^* \textit{ and } y \in \mathbf{R}^*$$

We now define the order relations on $\mathbf{R}$.

Definition 4 Let x and y be real numbers. We define

$$(2.13) \qquad x > y \ (\text{or } y < x) \qquad \text{if } x - y \in \mathbf{R}^+$$

and

$$(2.14) \qquad x \geq y \ (\text{or } y \leq x) \qquad \text{if } x - y \in \mathbf{R}^{0+}$$

A real number x is *negative* if $x < 0^*$, that is, if $-x$ is positive. Consider real numbers x, x', y, and y' with

$$(2.15) \qquad x = x' \qquad y = y' \qquad \text{and} \qquad x > y$$

Then
$$x' - y' = x - y \in \mathbf{R}^+$$

Therefore

$$(2.16) \qquad\qquad x' > y'$$

We express the fact that (2.16) holds whenever (2.15) is valid by saying that $>$ is a *relation* on $\mathbf{R}$. We express the fact that $x > y$ if and only if $y < x$ by saying that $>$ and $<$ are *transposed relations*. Similarly, $\leq$ and $\geq$ are transposed relations.

If $x < y$ or $x = y$ then $x \leq y$. The converse is not valid: As we shall see later, it is possible that we have $x \leq y$ without being able to prove that $x < y$ or $x = y$. For this reason it was necessary to define the relations $<$ and $\leq$ independently of each other.

The following rules for manipulating inequalities are easily proved from Proposition 4. We omit the proofs.

Proposition 5 *For all real numbers x, y, z, and t,*

$$(2.17)$$

(a) $x < z$ *whenever $x < y$ and $y \leq z$ or $x \leq y$ and $y < z$*
(b) $x \leq z$ *whenever $x \leq y$ and $y \leq z$*
(c) $x + y \leq z + t$ *whenever $x \leq z$ and $y \leq t$*
(d) $x + y < z + t$ *whenever $x \leq z$ and $y < t$*
(e) $xy \leq zy$ *whenever $x \leq z$ and $y \geq 0^*$*
(f) $xy < zy$ *whenever $x < z$ and $y > 0^*$*
(g) $-x > -y$ *if $x < y$*
(h) $-x \geq -y$ *if $x \leq y$*
(i) $\max \{x,y\} \geq x$
(j) $\min \{x,y\} \leq x$
(k) $x = y$ *if $x \leq y$ and $y \leq x$*
(l) $|x| \geq 0^*$
(m) $|x + y| \leq |x| + |y|$

An important property of the relation $<$, of which we shall make no use, is the antisymmetry property, which states that at most one of the relations $x < y$ and $y < x$ is valid for given real numbers x and y. This negative statement has no place in the affirmative mathematics we are trying to develop, except as motivation. Its place is taken by the affirmative statement $(2.17)(k)$. As a general principle, negative statements are only for counterexamples and motivation; they are not to be used in subsequent work.

Definition 5 For real numbers x and y we write $x \neq y$ if and only if $x < y$ or $x > y$.

Inequality $\neq$ is a relation because both $<$ and $>$ are relations. As motivation we have the negative statement that at most one of the relations $x \neq y$, $x = y$ can hold for given real numbers x and y. In other words, at most one of the relations $x < y$, $x > y$, $x = y$ can hold. This is clear from the definitions.

The following proposition defines the inverse x^{-1} of a real number $x \neq 0^*$, and derives the basic properties of the operation $x \to x^{-1}$.

Proposition 6 *Let x be a nonzero real number (so that $|x| \in \mathbf{R}^+$). There exists a positive integer N with $|x_m| \geq N^{-1}$ for $m \geq N$. Define*

$$(2.18) \qquad\qquad y_n \equiv (x_{N^3})^{-1} \qquad (n < N)$$

and

$$(2.19) \qquad\qquad y_n \equiv (x_{nN^2})^{-1} \qquad (n \geq N)$$

Then

$$(2.20) \qquad\qquad x^{-1} \equiv \{y_n\}$$

is a real number which is positive if x is positive and negative if x is negative, and $xx^{-1} = 1^*$.

If t is any real number for which $xt = 1^*$, then $t = x^{-1}$. The operation $x \to x^{-1}$ is a function. If $x \neq 0$ and $y \neq 0$, then $(xy)^{-1} = x^{-1}y^{-1}$. If $\alpha \neq 0$ is rational, then $(\alpha^*)^{-1} = (\alpha^{-1})^*$. If $x \neq 0$, then $(x^{-1})^{-1} = x$.

Proof   Our definitions guarantee that $|y_n| \leq N$ for all n.

Consider positive integers m and n. Write

$$j \equiv \max \{m, N\}, \; k \equiv \max \{n, N\}$$

Then
$$
\begin{aligned}
|y_m - y_n| &= |y_m| \, |y_n| \, |x_{jN^2} - x_{kN^2}| \\
&\leq N^2((jN^2)^{-1} + (kN^2)^{-1}) = j^{-1} + k^{-1} \leq m^{-1} + n^{-1}
\end{aligned}
$$

Therefore x^{-1} is a real number.

Assume now that $x > 0^*$. Then $x_n > 0$ for all n sufficiently large. Hence $y_n > K_x^{-1}$ (where K_x is the canonical bound for x) for all n sufficiently large. It follows that $x^{-1} > 0^*$. A similar proof shows that $x^{-1} < 0^*$ whenever $x < 0^*$.

Let k be the maximum of the canonical bounds for x and x^{-1}. Write $xx^{-1} \equiv \{z_n\}$. Then

$$z_n \equiv x_{2nk} y_{2nk} \equiv x_{2nk}(x_{2nN^2k})^{-1} \qquad (n \geq N)$$

Therefore
$$
\begin{aligned}
|z_n - 1^*| &= |x_{2nN^2k}|^{-1}|x_{2nk} - x_{2nN^2k}| \\
&\leq |y_{2nk}|((2nk)^{-1} + (2nN^2k)^{-1}) \leq n^{-1}
\end{aligned}
$$

for $n \geq N$. It follows that $xx^{-1} = 1^*$.

If t is any real number with $xt = 1^*$, then

$$x^{-1} = x^{-1}(xt) = (x^{-1}x)t = (xx^{-1})t = t$$

If $x = x'$, then

$$x'x^{-1} = xx^{-1} = 1^*$$

Therefore $x^{-1} = (x')^{-1}$. It follows that $x \to x^{-1}$ is a function.

If $x \neq 0$ and $y \neq 0$, then

$$(xy)x^{-1}y^{-1} = xx^{-1}yy^{-1} = 1^*$$

Therefore $x^{-1}y^{-1} = (xy)^{-1}$.

If $\alpha \neq 0$ is rational, then $\alpha^* \equiv \{\alpha, \alpha, \ldots\}$. Therefore

$$(\alpha^*)^{-1} = \{\alpha^{-1}, \alpha^{-1}, \ldots\} = (\alpha^{-1})^*$$

For each x in $\mathbf{R}^+$, x^{-1} is in $\mathbf{R}^+$, and thus $(x^{-1})^{-1}$ exists. Since $x^{-1}x = xx^{-1} = 1^*$, it follows that $(x^{-1})^{-1} = x$. Similarly $(x^{-1})^{-1} = x$ if x is negative. Therefore $(x^{-1})^{-1} = x$ whenever $x \neq 0$.

As the previous propositions show, $(\alpha\beta)^* = \alpha^*\beta^*$, $(\alpha + \beta)^* = \alpha^* + \beta^*$, $(-\alpha)^* = -\alpha^*$, $(|\alpha|)^* = |\alpha^*|$, and $(\alpha^{-1})^* = (\alpha^*)^{-1}$ for all rational numbers α and β. Also $\alpha \bigtriangleup \beta$ if and only if $\alpha^* \bigtriangleup \beta^*$, where $\bigtriangleup$ stands for any of the relations $=$, $<$, $>$, and $\neq$. This situation is expressed by saying that the map $\alpha \rightarrow \alpha^*$ is an *order isomorphism* from Q into $\mathbf{R}$. This justifies identifying Q with a subset of $\mathbf{R}$, as we previously identified Z with a subset of Q. Henceforth we make no distinction between a rational number α and the corresponding real number α^*.

The next lemma shows that the nth rational approximation x_n to a real number $x \equiv \{x_n\}$ actually approximates x to within n^{-1}.

Lemma 3   *For each real number $x \equiv \{x_n\}$, we have*

$$(2.21) \qquad\qquad |x - x_n| \leq n^{-1} \qquad (n \in Z^+)$$

Proof   By (2.8) (a), (c), (d), (e) and (2.9), the mth rational approximation to $n^{-1} - |x - x_n|$ is

$$n^{-1} - |x_{4m} - x_n| \geq n^{-1} - ((4m)^{-1} + n^{-1}) = -(4m)^{-1} > -m^{-1}$$

By Definition 3, we have $n^{-1} - |x - x_n| \in \mathbf{R}^{0+}$. Therefore $|x - x_n| \leq n^{-1}$.

Lemma 4   *If $x \equiv \{x_n\}$ and $y \equiv \{y_n\}$ are real numbers, with $x < y$, there exists a rational number α with $x < \alpha < y$.*

Proof   By Definition 2, we have $y - x \equiv \{y_{2n} - x_{2n}\}_{n=1}^{\infty}$. Since $y - x \in \mathbf{R}^+$, by Definition 3 there exists n in Z^+ with $y_{2n} - x_{2n} > n^{-1}$.

Write

$$\alpha \equiv \tfrac{1}{2}(x_{2n} + y_{2n})$$

Then $\alpha - x \geq \alpha - x_{2n} - |x_{2n} - x| \geq \tfrac{1}{2}(y_{2n} - x_{2n}) - (2n)^{-1} > 0$

Also, $y - \alpha \geq y_{2n} - \alpha - |y_{2n} - y| \geq \tfrac{1}{2}(y_{2n} - x_{2n}) - (2n)^{-1} > 0$

Therefore $x < \alpha < y$.

As a corollary, for each x in $\mathbf{R}$ and r in $\mathbf{R}^+$ there exists α in Q with $|x - \alpha| < r$. Here is another corollary.

Proposition 7 *If $x_1, \ldots, x_n$ are real numbers with $x_1 + \cdots + x_n > 0$, then $x_i > 0$ for some i $(1 \leq i \leq n)$.*

Proof By Lemma 4 there exists a rational number α with $0 < \alpha < x_1 + \cdots + x_n$. Let a_i $(1 \leq i \leq n)$ be a rational number with

$$|x_i - a_i| < (2n)^{-1}\alpha$$

Then $\displaystyle\sum_{i=1}^{n} a_i \geq \sum_{i=1}^{n} x_i - \sum_{i=1}^{n} |x_i - a_i| > \tfrac{1}{2}\alpha$

Therefore $a_i > (2n)^{-1}\alpha$ for some i. It follows that

$$x_i \geq a_i - |x_i - a_i| > 0$$

Corollary *If $x, y,$ and z are real numbers with $y < z$, then either $x < z$ or $x > y$.*

Proof Since $z - x + x - y = z - y > 0$, either $z - x > 0$ or $x - y > 0$, by Proposition 7.

The next lemma gives an extremely useful method for proving inequalities of the form $x \leq y$.

Lemma 5 *Let x and y be real numbers such that the assumption $x > y$ implies that $0 = 1$. Then $x \leq y$.*

Proof Without loss of generality, we take $y = 0$. For each n in Z^+, either $x_n \leq n^{-1}$ or $x_n > n^{-1}$. The case $x_n > n^{-1}$ is ruled out, since it implies $x > 0$. Therefore $-x_n \geq -n^{-1}$, for all n, so that $-x \geq 0$.

Theorem 1 *Let $\{a_n\}$ be a sequence of real numbers. Let x_0 and y_0 be real numbers, $x_0 < y_0$. Then there exists a real number x with*

$$(2.22) \qquad\qquad x_0 \le x \le y_0$$

and

$$(2.23) \qquad\qquad x \ne a_n \qquad (n \in \mathbf{Z}^+)$$

Proof *We construct by induction sequences $\{x_n\}$ and $\{y_n\}$ of rational numbers such that*

$$(2.24) \qquad \text{(i)} \quad x_0 \le x_n \le x_m < y_m \le y_n \le y_0 \qquad (m \ge n \ge 1)$$
$$\text{(ii)} \quad x_n > a_n \text{ or } y_n < a_n \qquad (n \ge 1)$$
$$\text{(iii)} \quad y_n - x_n < n^{-1} \qquad (n \ge 1)$$

Assume that $n \ge 1$ and that $x_0, \ldots, x_{n-1}, y_0, \ldots, y_{n-1}$ have been constructed. Either $a_n > x_{n-1}$ or $a_n < y_{n-1}$. In case $a_n > x_{n-1}$, let x_n be any rational number with $x_{n-1} < x_n < \min\{a_n, y_{n-1}\}$, and let y_n be any rational number with $x_n < y_n < \min\{a_n, y_{n-1}, x_n + n^{-1}\}$. Then the relevant inequalities are satisfied. In case $a_n < y_{n-1}$, let y_n be any rational number with $\max\{a_n, x_{n-1}\} < y_n < y_{n-1}$, and x_n any rational number with $\max\{a_n, x_{n-1}, y_n - n^{-1}\} < x_n < y_n$. Again, the relevant inequalities are satisfied. This completes the induction.

From (i) and (iii) it follows that

$$|x_m - x_n| = x_m - x_n < y_n - x_n < n^{-1} \qquad (m \ge n)$$

Similarly $|y_m - y_n| < n^{-1}$ for $m \ge n$. Therefore $x \equiv \{x_n\}$ and $y \equiv \{y_n\}$ are real numbers. By (iii), they are equal. By (i), $x_n \le x$ and $y_n \ge y$ for all n. If $a_n < x_n$ then $a_n < x$, so $a_n \ne x$. If $a_n > y_n$ then $a_n > y = x$, so $a_n \ne x$. Thus x satisfies (2.22) and (2.23).

Theorem 1 is the famous theorem of Cantor, that the real numbers are uncountable. The proof is essentially Cantor's "diagonal" proof. Both Cantor's theorem and his method of proof are of great importance.

The time has come to consider some counterexamples. Let $\{n_k\}$ be a sequence of integers, each of which is either 0 or 1, for which we are unable to prove either that $n_k = 1$ for some k or that $n_k = 0$ for all k. This corresponds to what Brouwer calls "a fugitive property of the natural numbers." Such a sequence can be defined, for example, as follows. Let n_k be 0 if $u^t + v^t \neq w^t$ for all integers u, v, w, t with $0 < u$, v, $w \leq k$ and $3 \leq t \leq k$. Otherwise let n_k be 1. Then we are unable to prove $n_k = 1$ for some k, because this would disprove Fermat's last theorem. We are unable to prove $n_k = 0$ for all k, because this would prove Fermat's last theorem.

Define now $x_k \equiv 0$ if $n_j = 0$ for all $j \leq k$, and $x_k \equiv 2^{-m}$ otherwise, where m is the least positive integer such that $n_m = 1$. Then $x \equiv \{x_k\}$ is a nonnegative real number, but we are unable to prove $x > 0$ or $x = 0$. Since nothing is true unless and until it has been proved, it is untrue that $x > 0$ or $x = 0$.

Of course, if Fermat's last theorem is proved tomorrow, we shall probably still be able to define a fugitive sequence $\{n_k\}$ of integers. Thus it is unlikely that there will ever exist a constructive proof that for every real number $x \geq 0$ either $x > 0$ or $x = 0$. We express this fact by saying that there exists a real number $x \geq 0$ such that it is *not* true that $x > 0$ or $x = 0$.

In much the same way we can construct a real number x such that it is *not* true that $x \geq 0$ or $x \leq 0$.

3. SEQUENCES AND SERIES OF REAL NUMBERS

We develop methods for defining a real number in terms of approximations by other real numbers.

Definition 6 A sequence $\{x_n\}$ of real numbers *converges* to a real number x_0 if for each k in Z^+ there exists N_k in Z^+ with

$$(3.1) \qquad\qquad |x_n - x_0| \leq k^{-1} \qquad (n \geq N_k)$$

To express that $\{x_n\}$ converges to x_0 we write

$$(3.2) \qquad\qquad \lim_{n \to \infty} x_n = x_0$$

or

$$(3.3) \qquad\qquad x_n \to x_0 \qquad \text{as } n \to \infty$$

A sequence of real numbers is said to *converge* when there exists a real number to which it converges.

It is easily seen that if $\{x_n\}$ converges to both x_0 and x_0', then $x_0 = x_0'$.

A convergent sequence is bounded: there exists r in R^+ such that $|x_n| \leq r$ for all n.

A convergent sequence of real numbers is not determined until the limit x_0 and the sequence $\{N_k\}$ are given, as well as the sequence $\{x_n\}$ itself. Even when they are not mentioned explicitly, these quantities are implicitly present. Similar comments apply to many subsequent definitions, including the following.

Definition 7   A sequence $\{x_n\}$ of real numbers is a *Cauchy sequence* if for each k in Z^+ there exists M_k in Z^+ such that

$$(3.4) \qquad\qquad |x_m - x_n| \le k^{-1} \qquad (m,\, n \ge M_k)$$

Theorem 2   *A sequence $\{x_n\}$ of real numbers converges if and only if it is a Cauchy sequence.*

Proof   Assume that $\{x_n\}$ converges to a real number x_0. Let the sequence $\{N_k\}$ satisfy (3.1). Write $M_k \equiv N_{2k}$. Then

$$|x_m - x_n| \le |x_m - x_0| + |x_n - x_0| \le (2k)^{-1} + (2k)^{-1} = k^{-1}$$

for $m,\, n \ge M_k$. Therefore $\{x_n\}$ is a Cauchy sequence.

Assume conversely that $\{x_n\}$ is a Cauchy sequence. Let the sequence $\{M_k\}$ satisfy (3.4). Write $N_k \equiv \max\,\{3k, M_{2k}\}$. Then

$$|x_m - x_n| \le (2k)^{-1} \qquad (m,\, n \ge N_k)$$

Let y_k be the $(2k)$th rational approximation to x_{N_k}. For $m \ge n$,

$$\begin{aligned}
|y_m - y_n| &\le |y_m - x_{N_m}| + |x_{N_m} - x_{N_n}| + |x_{N_n} - y_n| \\
&\le (2m)^{-1} + (N_m^{-1} + N_n^{-1}) + (2n)^{-1} < m^{-1} + n^{-1}
\end{aligned}$$

Therefore $y \equiv \{y_n\}$ is a real number. To see that $\{x_n\}$ converges to y we consider $n \ge N_k$ and compute

$$\begin{aligned}
|y - x_n| &\le |y - y_n| + |y_n - x_{N_n}| + |x_{N_n} - x_n \\
&\le n^{-1} + (2n)^{-1} + (2k)^{-1} \le (3k)^{-1} + (6k)^{-1} + (2k)^{-1} = k^{-1}
\end{aligned}$$

A *subsequence* of a sequence $\{x_n\}$ is a sequence of the form $\{x_{n_k}\}_{k=1}^{\infty}$, where $\{n_k\}_{k=1}^{\infty}$ is a sequence of positive integers with $n_j < n_k$ whenever $j < k$.

A subsequence of a convergent sequence converges to the same limit, and a convergent sequence converges to the same limit after modifications involving only finitely many terms, including possible insertions or deletions.

If $x \equiv \{x_n\}$ is a regular sequence of rational numbers, then $\{x_n^*\}$ converges to x.

A theorem of classical mathematics states that every bounded monotone sequence of real numbers converges. A counterexample to this statement is any monotone sequence $\{x_k\}$ such that $x_k = 0$ or $x_k = 1$ for all k, but it is not known whether $x_k = 0$ for all k.

It is useful to supplement Definition 6 by writing

$$\lim_{n \to \infty} x_n = \infty$$

to express the fact that for each k in Z^+ there exists N_k in Z^+ with

$$x_n > k \qquad (n \geq N_k)$$

We also define

$$\lim_{n \to \infty} x_n = -\infty$$

to mean $\lim_{n \to \infty} -x_n = \infty$.

The next proposition shows that we may work with real numbers constructed as limits by working with their approximations.

Proposition 8   *Assume that $x_n \to x_0$ as $n \to \infty$, and $y_n \to y_0$ as $n \to \infty$. Then*

(3.5) (a) $x_n + y_n \to x_0 + y_0$ *as $n \to \infty$*
      (b) $x_n y_n \to x_0 y_0$ *as $n \to \infty$*
      (c) $\max \{x_n, y_n\} \to \max \{x_0, y_0\}$ *as $n \to \infty$*
      (d) $x_0 = c$ *whenever $x_n = c$ for all n*
      (e) $x_n^{-1} \to x_0^{-1}$ *as $n \to \infty$ whenever $x_0 \neq 0$*
and $x_n \neq 0$ for all n
      (f) $x_0 \leq y_0$ *if $x_n \leq y_n$ for all n*

Proof
(a) For each k in Z^+ there exists N_k in Z^+ such that

$$|x_n - x_0| \leq (2k)^{-1}, \; |y_n - y_0| \leq (2k)^{-1} \qquad (n \geq N_k)$$

Then $|x_n + y_n - (x_0 + y_0)| \leq (2k)^{-1} + (2k)^{-1} = k^{-1} \qquad (n \geq N_k)$

Therefore $x_n + y_n \to x_0 + y_0$ as $n \to \infty$.

(b) Choose m in Z^+ with $|y_0| \le m$ and $|x_n| \le m$ for all n. For each k in Z^+ choose N_k in Z^+ with

$$|x_n - x_0| \le (2mk)^{-1}, \; |y_n - y_0| \le (2mk)^{-1} \qquad (n \ge N_k)$$

Then for $n \ge N_k$,

$$\begin{aligned}
|x_n y_n - x_0 y_0| &\le |x_n(y_n - y_0)| + |y_0(x_n - x_0)| \\
&\le m(|y_n - y_0| + |x_n - x_0|) \le k^{-1}
\end{aligned}$$

Therefore $x_n y_n \to x_0 y_0$ as $n \to \infty$.

(c) Since

$$|\max \{x_n, y_n\} - \max \{x_0, y_0\}| \le \max \{|x_n - x_0|, |y_n - y_0|\}$$

it follows that

$$\max \{x_n, y_n\} \to \max \{x_0, y_0\} \text{ as } n \to \infty$$

(d) If $x_n = c$ for all n, then $\{x_n\}$ converges to c. Therefore $x_0 = c$

(e) Since $|x_0| > 0$,

$$|x_n| \ge |x_0| - |x_n - x_0| > \tfrac{1}{2}|x_0|$$

whenever n is large enough, say for $n \ge n_0$. Let k and n be positive integers such that $n \ge n_0$ and $|x_n - x_0| < (2k)^{-1}|x_0|^2$. Then

$$|x_n^{-1} - x_0^{-1}| = |x_n|^{-1}|x_0|^{-1}|x_n - x_0| \le 2|x_0|^{-2}(2k)^{-1}|x_0|^2 = k^{-1}$$

Therefore $x_n^{-1} \to x_0^{-1}$ as $n \to \infty$.

(f) We compute

$$\begin{aligned}
y_0 - x_0 &= \lim_{n \to \infty} y_n - \lim_{n \to \infty} x_n = \lim_{n \to \infty} (y_n - x_n) = \lim_{n \to \infty} |y_n - x_n| \\
&= \lim_{n \to \infty} \max \{y_n - x_n, x_n - y_n\} = \max \{y_0 - x_0, x_0 - y_0\} \ge 0
\end{aligned}$$

by (a), (b), (c), and (d).

For each sequence $\{x_n\}$ of real numbers the number

$$s_n \equiv \sum_{k=1}^{n} x_k$$

is called the nth *partial sum* of $\{x_n\}$, and $\{s_n\}$ is called the sequence of partial sums of the sequence $\{x_n\}$. A *sum* s_0 of $\{x_n\}$ is a limit of the sequence $\{s_n\}$ of partial sums, and we write

$$s_0 = \sum_{n=1}^{\infty} x_n$$

to indicate that s_0 is a sum of $\{x_n\}$. A sequence which is meant to be summed is called a *series*. A series is said to *converge* to its sum. Thus the sequence $\{2^{-n}\}_{n=1}^{\infty}$ converges to 0 as a sequence, but as a series it converges to $\sum_{n=1}^{\infty} 2^{-n} = 1$.

A convergent series remains convergent, but not necessarily to the same sum, after modification of finitely many of its terms.

The series $\{x_n\}$ is often loosely referred to as *the series* $\sum_{n=1}^{\infty} x_n$.

If the series $\sum_{n=1}^{\infty} x_n$ converges, then $x_n \to 0$ as $n \to \infty$.

A series $\sum_{n=1}^{\infty} x_n$ is said to *converge absolutely* when the series $\sum_{n=1}^{\infty} |x_n|$ converges.

In classical analysis a series of nonnegative terms converges if the partial sums are bounded. This is not true in constructive analysis. However, we have the following result.

Proposition 9   *If $\sum_{n=1}^{\infty} y_n$ is a convergent series of nonnegative terms, and if $|x_n| \leq y_n$ for each n, then $\sum_{n=1}^{\infty} x_n$ converges.*

Proof   Since $\sum_{n=1}^{\infty} y_n$ is convergent, the sequence of partial sums is a Cauchy sequence. Therefore there exists for each k in Z^+ an N_k in Z^+ with

$$\sum_{j=n+1}^{m} y_j \leq k^{-1} \qquad (m \geq n \geq N_k)$$

Then
$$\left| \sum_{j=n+1}^{m} x_j \right| \leq \sum_{j=n+1}^{m} y_j \leq k^{-1} \qquad (m \geq n \geq N_k)$$

Therefore the sequence of partial sums of the series $\sum_{n=1}^{\infty} x_n$ is a Cauchy sequence. By Theorem 2, the series converges.

The criterion of Proposition 9 is known as the *comparison test*. It follows from the comparison test that every absolutely convergent series is convergent.

The terms of an absolutely convergent series $\sum_{n=1}^{\infty} x_n$ may be re-ordered without affecting the sum s_0 of the series. More precisely, if $\lambda: Z^+ \to Z^+$ is a bijection, then $\sum_{n=1}^{\infty} x_{\lambda(n)}$ exists and equals s_0. This may not be true if the series $\sum_{n=1}^{\infty} x_n$ is merely convergent.

A sequence $\{x_n\}$ is said to *diverge* if there exists ϵ in $\mathbf{R}^+$ such that for each k in Z^+ there exist m and n in Z^+ with $m, n \geq k$ and $|x_m - x_n| \geq \epsilon$. The motivation for this definition is, of course, that a sequence cannot be both convergent and divergent. A series is said to *diverge* if the sequence of its partial sums diverges. The series $\sum_{n=1}^{\infty} x_n$ diverges whenever there exists r in $\mathbf{R}^+$ such that $|x_n| \geq r$ for infinitely many values of n.

The following very useful test for convergence and divergence is called the *ratio test*.

Proposition 10 *Let* $\sum_{n=1}^{\infty} x_n$ *be a series, c a positive real number, and N a positive integer. Then* $\sum_{n=1}^{\infty} x_n$ *converges if* $c < 1$ *and*

$$(3.6) \qquad |x_{n+1}| \leq c|x_n| \qquad (n \geq N)$$

and diverges if $c > 1$ *and*

$$(3.7) \qquad |x_{n+1}| > c|x_n| \qquad (n \geq N)$$

Proof Assume that $c < 1$ and that (3.6) is valid. Then $|x_n| \leq c^{n-N}|x_N|$ for $n \geq N$. By the comparison test, $\sum_{n=1}^{\infty} x_n$ converges.

Next, assume that $c > 1$ and that (3.7) holds. Then

$$|x_n| \geq c^{n-N-1}|x_{N+1}| \geq |x_{N+1}| \qquad (n \geq N + 1)$$

and

$$|x_{N+1}| > c|x_N| \geq 0$$

Therefore $\sum_{n=1}^{\infty} x_n$ diverges.

A corollary of the ratio test is that if the limit

$$L \equiv \lim_{n \to \infty} |x_{n+1}x_n^{-1}|$$

exists, then $\sum_{n=1}^{\infty} x_n$ converges whenever $L < 1$ and diverges whenever $L > 1$.

Important real numbers represented by series are

$$e = 1 + \sum_{n=1}^{\infty} (n!)^{-1}$$

and

$$\pi = 4 \sum_{n=0}^{\infty} (-1)^n (2n + 1)^{-1}$$

The series for e converges by the ratio test. The convergence of the series for π is a consequence of the general result that a series $\sum_{n=1}^{\infty} (-1)^n x_n$ converges whenever (i) $x_n \geq 0$ for all n and (ii) the sequence $\{x_n\}$ converges monotonely to 0. To see this, consider positive integers m and n, with $m \geq n$. Then

$$0 \leq (x_n - x_{n+1}) + (x_{n+2} - x_{n+3}) + \cdots + (-1)^{m+n}x_m$$
$$= (-1)^n \sum_{k=n}^{m} (-1)^k x_k$$
$$= x_n - (x_{n+1} - x_{n+2}) - \cdots + (-1)^{m+n}x_m \leq x_n$$

It follows that the sequence of partial sums of the series is a Cauchy sequence. Therefore the series converges.

4. CONTINUOUS FUNCTIONS

A property P which is applicable to the elements of a set S is defined by a statement of the requirements that an element of S must satisfy in order to have property P. To construct an element of S with property P we must (a) construct an element of S, (b) perform certain additional constructions which depend on the property P, and (c) prove that the entities constructed satisfy certain requirements that are characteristic of the property P. Each property P applicable to elements of S determines a subset of S, which is denoted by

$$\{x : x \in S, \ x \text{ has property } P\}$$

Properties P applicable to S and subsets A of S are essentially the same things regarded from different points of view.

The intersection

$$\{x: x \in S,\ x \text{ has property } P\} \cap \{x: x \in S,\ x \text{ has property } Q\}$$

is defined to be the set

$$\{x: x \in S,\ x \text{ has properties } P \text{ and } Q\}$$

For certain purposes it is convenient to extend the real numbers $\mathbf{R}$ by adjoining elements $-\infty$ and ∞. To construct an element of the set $\mathbf{R}_\infty$ of extended real numbers we either construct a real number, pick the element $-\infty$, or pick the element $+\infty$. The order relation $<$ is extended to $\mathbf{R}_\infty$ by defining $-\infty < \infty$, $-\infty < x$, and $x < \infty$ for all real numbers x. The order relation $\leq$ is extended in the same way. It is often convenient to phrase the order relations on $\mathbf{R}$ in terms of intervals.

Definition 8   For each a in $\mathbf{R}_\infty$ we define

$$(4.1) \qquad \begin{aligned}
(-\infty, a) &\equiv \{x: x \in \mathbf{R},\ x < a\} \\
(-\infty, a] &\equiv \{x: x \in \mathbf{R},\ x \leq a\} \\
(a, \infty) &\equiv \{x: x \in \mathbf{R},\ a < x\} \\
[a, \infty) &\equiv \{x: x \in \mathbf{R},\ a \leq x\}
\end{aligned}$$

For each a and b in $\mathbf{R}_\infty$ we define

$$(4.2) \qquad \begin{aligned}
(a,b) &\equiv (a, \infty) \cap (-\infty, b) \\
(a,b] &\equiv (a, \infty) \cap (-\infty, b] \\
[a,b) &\equiv [a, \infty) \cap (-\infty, b) \\
[a,b] &\equiv [a, \infty) \cap (-\infty, b]
\end{aligned}$$

For example, $x \in (a,b)$ means that $a < x < b$. Each of these sets is an *interval*, whose *end points* are a and b. The interval (a,b) is *open*, $[a,b]$ is *closed*, $(a,b]$ is *half-open* on the left, and $[a,b)$ is *half-open* on the right. An interval is *finite* if both its end points a and b are finite, and *proper* if $a < b$. An interval I is *nonvoid* if $x \in I$ for some x in $\mathbf{R}$. A closed finite nonvoid interval is *compact*.

The rules for manipulating with intervals, which we use in the sequel without mention or proof, are implicit in Proposition 5.

Often when one real number depends on another the dependence is

smooth, or continuous. An exact description of what this means is given
in the following definition.

Definition 9 A real-valued function f defined on a compact interval
I is *continuous* on I if for each $\epsilon > 0$ there exists $\omega(\epsilon) > 0$ such that

$$(4.3) \qquad\qquad |f(x) - f(y)| \leq \epsilon$$

whenever $x, y \in I$ and $|x - y| \leq \omega(\epsilon)$. The operation $\epsilon \to \omega(\epsilon)$ is called
a *modulus of continuity* for f.

A real-valued function f on an arbitrary interval J is *continuous* on J
if it is continuous on every compact subinterval I of J. For example,
when a and b are finite and $a < b$, then f is continuous on (a,b) if and
only if it is continuous on $[a + \delta, b - \delta]$ for each δ with $0 < \delta <$
$\frac{1}{2}(b - a)$.

The modulus of continuity ω is an indispensable part of the definition
of a continuous function on a compact interval, although sometimes it
is not mentioned explicitly. In the same way, the moduli of continuity
of the restrictions of f to each compact subinterval are indispensable
parts of the definition of a continuous function f on a general interval.
The constant functions $x \to c$ and the identity function $x \to x$ are continuous on every interval.

Definition 10 A real number c is called the *least upper bound* (respectively, *greatest lower bound*) of a subset A of $\mathbf{R}$ and written

$$(4.4) \qquad c \equiv \text{l.u.b. } A \qquad (\text{respectively, } c \equiv \text{g.l.b. } A)$$

if

$$(4.5) \qquad\qquad x \leq c \qquad (\text{respectively, } x \geq c)$$

for all x in A and if for each $\epsilon > 0$ there exists x in A with $c - x < \epsilon$
(respectively, $x - c < \epsilon$).

Theorem 3 *Let the subset A of $\mathbf{R}$ have the property that for each $\epsilon > 0$
there exist finitely many points $y_1, \ldots, y_n$ in A such that for each x in A
at least one of the numbers $|x - y_1|, \ldots, |x - y_n|$ is less than ϵ. (Such
a set is called* totally bounded.) *Then* l.u.b. A *and* g.l.b. A *exist.*

Proof For each positive integer k choose points $y_1, \ldots, y_n$ in A such
that for each x in A at least one of the numbers $|x - y_1|, \ldots, |x - y_n|$

is less than k^{-1}. For some value of m $(1 \leq m \leq n)$ we have

$$y_m \geq \max \{y_1, \ldots, y_n\} - k^{-1}$$

Write $x_k \equiv y_m$. If x is any point in A, we have $|x - y_i| < k^{-1}$ for some value of i $(1 \leq i \leq n)$. Hence

$$x - x_k = x - y_i + y_i - y_m < k^{-1} + k^{-1} = 2k^{-1}$$

Thus the sequence $\{x_k\}$ of elements of A has the property that $x - x_k < 2k^{-1}$ for all x in A and all k in Z^+. In particular, $|x_j - x_k| < 2j^{-1} + 2k^{-1}$ for all j and k. Therefore $\{x_k\}$ is a Cauchy sequence, whose limit we call x_0. For all x in A,

$$x - x_0 = \lim_{k \to \infty} (x - x_k) \leq \lim_{k \to \infty} 2k^{-1} = 0$$

Therefore x_0 is an upper bound for A, and since $x_0 = \lim_{k \to \infty} x_k$ it is a least upper bound. The proof that g.l.b. A exists is similar.

Corollary   If $f: [a,b] \to \mathbf{R}$ is a continuous function on a compact interval, then the quantities

$$c = \text{l.u.b. } \{f(x) : x \in [a,b]\}$$
and
$$d \equiv \text{g.l.b. } \{f(x) : x \in a,b]\}$$

(called, respectively, the *supremum* and the *infimum* of f on the interval $[a,b]$) exist.

Proof   Consider $\epsilon > 0$. Choose real numbers

$$a = a_0 < a_1 < \cdots < a_n = b$$

such that $a_{i+1} - a_i < \omega(\epsilon)$ $(0 \leq i \leq n - 1)$, where ω is the modulus of continuity of f. Then for each x in $[a,b]$ we have $|x - a_i| < \omega(\epsilon)$ for some i. Therefore $|f(x) - f(a_i)| \leq \epsilon$ for some i. Since ϵ is arbitrary, it follows that the set $A \equiv \{f(x) : x \in [a,b]\}$ is totally bounded. Therefore c and d exist.

The proof of the following proposition, which resembles the proof of Proposition 8, is left to the reader.

Proposition 11   *Let f and g be continuous real-valued functions defined on an interval I. Then the functions $f + g$, fg, and $\max \{f,g\}$ are con-*

tinuous on I. If f is bounded away from 0 on every compact subinterval J of I, that is, if $|f(x)| \geq c$ for all x in J and some $c > 0$ (depending on J), then f^{-1} is continuous on I.

Proposition 11 implies that the quotient of continuous functions is continuous, provided that the denominator is bounded away from 0 on every compact subinterval. It also implies that a polynomial function

$$x \to c_0 x^n + c_1 x^{n-1} + \cdots + c_n$$

is continuous on every interval, and that $|f|$ is continuous on each interval where f is continuous.

The composition of continuous functions is continuous, in the sense that if $f: I \to J$ and $g: J \to \mathbf{R}$ are continuous, then $g \circ f$ is continuous, provided that f maps every compact subinterval of I into a compact subinterval of J. To prove this it is sufficient to consider the case in which I and J are both compact. Let ω be the modulus of continuity of f and σ the modulus of continuity of g. Then if x, $y \in I$, $\epsilon > 0$, and $|x - y| \leq \omega(\sigma(\epsilon))$, we have $|f(x) - f(y)| \leq \sigma(\epsilon)$. Therefore

$$|g(f(x)) - g(f(y))| \leq \epsilon$$

It follows that $g \circ f$ is continuous, with modulus of continuity $\omega \circ \sigma$.

Just as sequences of real numbers can converge to real numbers, sequences of continuous functions can converge to continuous functions. In fact, most of the important functions of analysis are defined in this way.

Definition 11 A sequence $\{f_n\}$ of continuous functions on a compact interval I *converges* on I to a continuous function f if for each $\epsilon > 0$ there exists N_ϵ in Z^+ such that

$$(4.6) \qquad |f_n(x) - f(x)| \leq \epsilon \qquad (x \in I, n \geq N_\epsilon)$$

A sequence $\{f_n\}$ of continuous functions on an arbitrary interval J *converges* on J to a continuous function f if it converges to f on every compact subinterval I of J.

Definition 11 can be recast to bear a closer resemblance to Definition 6. To this end we define the *norm* $\|f\|_I$ of a continuous function f on a compact interval I to be the supremum of $|f|$ on I. Then $\{f_n\}$ converges to f on I if and only if for each k in Z^+ there exists N_k in Z^+ with

$$(4.7) \qquad \|f_n - f\|_I \leq k^{-1} \qquad (n \geq N_k)$$

Definition 12   A sequence $\{f_n\}$ of continuous functions on a compact interval I is a *Cauchy sequence* on I if for each $\epsilon > 0$ there exists M_ϵ in Z^+ such that

$$(4.8) \qquad |f_m(x) - f_n(x)| \leq \epsilon \qquad (x \in I,\ m,\ n \geq M_\epsilon)$$

A sequence $\{f_n\}$ of continuous functions on an arbitrary interval J is a *Cauchy sequence* on J if it is a Cauchy sequence on every compact subinterval I of J.

The sequence $\{f_n\}$ is a Cauchy sequence on the compact interval I if and only if for every k in Z^+ there exists M_k in Z^+ such that

$$(4.9) \qquad \|f_m - f_n\|_I \leq k^{-1} \qquad (m,\ n \geq M_k)$$

Notice that a sequence $\{c_n\}$ of real numbers converges if and only if the corresponding sequence of constant functions, which we also denote by $\{c_n\}$, converges on a given nonvoid interval I, and that a sequence of real numbers is a Cauchy sequence if and only if the corresponding sequence of constant functions is a Cauchy sequence on I. Because of these remarks the following theorem is a generalization of Theorem 2.

Theorem 4   *A sequence $\{f_n\}$ of continuous functions on an interval J converges on J if and only if it is a Cauchy sequence on J.*

Proof   Assume that $\{f_n\}$ converges to f on J. Let I be any compact subinterval of J. For each $\epsilon > 0$ choose N_ϵ in Z^+ satisfying (4.6), and write $M_\epsilon \equiv N_{\epsilon/2}$. Then, whenever $m,\ n \geq M_\epsilon$ and $x \in I$, we have

$$|f_m(x) - f_n(x)| \leq |f_m(x) - f(x)| + |f_n(x) - f(x)| \leq \frac{\epsilon}{2} + \frac{\epsilon}{2} = \epsilon$$

Therefore $\{f_n\}$ is a Cauchy sequence on I. It follows that $\{f_n\}$ is a Cauchy sequence on J.

Assume conversely that $\{f_n\}$ is a Cauchy sequence on J. Then $\{f_n(x)\}$ is a Cauchy sequence of real numbers, for each x in J, whose limit we denote by $f(x)$. We shall show that $f: J \to \mathbf{R}$ is a continuous function and that $\{f_n\}$ converges to f on J. It is enough to show that f is continuous on each compact subinterval I of J and that $\{f_n\}$ converges to f on I. To this end, choose the positive integers M_ϵ such that (4.8) is valid, and for each n in Z^+ let ω_n be the modulus of continuity of f_n on

I. For each $\epsilon > 0$ write

$$\omega(\epsilon) \equiv \omega_M(\epsilon/3)$$

where $M \equiv M_{\epsilon/3}$. Then whenever $x, y \in I$ and $|x - y| \leq \omega(\epsilon)$, we have

$$
\begin{aligned}
|f(x) - f(y)| &\leq |f(x) - f_M(x)| + |f_M(x) - f_M(y)| + |f_M(y) - f(y)| \\
&= \lim_{n \to \infty} |f_n(x) - f_M(x)| + |f_M(x) - f_M(y)| \\
&\qquad\qquad\qquad\qquad + \lim_{n \to \infty} |f_M(y) - f_n(y)| \\
&\leq \frac{\epsilon}{3} + \frac{\epsilon}{3} + \frac{\epsilon}{3} = \epsilon
\end{aligned}
$$

Therefore f is continuous with modulus of continuity ω. Finally, if $x \in I$, $\epsilon > 0$, and $n \geq M_\epsilon$, then

$$|f_n(x) - f(x)| = \lim_{m \to \infty} |f_n(x) - f_m(x)| \leq \epsilon$$

Therefore $\{f_n\}$ converges to f on I.

Notations to express the fact that $\{f_n\}$ converges to f are $\lim\limits_{n \to \infty} f_n = f$, and $f_n \to f$ as $n \to \infty$.

To each sequence $\{f_n\}$ of continuous functions on an interval I corresponds a sequence $\{g_n\}$ of *partial sums*, defined by

$$g_n \equiv \sum_{k=1}^{n} f_k$$

If $\{g_n\}$ converges to a continuous function g on I, then g is the *sum* of the series $\sum\limits_{n=1}^{\infty} f_n$:

$$g \equiv \sum_{n=1}^{\infty} f_n$$

and the series is said to *converge* to g on I. If $\sum\limits_{n=1}^{\infty} |f_n|$ converges on I, then $\sum\limits_{n=1}^{\infty} f_n$ is said to *converge absolutely* on I.

The comparison test and the ratio test carry over to series of functions.

The comparison test states that if $\sum_{n=1}^{\infty} g_n$ is a convergent series of nonnegative continuous functions on an interval I, then the series $\sum_{n=1}^{\infty} f_n$ of continuous functions on I converges on I whenever $|f_n(x)| \leq g_n(x)$ for all n in Z^+ and all x in I.

The ratio test states that if $\sum_{n=1}^{\infty} f_n$ is a series of continuous functions on an interval J such that for each compact subinterval I of J there exists a constant c_I, $0 < c_I < 1$, and a positive integer N_I with

$$|f_{n+1}(x)| \leq c_I|f_n(x)| \qquad (x \in I,\ n \geq N_I)$$

then $\sum_{n=1}^{\infty} f_n$ converges absolutely on J.

A *power series* is a series of the form $\sum_{n=0}^{\infty} a_n(x - x_0)^n$, where $a_n(x - x_0)^n$ represents the function $x \to a_n(x - x_0)^n$ and where $a_0(x - x_0)^0 \equiv a_0$ for all x. The ratio test has the following corollary.

Proposition 12 *Let the power series*

$$(4.10) \qquad \sum_{n=0}^{\infty} a_n(x - x_0)^n$$

have the property that there exists $r > 0$ and N in Z^+ such that $|a_{n+1}| \leq r^{-1}|a_n|$ for all $n \geq N$. Then (4.10) converges absolutely on the interval $(x_0 - r, x_0 + r)$.

Proof If I is a compact subinterval of $(x_0 - r,\ x_0 + r)$, then there exists r_0 with $0 < r_0 < r$ such that $|x - x_0| \leq r_0$ for all x in I. Then

$$|a_{n+1}(x - x_0)^{n+1}| \leq r^{-1}r_0|a_n(x - x_0)^n| \qquad (n \geq N)$$

By the ratio test, (4.10) therefore converges absolutely on I.

5. DIFFERENTIATION

The rate at which a function is changing is a fundamental property of the function. Here is the precise definition of this concept.

Definition 13 Let f and g be continuous functions on a compact proper interval I such that for each $\epsilon > 0$ there exists $\delta(\epsilon) > 0$ with

(5.1)
$$|f(y) - f(x - g(x)(y - x)| \leq \epsilon|y - x| \qquad (x, y \in I, |y - x| \leq \delta(\epsilon))$$

Then f is said to be *differentiable* on I, g is called a *derivative* of f on I, and δ is called a *modulus of differentiability* for f on I. If f and g are continuous on the proper interval J, then g is a *derivative* of f on J if it is a derivative of f on every compact proper subinterval I of J, and f is said to be *differentiable* on J.

To express that g is a derivative of f we write

$$(5.2) \qquad g = f' \qquad \text{or} \qquad g = Df \qquad \text{or} \qquad g(x) = \frac{df(x)}{dx}$$

One way to interpret Definition 13 is that the *difference quotient*

$$(5.3) \qquad\qquad (f(y) - f(x))(y - x)^{-1}$$

approaches $g(x)$ as y approaches x. In other words, g is the rate of change of f.

If f has two derivatives g_1 and g_2 on I, they are equal.

Theorem 5 *Let f_1 and f_2 be differentiable functions on an interval I. Then $f_1 + f_2$ and $f_1 f_2$ are differentiable on I. In case f_1 is bounded away from 0 on every compact subinterval of I, then f_1^{-1} is differentiable on I. The function $x \to x$ is differentiable on $\mathbf{R}$. For each c in $\mathbf{R}$ the function $x \to c$ is differentiable on $\mathbf{R}$. The derivatives in question are given by the following equations.*

$$(5.4) \qquad\qquad$$

$$(a) \quad D(f_1 + f_2) = Df_1 + Df_2$$

$$(b) \quad D(f_1 f_2) = f_1\, Df_2 + f_2\, Df_1$$

$$(c) \quad Df_1^{-1} = -f_1^{-2}Df_1$$

$$(d) \quad \frac{dx}{dx} = 1$$

$$(e) \quad \frac{dc}{dx} = 0$$

Proof It is enough to consider the case in which I is compact. Let δ_1 and δ_2 be moduli of differentiability for f_1 and f_2, respectively, on I, and ω_1 a modulus of continuity for f_1 on I.

(a) Whenever $x, y \in I$ and $|y - x| \le \delta(\epsilon) \equiv \min \{\delta_1(\epsilon/2), \delta_2(\epsilon/2)\}$, we have

$$|f_1(y) + f_2(y) - (f_1(x) + f_2(x)) - (f_1'(x) + f_2'(x))(y - x)|$$
$$\le |f_1(y) - f_1(x) - f_1'(x)(y - x)| + |f_2(y) - f_2(x) - f_2'(x)(y - x)|$$
$$\le \frac{\epsilon}{2} + \frac{\epsilon}{2} = \epsilon$$

Thus $f_1 + f_2$ is differentiable on I with derivative $f_1' + f_2'$ and modulus of differentiability δ.

(b) Let M be a common bound for $|f_1|$, $|f_2|$, and $|f_2'|$ on the interval I. (For instance, define $M \equiv \max \{\|f_1\|_I, \|f_2\|_I, \|f_2'\|_I\}$.) Then whenever $x, y \in I$ and

$$|y - x| \le \delta(\epsilon) \equiv \min \{\delta_1((3M)^{-1}\epsilon), \delta_2((3M)^{-1}\epsilon), \omega_1((3M)^{-1}\epsilon)\}$$

we have

$$|f_1(y)f_2(y) - f_1(x)f_2(x) - (f_1(x)f_2'(x) + f_2(x)f_1'(x))(y - x)|$$
$$\le |f_1(y)| \, |f_2(y) - f_2(x) - f_2'(x)(y - x)|$$
$$+ |f_1(y) - f_1(x)| \, |f_2'(x)| \, |y - x|$$
$$+ |f_2(x)| \, |f_1(y) - f_1(x) - f_1'(x)(y - x)|$$
$$\le 3M(3M)^{-1}\epsilon|y - x| = \epsilon|y - x|$$

Therefore $f_1 f_2$ is differentiable on I with derivative $f_1 f_2' + f_2 f_1'$ and modulus of differentiability δ.

(c) For each $\epsilon > 0$ write

$$\delta(\epsilon) \equiv \min \{\delta_1(\tfrac{1}{2}M^{-2}\epsilon), \omega_1(\tfrac{1}{2}M^{-4}\epsilon)\}$$

where $M \equiv \max \{\|f_1^{-1}\|_I, \|f_1'\|_I\}$. Then whenever $x, y \in I$ and $|y - x| \le \delta(\epsilon)$, we have

$$|f_1^{-1}(y) - f_1^{-1}(x) + f_1^{-2}(x)f_1'(x)(y - x)|$$
$$= |f_1^{-1}(x)f_1^{-1}(y)| \, |f_1(y) - f_1(x) - f_1(y)f_1^{-1}(x)f_1'(x)(y - x)|$$
$$\le M^2|f_1(y) - f_1(x) - f_1'(x)(y - x)|$$
$$+ M^2|f_1'(x)f_1(x)^{-1}| \, |f_1(y) - f_1(x)| \, |y - x|$$
$$\le M^2(\tfrac{1}{2}M^{-2}\epsilon)|y - x| + M^4(\tfrac{1}{2}M^{-4}\epsilon)|y - x| = \epsilon|y - x|$$

Therefore f_1^{-1} is differentiable on I with derivative $-f_1^{-2}f_1'$ and modulus of differentiability δ.

(d) This is obvious.

(e) This is obvious too.

Corollary *For all positive integers n,*

$$(5.5) \qquad \frac{dx^n}{dx} = nx^{n-1}$$

Proof The proof is by induction on n. When $n = 1$ (5.5) is just (5.4) (d). If (5.5) is true for a given value of n, then

$$\frac{dx^{n+1}}{dx} = \frac{d(x \cdot x^n)}{dx} = x^n + x(nx^{n-1}) = (n+1)x^n$$

by (5.4)(b). Therefore (5.5) is true for all n.

Theorem 5 and its corollary imply the formula

$$D(f_1 f_2^{-1}) = f_2^{-2}(f_2\, Df_1 - f_1\, Df_2)$$

for the derivative of a quotient, and the formula

$$D\left(\sum_{k=0}^{n} a_{n-k}x^k\right) = \sum_{k=1}^{n} ka_{n-k}x^{k-1}$$

for the derivative of a polynomial.

The next theorem is the so-called chain rule for the derivative of a composite function. Its intuitive meaning is that the rate of change of quantity C with respect to quantity A is the product of the rate of change of C with respect to some third quantity B by the rate of change of B with respect to A.

Theorem 6 *Let $f\colon I \to \mathbf{R}$ and $g\colon J \to \mathbf{R}$ be differentiable functions such that f maps every compact subinterval of I into some compact subinterval of J. Then $g \circ f$ is differentiable, and*

$$(5.6) \qquad (g \circ f)' = (g' \circ f)f'$$

Proof It is no loss of generality to assume that I and J are compact. Let δ_f be the modulus of differentiability and ω_f the modulus of continuity of f on I. Let δ_g be the modulus of differentiability of g on J. For each $\epsilon > 0$ write

$$\delta(\epsilon) \equiv \min\{\omega_f(\delta_g(\alpha)),\, \delta_f(\beta)\}$$

where $\quad \alpha \equiv (1 + \|f'\|_I)^{-1}\dfrac{\epsilon}{2} \quad$ and $\quad \beta \equiv (\alpha + \|g'\|_J)^{-1}\dfrac{\epsilon}{2}$

Then for $x,\ y \in I$ and $|y - x| \le \delta(\epsilon)$ we have $|f(y) - f(x)| \le \delta_g(\alpha)$, so that

$$|g(f(y)) - g(f(x)) - g'(f(x))(f(y) - f(x))| \le \alpha|f(y) - f(x)|$$

Also $\quad |f(y) - f(x)| \le \|f'\|_I |y - x| + |f(y) - f(x) - f'(x)(y - x)|$

and $\qquad\qquad |f(y) - f(x) - f'(x)(y - x)| \le \beta|y - x|$

Using these inequalities and noting that $\alpha\|f'\|_I < \epsilon/2$, we compute

$$
\begin{aligned}
|g(f(y)) &- g(f(x)) - g'(f(x))f'(x)(y - x)| \\
&\le |g(f(y)) - g(f(x)) - g'(f(x))(f(y) - f(x))| \\
&\qquad\qquad + |g'(f(x))|\,|f(y) - f(x) - f'(x)(y - x)| \\
&\le \alpha|f(y) - f(x)| + \|g'\|_J |f(y) - f(x) - f'(x)(y - x)| \\
&\le \alpha\|f'\|_I |y - x| + (\alpha + \|g'\|_J)|f(y) - f(x) - f'(x)(y - x)| \\
&< \frac{\epsilon}{2}|y - x| + \frac{\epsilon}{2}|y - x| = \epsilon|y - x|
\end{aligned}
$$

It follows that $g \circ f$ is differentiable on I with derivative $(g' \circ f)f'$ and modulus of differentiability δ.

The next lemma is called *Rolle's theorem*.

Lemma 6 *Let f be differentiable on the interval $[a,b]$, and let $f(a) = f(b)$. Then for each $\epsilon > 0$ there exists x in $[a,b]$ with*

$$(5.7) \qquad\qquad |f'(x)| \le \epsilon$$

Proof Let ω be the modulus of continuity of f' on $[a,b]$ and δ the modulus of differentiability of f on $[a,b]$. Choose real numbers

$$a = x_0 < x_1 < \cdots < x_n = b$$

for which

$$x_{i+1} - x_i \le \min\left\{\delta\left(\frac{\epsilon}{2}\right),\ \omega\left(\frac{\epsilon}{2}\right)\right\} \qquad (0 \le i \le n - 1)$$

Then for $0 \le i \le n - 1$, we have

$$
\begin{aligned}
f(x_{i+1}) - f(x_i) &= f'(x_i)(x_{i+1} - x_i) + f(x_{i+1}) - f(x_i) - f'(x_i)(x_{i+1} - x_i) \\
&\le \left(f'(x_i) + \frac{\epsilon}{2}\right)(x_{i+1} - x_i) < (f'(x_i) + \epsilon)(x_{i+1} - x_i)
\end{aligned}
$$

Therefore

$$0 = f(b) - f(a) = \sum_{i=0}^{n-1} f(x_{i+1}) - f(x_i) < \sum_{i=0}^{n-1} (f'(x_i) + \epsilon)(x_{i+1} - x_i)$$

By Proposition 7 it follows that $f'(x_i) > -\epsilon$ for at least one value of i, say for $i = j$. Similarly, $f'(x_i) < \epsilon$ for at least one value of i, say for $i = k$.

By the Corollary to Proposition 7, either (5.7) is valid or

$$(5.8) \qquad |f'(x_i)| > \frac{\epsilon}{2} \qquad (0 \leq i \leq n - 1)$$

We may therefore assume (5.8). Since $|f'(x_{i+1}) - f'(x_i)| \leq \epsilon/2$, the quantities $f'(x_i)$ and $f'(x_{i+1})$ are either both positive or both negative. It follows that the quantities $f'(x_i)$ $(0 \leq i \leq n - 1)$ are either all positive or all negative. In the former case, $0 < f'(x_k) < \epsilon$, so that (5.7) holds with $x \equiv x_k$; and in the latter case, $-\epsilon < f'(x_j) < 0$, so that (5.7) holds with $x \equiv x_j$.

Rolle's theorem implies the *law of the mean*, which gives a basic estimate for the difference of two values of a differentiable function.

Theorem 7 Let f be *differentiable on the interval* $[a,b]$. *Then for each* $\epsilon > 0$ *there exists* x *in* $[a,b]$ *with*

$$(5.9) \qquad |f(b) - f(a) - f'(x)(b - a)| \leq \epsilon$$

Proof Define the function h by

$$h(x) \equiv (x - a)(f(b) - f(a)) - f(x)(b - a)$$

Then $h(b) = h(a) = -f(a)(b - a)$. By Lemma 6 there exists x in $[a,b]$ with

$$|h'(x)| = |f(b) - f(a) - f'(x)(b - a)| \leq \epsilon$$

Definition 14 Let f, $f^{(1)}$, $f^{(2)}$, $\ldots$, $f^{(n)}$ be differentiable functions on an interval I with

$$Df = f^{(1)}, \ Df^{(1)} = f^{(2)}, \ \ldots \ , Df^{(n-1)} = f^{(n)}$$

on I. Then $f^{(n)}$ is called the *nth derivative* of f on I and written $D^n f$, or simply $f^{(n)}$, and f is said to be *n times differentiable* on I. The function f itself may be written $f^{(0)}$ or $D^0 f$.

A natural way to simplify a continuous function and set it up for computation is to replace it by a polynomial approximation. The basic result on polynomial approximation of differentiable functions is *Taylor's theorem* (Theorem 8 below). The motivation for Taylor's theorem goes as follows. Consider an n times differentiable function f on an interval I, and a point a in I. It is natural to approximate f by a polynomial of degree n whose derivatives of orders 0, 1, . . . , n at a have the same values as the corresponding derivatives of f at a. The unique such polynomial is obviously

$$(5.10) \qquad \sum_{k=0}^{n} \frac{f^{(k)}(a)}{k!} (x - a)^k$$

This approximation is useful for a given value b of x only when there exists a good estimate for the remainder

$$(5.11) \qquad R \equiv f(b) - \sum_{k=0}^{n} \frac{f^{(k)}(a)}{k!} (b - a)^k$$

This is the purpose of Taylor's theorem.

Theorem 8 *Let f be an $(n + 1)$ times differentiable function on an interval I. Let ϵ be a positive constant, and let a and b be points of I. Then there exists c, with* min $\{a,b\} \leq c \leq$ max $\{a,b\}$, *such that*

$$(5.12) \qquad \left| R - \frac{f^{(n+1)}(c)}{n!} (b - c)^n (b - a) \right| \leq \epsilon$$

where R is given by (5.11).

Proof If $|a - b|$ is sufficiently small, (5.12) holds with $c = b$. Hence it is enough to consider the case $a \neq b$. Define the differentiable function g on I by

$$g(x) \equiv f(b) - f(x) - \frac{f'(x)}{1!} (b - x) - \cdots - \frac{f^n(x)}{n!} (b - x)^n$$
$$- R \frac{b - x}{b - a}$$

Then $g(a) = g(b) = 0$, and

$$g'(x) = -f'(x) + f'(x) - f''(x)(b - x) + f''(x)(b - x)$$
$$- \frac{f'''(x)}{2} (b - x)^2 + \cdots - \frac{f^{(n+1)}(x)}{n!} (b - x)^n + R(b - a)^{-1}$$
$$= -\frac{f^{(n+1)}(x)}{n!} (b - x)^n + R(b - a)^{-1}$$

since the terms cancel in pairs except for the last two. By Rolle's theorem there exists c with $a \le c \le b$ or $b \le c \le a$ and $|g'(c)| \le \epsilon(b - a)^{-1}$. This is equivalent to (5.12).

An important special case occurs when f is an infinitely differentiable function on an open interval $I \equiv (a - c, a + c)$ such that for each r in $(0,c)$,

$$(5.13) \qquad \frac{r^n f^{(n)}}{n!} \to 0 \qquad \text{as } n \to \infty$$

on I. It follows from Theorem 8 that the *Taylor's series*

$$(5.14) \qquad \sum_{n=0}^{\infty} \frac{f^n(a)}{n!} (x - a)^n$$

converges to f on I. While the convergence of (5.14) already follows from the comparison test, Theorem 8 is needed to show that the sum is f.

Examples of (5.14) will occur later, when some of the common functions of analysis are expanded in Taylor's series.

6. INTEGRATION

Differentiation has to do with the instantaneous behavior of a function; we now look at a process, called *integration*, which has to do with the behavior of a function on the average—in fact it is essentially a rigorous formulation of the averaging process. Differentiation and integration will turn out to be inverse processes in a certain precise sense.

If $a = a_0 \le a_1 \le \cdots \le a_n = b$ are real numbers, the finite sequence $P \equiv \{a_0, a_1, \ldots, a_n\}$ is called a *partition* of the interval $[a,b]$, n is called the *length* of P, and

$$\text{mesh } P \equiv \max \{|a_{i+1} - a_i| : 0 \le i \le n - 1\}$$

is called the *mesh* of P. A partition Q of $[a,b]$ is a *refinement* of P if P is a subsequence of Q.

For each partition $P \equiv \{a_0, \ldots, a_n\}$ of $[a,b]$ and each continuous function f on $[a,b]$, $S(f,P)$ will denote an arbitrary sum of the type

$$(6.1) \qquad \sum_{i=0}^{n-1} f(x_i)(a_{i+1} - a_i)$$

where $x_i \in [a_i, a_{i+1}]$ $(0 \le i \le n - 1)$. In particular, if P is the partition $\{a = a_0, a + (b - a)/n = a_1, \ldots, b = a_n\}$, the quantity

$$(6.2) \qquad S(f,a,b,n) \equiv \frac{b - a}{n} \sum_{i=0}^{n-1} f\left(a + i\,\frac{b - a}{n}\right)$$

is one of the numbers $S(f,P)$.

Theorem 9 *Let f be a continuous function on a compact interval $[a,b]$ with modulus of continuity ω. Then the sequence*

$$(6.3) \qquad \{S(f,a,b,n)\}_{n=1}^{\infty}$$

converges, to a limit which is called the integral of f from a to b, *and is written*

$$(6.4) \qquad \int_a^b f(x)\,dx$$

If P is any partition of $[a,b]$, if $\epsilon > 0$, and if mesh $P \le \omega(\epsilon)$, then

$$(6.5) \qquad \left| S(f,P) - \int_a^b f(x)\,dx \right| \le \epsilon(b - a)$$

Proof Consider any partition $P \equiv \{a_0, \ldots, a_n\}$ of $[a,b]$ with mesh $P \le \omega(\epsilon)$ and any refinement $Q \equiv \{b_0, \ldots, b_m\}$ of P. For each i, $0 \le i \le n - 1$, let $\sum_i$ denote summation over those indices j for which $a_i \le b_j < a_{i+1}$. Then

$$(6.6)$$

$$
\begin{aligned}
|S(f,P) - S(f,Q)| &= \left| \sum_{i=0}^{n-1} f(x_i)(a_{i+1} - a_i) - \sum_{j=0}^{m-1} f(y_j)(b_{j+1} - b_j) \right| \\
&= \left| \sum_{i=0}^{n-1} f(x_i) \sum_i (b_{j+1} - b_j) - \sum_{i=0}^{n-1} \sum_i f(y_j)(b_{j+1} - b_j) \right| \\
&\le \sum_{i=0}^{n-1} \sum_i |f(x_i) - f(y_j)|(b_{j+1} - b_j) \\
&\le \sum_{i=0}^{n-1} \sum_i \epsilon(b_{j+1} - b_j) = \epsilon(b - a)
\end{aligned}
$$

Let R be another partition of $[a,b]$, with mesh $R \le \omega(\delta)$. Applying (6.6) to any common refinement Q of P and R, we get

$$(6.7) \quad |S(f,P) - S(f,R)| \leq |S(f,P) - S(f,Q)| + |S(f,Q) - S(f,R)|$$
$$\leq \epsilon(b - a) + \delta(b - a)$$

In particular,

$$(6.8) \qquad |S(f,a,b,j) - S(f,a,b,k)| \leq 2\epsilon(b-a)$$

for all positive integers j and k with $j,\ k \geq \omega(\epsilon)^{-1}(b - a)$. Therefore (6.3) is a Cauchy sequence. For each positive integer $k \geq \omega(\delta)^{-1}(b - a)$, inequality (6.7) becomes

$$|S(f,P) - S(f,a,b,k)| \leq (\epsilon + \delta)(b - a)$$

Passing to the limit as $k \to \infty$ and $\delta \to 0$ gives (6.5).

Proposition 13 *Let f and g be continuous functions on a compact interval $[a,b]$. Let α and β be real numbers. Then*

$$(6.9) \qquad \int_a^b (\alpha f(x) + \beta g(x))\, dx = \alpha \int_a^b f(x)\, dx + \beta \int_a^b g(x)\, dx$$

and

$$(6.10) \qquad \int_a^b f(x)\, dx \leq \int_a^b g(x)\, dx$$

if $f \leq g$.

Proof Equation (6.9) follows from the fact that

$$S(\alpha f + \beta g,\, a,\, b,\, n) = \alpha S(f,a,b,n) + \beta S(g,a,b,n)$$

for all n, and (6.10) from the inequalities

$$S(f,a,b,n) \leq S(g,a,b,n)$$

As a corollary we have

$$(6.11) \qquad c_1(b - a) \leq \int_a^b f(x)\, dx \leq c_2(b - a)$$

whenever $c_1 \leq f(x) \leq c_2$ for all x in $[a,b]$. By setting $-c_1 = c_2 = \|f\|_{[a,b]}$

we obtain

$$(6.12) \qquad \left| \int_a^b f(x)\, dx \right| \leq \|f\|_{[a,b]} (b - a)$$

Proposition 14 *If f is a continuous function on a compact interval $[a,b]$, then*

$$(6.13) \qquad \int_a^b f(x)\, dx = \int_a^c f(x)\, dx + \int_c^b f(x)\, dx \qquad (a \leq c \leq b)$$

Proof For arbitrary partitions $P \equiv \{a, \ldots, c\}$ of $[a,c]$ and $Q \equiv \{c, \ldots, b\}$ of $[c,b]$ we have

$$S(f,P) + S(f,Q) = S(f,R)$$

with appropriate choices of the approximating sums, where R is the partition $\{a, \ldots, c, \ldots, b\}$ of $[a,b]$. Since

$$\text{mesh } R = \max \{\text{mesh } P,\ \text{mesh } Q\}$$

equality (6.13) follows by passage to the limit as mesh $P \to 0$, mesh $Q \to 0$.

For arbitrary real numbers a and b and each continuous function f on $[\min \{a,b\},\ \max \{a,b\}]$ we define

$$(6.14) \qquad \int_a^b f(x)\, dx \equiv \int_\lambda^b f(x)\, dx - \int_\lambda^a f(x)\, dx$$

where $\lambda = \min \{a,b\}$. In case $a \leq b$ this agrees with the definition already given.

Equality (6.9) is valid for arbitrary real numbers a and b, provided that the integrals are defined (which means that f and g are continuous on the interval $[\min \{a,b\},\ \max \{a,b\}]$). Equality (6.13) is valid whenever f is continuous on the interval $[\min \{a,b,c\},\ \max \{a,b,c\}]$. In particular, setting $b = a$ in (6.13) gives

$$\int_a^c f(x)\, dx = - \int_c^a f(x)\, dx$$

whenever the integrals are defined. Equality (6.12) holds for arbitrary real numbers a and b, with the norm $\|f\|_{[a,b]}$ of f on the interval $[a,b]$ replaced by the norm of f on $[\min \{a,b\},\ \max \{a,b\}]$.

We now prove that differentiation and integration are inverse processes, a result whose imposing title is *The fundamental theorem of calculus*.

Theorem 10 *Let f be a continuous function on a proper interval I. Let a be a point of I. Write*

$$g(x) \equiv \int_a^x f(t)\,dt$$

for all x in I. Then $Dg = f$ on I. If g_0 is any differentiable function for which $Dg_0 = f$, the difference $g - g_0$ is a constant function.

Proof There is no loss of generality in taking I to be compact. Let ω be the modulus of continuity of f on I. For each $\epsilon > 0$ and each pair of points x, y in I with $|x - y| \leq \omega(\epsilon)$, we have

$$|g(y) - g(x) - f(x)(y - x)| = \left| \int_x^y f(t)\,dt - \int_x^y f(x)\,dt \right|$$
$$= \left| \int_x^y (f(t) - f(x))\,dt \right| \leq \epsilon|y - x|$$

Therefore g is differentiable on I with derivative f and modulus of differentiability ω. If also $Dg_0 = f$ on I, then $D(g - g_0) = 0$ on I. By the mean-value theorem, $(g - g_0)(a) = (g - g_0)(b)$ whenever a, $b \in I$ and $a < b$. Therefore $g - g_0$ is a constant function.

Our next lemma has an interest of its own.

Lemma 7 *Let $\{f_n\}$ be a sequence of continuous functions converging to a function f on a nonvoid interval I. Let a be any point of I. Write*

$$g(x) \equiv \int_a^x f(t)\,dt \qquad (x \in I)$$

and
$$g_n(x) \equiv \int_a^x f_n(t)\,dt \qquad (x \in I)$$

for each positive integer n. Then $g_n \to g$ on I as $n \to \infty$.

Proof It is no loss of generality to assume that I is compact. Let r be the length of I. Then for each x in I, we have

$$|g_n(x) - g(x)| = \left| \int_a^x (f_n(t) - f(t))\,dt \right|$$
$$\leq r\|f_n - f\|_I \to 0 \qquad \text{as } n \to \infty$$

Therefore $\{g_n\}$ converges to g on I.

Theorem 11 *Let $\{f_n\}$ be a sequence of differentiable functions which converges to a continuous function f on a proper interval I. Assume that the sequence $\{f_n'\}$ of derivatives converges to a continuous function g on I. Then $f' = g$ on I.*

Proof It is no loss of generality to assume that I is compact. Let a be any point of I. Write

$$h(x) \equiv \int_a^x g(t)\, dt \qquad (x \in I)$$

and
$$h_n(x) \equiv \int_a^x f_n'(t)\, dt$$

By Theorem 10, we have $h' = g$ and $h_n' = f_n'$ $(n \in Z^+)$. Hence $f_n - h_n$ is constant on I. By Lemma 7, we have $h_n \to h$ on I as $n \to \infty$. Therefore

$$f - h = \lim_{n \to \infty} (f_n - h_n)$$

is constant on I. It follows that $f' = h' = g$ on I, as was to be proved.

7. CERTAIN IMPORTANT FUNCTIONS

A good point of departure for the study of the exponential function is obtained from an analysis of its rate of change. Intuitively the exponential function is that differentiable function on **R**, normalized to have the value 1 at 0, which increases at a rate proportional to its size. In other words, it is the solution of the differential equation

$$(7.1) \qquad\qquad\qquad Df = f$$

normalized by the condition $f(0) = 1$. Clearly such a function f, if it exists, is infinitely differentiable, and $f^{(n)}(0) = 1$ for all nonnegative integers n. Thus the Taylor's series for f about the point 0 is

$$(7.2) \qquad\qquad\qquad \sum_{n=0}^{\infty} \frac{x^n}{n!}$$

This series converges to f on every finite interval I because the derivatives of f are all bounded by the same bound on I. Therefore f, if it exists, is unique, and is given by (7.2). On the other hand, (7.2) converges on **R**, by the comparison test. We therefore define the exponential function exp to be the sum of the series (7.2):

$$(7.3) \qquad \exp (x) \equiv \sum_{n=0}^{\infty} \frac{x^n}{n!}$$

Clearly exp $(0) = 1$. Also,

$$(7.4) \qquad \frac{d \exp (x)}{dx} = \sum_{n=0}^{\infty} \frac{nx^{n-1}}{n!} = \sum_{n=1}^{\infty} \frac{x^{n-1}}{(n-1)!} = \exp (x)$$

by Theorem 11. The function exp has the desired properties, and is unique.

Proposition 15   *The function* exp *satisfies the* functional equation

$$(7.5) \qquad \exp (x + y) = \exp (x) \exp (y) \qquad (x, y \in \mathbf{R})$$

Proof   Let z be any positive real number. Then $\exp (z) \neq 0$, by (7.3). Define the differentiable function $f : \mathbf{R} \to \mathbf{R}$ by

$$f(x) = (\exp (z))^{-1} \exp (x + z)$$

The derivative of f is

$$\frac{df(x)}{dx} = (\exp (z))^{-1} \exp (x + z) \frac{d(x + z)}{dx} = f(x)$$

by (5.6). Also, $f(0) = 1$. By the uniqueness of the exponential function, $f(x) = \exp (x)$, or

$$(7.6) \qquad \exp (x + z) = \exp (x) \exp (z)$$

Consider now arbitrary real numbers x and y. Choose z with $z > 0$ and $z + y > 0$. Then, by (7.6),

$$\begin{aligned}
\exp (x + y) \exp (z) &= \exp (x + y + z) \\
&= \exp (x) \exp (y + z) = \exp (x) \exp (y) \exp (z)
\end{aligned}$$

which is equivalent to (7.5).

It can be shown that exp is the unique continuous function from $\mathbf{R}$ to $\mathbf{R}$ that has the value 1 at 0 and satisfies (7.5).

Equality (7.5) implies that $\exp(-x) = (\exp(x))^{-1}$. Since $\exp(x) \geq 1$ for $x \geq 0$, it follows that

$$(7.7) \qquad\qquad \exp(x) > 0 \qquad (x \in \mathbf{R})$$

The theory of the exponential function can also be approached as follows. To construct a function f with $f(0) = 1$ and $f' = f$, notice that $f' = f$ implies, by the mean-value theorem, that $f(x + \delta)$ is approximately equal to $f(x) + \delta f'(x) = (1 + \delta)f(x)$ when δ is small. Therefore, if t is any real number and n is a large positive integer, $f(t/n)$ is approximately $1 + (t/n), f(2t/n) = f(t/n + t/n)$ is approximately $(1 + (t/n))^2$, and so forth, and thus $f(t) = f(n(t/n))$ is approximately $(1 + (t/n))^n$. This heuristic argument suggests that we define

$$\exp(t) = \lim_{n \to \infty} \left(1 + \frac{t}{n}\right)^n$$

It is not hard to see that

$$\left(1 + \frac{t}{n}\right)^n = 1 + n\frac{t}{n} + \frac{n(n-1)}{2!}\left(\frac{t}{n}\right)^2 + \cdots$$

converges on $\mathbf{R}$ to $\sum_{n=0}^{\infty} t^n/n!$ as $n \to \infty$. This approach to the theory of the exponential function would therefore lead to the same series representation.

The logarithmic function ln is the inverse function to the exponential function, but we shall not define it that way. Arguing heuristically, on the basis of the chain rule (5.6), we say that

$$1 = \frac{dx}{dx} = \frac{d}{dx}(\exp(\ln(x))) = \exp(\ln(x))\frac{d}{dx}(\ln(x)) = x\frac{d}{dx}\ln(x)$$

or

$$(7.8) \qquad\qquad \frac{d}{dx}(\ln(x)) = x^{-1} \qquad (x > 0)$$

We therefore define ln (x) to be the integral of x^{-1}, specifically

$$(7.9) \qquad\qquad \ln(x) = \int_1^x t^{-1}\, dt \qquad (x > 0)$$

By Theorem 10, ln is differentiable on $(0, \infty)$, and (7.8) is valid.

Let y be any positive real number. By (5.6), the derivative of the function

$$(7.10) \qquad x \rightarrow \ln(xy) - \ln(y)$$

is x^{-1}. Also, (7.10) vanishes at 1. By the last statement of Theorem 10, $\ln(x) = \ln(xy) - \ln(y)$, or

$$(7.11) \qquad \ln(xy) = \ln(x) + \ln(y) \qquad (x, y > 0)$$

This is the functional equation for the logarithmic function, corresponding to the functional equation (7.5) for the exponential function.

By (7.7), the composite function $\ln \circ \exp$ exists and is differentiable everywhere on $\mathbf{R}$. By (5.6), we have

$$\frac{d}{dx}(\ln(\exp(x))) = 1 \qquad (x \in \mathbf{R})$$

Since also $\ln(\exp(0)) = \ln(1) = 0$, we have

$$(7.12) \qquad \ln(\exp(x)) = x \qquad (x \in \mathbf{R})$$

by the last part of Theorem 10. Consider $x > 0$ and write $y \equiv \exp(\ln(x))$. Then

$$\ln(y) = \ln(\exp(\ln(x))) = \ln(x)$$

by (7.12), or

$$0 = \ln(y) - \ln(x) = \int_x^y t^{-1}\,dt$$

If $y > x$, this gives $0 \geq y^{-1}(y - x)$, a contradiction. By Lemma 5, we have $y \leq x$. Similarly, $x \leq y$. Hence $x = y$. Thus $\exp(\ln(x)) = x$ for all $x > 0$. It follows that the functions $\exp: \mathbf{R} \rightarrow \mathbf{R}^+$ and $\ln: \mathbf{R}^+ \rightarrow \mathbf{R}$ are inverse to each other.

The trigonometric functions sin and cos also can be approached via an intuitive analysis of their rates of change. From this analysis we are led to believe that

$$(7.13) \qquad \frac{d}{dx}\cos x = -\sin x \qquad \frac{d}{dx}\sin x = \cos x \qquad (x \in \mathbf{R})$$

We therefore define these functions by power series constructed in such a way that (7.13) will hold. Remembering that cos (0) = 1, sin (0) = 0, we are forced to define

$$
(7.14) \qquad \cos x \equiv \sum_{n=0}^{\infty} (-1)^n \frac{x^{2n}}{(2n)!}
$$
$$
\sin x \equiv \sum_{n=0}^{\infty} (-1)^n \frac{x^{2n+1}}{(2n+1)!} \qquad (x \in \mathbf{R})
$$

Theorem 11 implies that (7.13) is valid.

Proposition 16 *The functions sin and cos satisfy the functional equations*

$$
(7.15) \qquad
\begin{aligned}
(a) \quad & \sin (x + y) = \sin x \cos y + \cos x \sin y \\
(b) \quad & \cos (x + y) = \cos x \cos y - \sin x \sin y
\end{aligned}
$$

for all x, y in **R**.

Proof Consider both sides of (7.15)(a) as functions of x, say $f(x)$ and $g(x)$. The functions f and g have Taylor's series about $x = 0$ which converge on **R**. The coefficients of these series are expressed in terms of the various derivatives of f and g at $x = 0$. Therefore, to show that $f = g$ on **R**, it is enough to show $f^{(n)}(0) = g^{(n)}(0)$ for all $n \geq 0$. Since $f^{(2)} = -f$ and $g^{(2)} = -g$, it is enough to notice that

$$
f(0) = \sin y = \sin 0 \cos y + \cos 0 \sin y = g(0)
$$
$$
\text{and} \qquad f'(0) = \cos y = \cos 0 \cos y + (-\sin 0) \sin y = g'(0)
$$

Therefore (7.15)(a) is valid. Equation (7.15)(b) follows from (7.15)(a) by differentiation with respect to x.

Putting $y = -x$ in (7.15)(b) gives the useful identity

$$
(7.16) \qquad \cos^2 x + \sin^2 x = 1
$$

Further study of the trigonometric functions depends on properties of the number π, which we construct as twice the first positive zero of the cosine function.

Theorem 12   *The sequence $\{x_n\}$ of real numbers defined inductively by the equations*

$$(7.17) \qquad (a) \quad x_1 \equiv 1$$
$$(b) \quad x_{n+1} \equiv x_n + \cos x_n \qquad (n \geq 1)$$

is a monotone-increasing converging sequence, whose limit $\pi/2$ satisfies the conditions

$$(7.18) \qquad (a) \quad \cos \frac{\pi}{2} = 0$$

$$(b) \quad \cos x > 0 \qquad (0 \leq x < \pi/2)$$

Proof   As a start we prove that

$$(7.19) \qquad \cos t > 0 \qquad (0 \leq t \leq x_n,\ n \in Z^+)$$

Consider first the case $n = 1$, so that $0 \leq t \leq x_1 = 1$. Then

$$\cos t = \left(1 - \frac{t^2}{2!}\right) + \left(\frac{t^4}{4!} - \frac{t^6}{6!}\right) + \cdots \geq 1 - \frac{1}{2} > 0$$

Next assume that (7.19) is true for a particular value of n, and consider a value of t with $x_n < t \leq x_{n+1}$. Now $|\sin x| \leq 1$ for all x in $\mathbf{R}$, by (7.16), and $|\sin x_n| < 1$, because $\cos x_n > 0$. Hence

$$\cos t = \cos x_n - \int_{x_n}^{t} \sin x \, dx > \cos x_n - (x_{n+1} - x_n) = 0$$

Since $\cos$ is a continuous function and $\cos t > 0$ whenever $0 \leq t \leq x_n$ or $x_n < t \leq x_{n+1}$, $\cos t > 0$ whenever $0 \leq t \leq x_{n+1}$. By induction, (7.19) holds for all n. Since $\cos x_n > 0$ for all n, the sequence $\{x_n\}$ is monotone-increasing.

From the mean-value theorem and (7.19) we obtain

$$\sin x_n \geq \sin 0 = 0$$

for all n, and $\sin t \geq \sin 1$ whenever $1 \leq t \leq x_n$ for some n. Also by the mean-value theorem, for each $n \geq 2$ there exists t in $[x_{n-1}, x_n]$ such that

$$x_{n+1} - x_n = x_n + \cos x_n - (x_{n-1} + \cos x_{n-1})$$
$$= x_n - x_{n-1} + \cos x_n - \cos x_{n-1}$$

is arbitrarily near to

$$x_n - x_{n-1} - (\sin t)(x_n - x_{n-1}) \leq (x_n - x_{n-1})(1 - \sin 1)$$

Therefore

$$x_{n+1} - x_n \leq (1 - \sin 1)(x_n - x_{n-1})$$

This gives

$$x_{n+1} - x_n \leq (1 - \sin 1)^{n-1}(x_2 - x_1)$$

for all $n \geq 1$. Thus $\{x_n\}$ is a Cauchy sequence, whose limit we denote by $\pi/2$.

If $0 \leq x < \pi/2$, then $x < x_n$ for some n, and thus $\cos x > 0$, by (7.19). Taking limits on both sides of (7.17)(b) as $n \to \infty$ gives (7.18)(a).

From (7.15) and (7.18)(a) we get $\sin (x + \pi/2) = \cos x$, and thus $\sin \pi/2 = 1$. Similarly $\sin (\pi/2 - x) = \cos x$ and $\cos (x + \pi/2) = - \sin x$. Hence $\sin (x + \pi) = \cos (x + \pi/2) = - \sin x$, and thus $\sin (x + 2\pi) = \sin x$. Similarly $\cos (x + \pi) = - \cos x$ and $\cos (x + 2\pi) = \cos x$. The functions sin and cos therefore are periodic of period 2π.

Next we study the function arc sin, which is inverse to the function sin. For motivation we take the derivative of the equation

$$\sin (\text{arc sin } x) = x$$

and get

$$\cos (\text{arc sin } x) \frac{d}{dx} (\text{arc sin } x) = 1$$

Since $\quad 1 = \cos^2 (\text{arc sin } x) + \sin^2 (\text{arc sin } x) = \cos^2 (\text{arc sin } x) + x^2$

it follows that

$$\frac{d}{dx} (\text{arc sin } x) = \pm (1 - x^2)^{-1/2}$$

With this motivation, we define

$$\text{arc sin } x = \int_0^x (1 - t^2)^{-1/2} \, dt \qquad (-1 < x < 1)$$

Then arc sin is differentiable on $(-1,1)$ and

$$\frac{d}{dx} (\text{arc sin } x) = (1 - x^2)^{-1/2}$$

If $-\pi/2 < x < \pi/2$, then $-1 < \sin x < 1$, and

$$\frac{d}{dx} \text{arc sin } (\sin x) = (1 - \sin^2 x)^{-1/2} \cos x = 1$$

Hence

$$\text{arc sin } (\sin x) = x \qquad (-\pi/2 < x < \pi/2)$$

Therefore the function arc sin maps the interval $(-1,1)$ into the interval $(-\pi/2,\pi/2)$. For $-1 < x < 1$ let $y = \sin (\text{arc sin } x)$. Then

$$\text{arc sin } y = \text{arc sin } x$$

or

$$\int_x^y (1 - t^2)^{-1/2} \, dt = 0$$

Therefore $x = y = \sin (\text{arc sin } x)$. It follows that the functions $\sin: (-\pi/2,\pi/2) \to (-1,1)$ and arc sin: $(-1,1) \to (-\pi/2,\pi/2)$ are inverses.

PROBLEMS

The word *"not"* is to be understood in the sense of the discussion at the end of Sec. 2.

1. Construct a set A such that $x = y$ for all elements x and y of A, but A is *not* subfinite or void.

2. Construct a surjection which is *not* onto.

3. Prove that $(x,y) \to xy$ is a function from $\mathbf{R} \times \mathbf{R}$ into $\mathbf{R}$.

4. Let x and y be real numbers such that $x \neq y$ implies $0 = 1$. Show that $x = y$.

5. If $x_1, \ldots, x_n$ are real numbers with $x_1 \cdots x_n < 0$, show that $x_i < 0$ for some i.

6. Call a pair (S,T) of nonvoid subsets of the set Q of rational numbers a *Dedekind cut* if $s < t$ whenever $s \in S$ and $t \in T$, and if for arbitrary rational numbers x and y with $x < y$ either $x \in S$ or $y \in T$. Show that for each Dedekind cut (S,T) there exists a unique real number $x(S,T)$, such that $s \leq x(S,T)$ for all s in S and $t \geq x(S,T)$ for all t in T. Show that for each real number x there exists a Dedekind cut (S,T) with $x(S,T) = x$.

7. Show that the real number system can be constructed from the Dedekind cuts. In other words, define directly equality, order, addition, and multiplication of Dedekind cuts in such a way that the map $(S,T) \to x(S,T)$ from the set of all Dedekind cuts into $\mathbf{R}$ is a bijection which preserves order and the operations of addition and multiplication.

8. Construct a real number a such that a is *not* positive or negative or equal to 0.

9. Construct a continuous function $f\colon [0,1] \to \mathbf{R}$ such that there is *not* a point x in $[0,1]$ with $f(x)$ equal to the supremum of f on $[0,1]$.

10. Show that a continuous function f on a compact interval admits a modulus of continuity ω which is a continuous function.

11. Let $f\colon [0,1] \to \mathbf{R}$ be a continuous function, with $f(0) < 0$ and $f(1) > 0$. Show that for each $\epsilon > 0$ there exists x in $[0,1]$ with $|f(x)| < \epsilon$.

12. Construct a continuous function $f\colon [0,1] \to \mathbf{R}$ with $f(0) < 0$ and $f(1) > 0$ such that there does *not* exist a point x in $[0,1]$ with $f(x) = 0$.

13. Let $f\colon [0,1] \to \mathbf{R}$ be a continuous function, with $f(0) < 0$ and $f(1) > 0$, such that for arbitrary real numbers a and b with $0 \leq a < b \leq 1$ there exists x in $[a,b]$ with $f(x) \neq 0$. Show that there exists x in $[0,1]$ with $f(x) = 0$.

14. Let $f\colon [0,1] \to \mathbf{R}$ be a continuous function, with $f(0) < 0$ and $f(1) > 0$, such that there exists a positive integer n and a positive constant c with $|f(x)| + |f'(x)| + \cdots + |f^{(n)}(x)| \geq c$ for all x in $[0,1]$. Show that there exists x in $[0,1]$ with $f(x) = 0$.

15. Let $f\colon [0,1] \to \mathbf{R}$ be a polynomial function, with $f(0) < 0$ and $f(1) > 0$. Show that there exists x in $[0,1]$ with $f(x) = 0$.

16. Show that if $f\colon [0,1] \to [0,1]$ is a continuous function, then for each $\epsilon > 0$ there exists x in $[0,1]$ with $|f(x) - x| < \epsilon$.

17. Construct a continuous function $f\colon [0,1] \to \mathbf{R}$ such that f' does not exist on any compact proper subinterval of $[0,1]$. (In other words, the assumption that f' exists on a compact proper subinterval leads to a contradiction.)

NOTES

As the definitions of Sec. 1 indicate, a single concept of classical mathematics may split into two or more distinct concepts when looked at constructively. For example, the distinction between a function which is onto and a surjection does not exist classically, nor does the distinction between a finite set and a subfinite set.

An alternative to Definition 1 would be to define real numbers by the method of Dedekind cuts (see Prob. 6). Another possibility would be to define a real number as a sequence $\{x_n\}$ of rational numbers *and* a sequence $\{N_k\}$ of integers, such that $|x_m - x_n| \leq k^{-1}$ whenever $m, n \geq N_k$.

Notice that a real number *is* a regular sequence of rational numbers, not an equivalence class of regular sequences of rational numbers. To define a real number to be such an equivalence class would be either incorrect or pointless; pointless (but correct) if the equivalence class is required to be specified by giving some particular regular sequence of rational numbers that belongs to it, and incorrect otherwise.

Precisely what is meant by a subset (in particular, what it means for $\mathbf{R}^+$ and $\mathbf{R}^{0+}$ to be subsets of $\mathbf{R}$) will be discussed in Chap. 3.

In classical mathematics, to assume an inequality $x > y$, derive a contradiction, and conclude that therefore $x \leq y$ (as in Lemma 5) is called an indirect proof. It is necessary to use caution in thinking of Lemma 5 in this way. Although $x \leq y$ is equivalent to the negation of $x > y$, it is *not* true that $x > y$ is equivalent to the negation of $x \leq y$.

There is a paradox growing out of Theorem 1 which the reader should resolve. Since every regular sequence of rational numbers can presumably be described by a phrase in the English language, and since the phrases in the English language can be sequentially ordered, the regular sequences of rational numbers can be sequentially ordered, in contradiction to Theorem 1.

The counterexamples following Theorem 1 are deceitful. The reader is asked not to form the impression that the purpose of constructive mathematics is to consider pathological numbers like the ones discussed here. The only reason for discussing such numbers is to show that certain statements (in this case, the statement that for each x in $\mathbf{R}^{0+}$ either $x > 0$ or $x = 0$) are not constructively valid. To appreciate this point better, the reader should examine the counterexamples given in Chap. 1, where the treatment is more detailed.

Sometimes a mathematician, when confronted with a counterexample of this sort, will say he is not interested in real numbers that have been artificially twisted out of shape. One reply to this is to point out that the set of real numbers x such that either $x > 0$ or $x \leq 0$ is *not* complete (nor is it closed under addition).

The "choice" involved in the proof of (*b*) of Proposition 8 is performed according to a definite rule.

Notice that divergence is a positive notion.

The concept defined in Definition 9 is classically called uniform continuity. We abbreviate this to continuity, since the classical version of continuity (i.e., pointwise continuity) is not used. Classically, one proves that pointwise continuity (on a compact interval) implies uniform continuity. Constructively, there is no proof of this known, so we assume uniform continuity from the start. Nothing essential is lost, and a certain simplicity is gained. Similar remarks apply to Definitions 11 and 13.

As discussed in Chap. 1, a nonvoid bounded set of real numbers need *not* have a least upper bound. This, of course, makes some parts of constructive mathematics harder than the corresponding classical material, but it also makes some parts more interesting.

It would be more correct to call Lemma 6 a constructive substitute for Rolle's theorem—the classical result states that $f'(x) = 0$ for some x in (a,b)

The proofs of Lemma 6 and Theorem 8 illustrate the minor inconveniences caused by not being able to compare arbitrary real numbers.

For Riemann integration of functions that are possibly discontinuous, see Prob. 4 of Chap. 7.

CHAPTER 3

SET THEORY

The chapter begins with a fuller discussion of sets and functions, including the constructive meaning of such terms as subset, union, and inclusion. Certain classical laws of the algebra of sets carry over, and others do not. In Sec. 2 we introduce the basic notion of a complemented set, which is used later (in Chaps. 6 and 7) to facilitate the development of the theory of measure and integration. The chapter closes with some remarks on general topology which support the conclusion that in most cases of interest it is not possible to define continuous functions in terms of the family of open sets.

Very little is left of general topology after that vehicle of classical mathematics has been taken apart and reassembled constructively. With some regret, plus a large measure of relief, we see this flamboyant engine collapse to constructive size. First, however, we come to terms with another old friend, the reverend theory of sets.

1. SOME BASIC NOTIONS OF THE THEORY OF SETS

As discussed previously, a *set* is defined by describing what must be done to construct an *element* of the set, and what must be done to show that two elements of the set are *equal*.

A *function* $f: A \to B$ is a rule which associates an element $b \equiv f(a)$ of a set B to each element a of a set A, in such a way that b can be found by a finite routine when a is given. Equal elements of B must be associated to equal elements of A. The set of all functions from a set A to a set B will be written $F(A,B)$.

When A is not countable the set $F(A,B)$ seems to have little practical interest, because to get a hold on its structure is too hard. For instance, it has been asserted by Brouwer that all functions in $F(\mathbf{R},\mathbf{R})$ are continuous, but no acceptable proof is known.

The developments of Sec. 1 of Chap. 2 remain in force, and the reader should review them at this time.

Definition 1 A *subset* (A,i) of a set B consists of a set A and a function $i: A \to B$, called the *inclusion map*, such that

$$(1.1) \qquad a_1 = a_2 \qquad \text{if and only if} \qquad i(a_1) = i(a_2)$$

for all a_1 and a_2 in A.

Although both the set A and the inclusion map i must be given for the subset (A,i) of B to be described, we shall speak somewhat loosely of "a subset A of a set B," the inclusion map being understood.

Definition 1 can be viewed as follows. The requirements for constructing an element a of A are (i) to construct an element $b = i(a)$ of B, (ii) to perform certain additional constructions, and (iii) to verify that certain conditions are satisfied.

It is convenient at times to identify an element a of a subset A of a set B with its *image* $i(a)$ in B. Also, we speak on occasion of an element a of A as *belonging* to B, and an element b of B as *belonging* to A in case $b = i(a)$ for some a in A.

Sometimes the construction of an element of a set A implicitly entails the construction of an element of a set B, without this being an explicit part of the definition. In such a case we are at liberty to bring the inclusion map to the fore and realize A as a subset of B, or not, as we see fit, providing, of course, that the equality relation on A agrees with the equality relation on B. In this spirit we regard Z as a subset of Q and Q as a subset of $\mathbf{R}$. If A is a subset of B and B is a subset of C, there is an obvious inclusion map realizing A as a subset of C.

Definition 2 If f is a function from a set A to a set B, the *image* $f(A)$ of A is the subset of B whose elements are the elements of A, whose equality relation is defined by taking $a_1 = a_2$ (in the new sense) to mean $f(a_1) = f(a_2)$, and whose inclusion map is f. If C is a subset of B, then $f^{-1}(C)$ is that subset of A consisting of all (a,c) in $A \times C$ such that $f(a) = i(c)$, with inclusion map $(a,c) \to a$.

Definition 3 Let A and B be subsets of a set S, with inclusion maps i_A and i_B. The subsets $A \cup B$ and $A \cap B$ of S, and the relations $A \subset B$ and $A = B$ are defined as follows.

(a) To construct an element c of $A \cup B$, either construct an element a of A or construct an element b of B. In the former case define $i(c) \equiv i_A(a)$, and in the latter case $i(c) \equiv i_B(b)$. Define elements c_1 and c_2 of $A \cup B$ to be *equal* if $i(c_1) = i(c_2)$.

(b) To construct an element c of $A \cap B$, construct elements a of A and b of B with $i_A(a) = i_B(b)$. Write $i(c) \equiv i_A(a)$, and define $c_1 = c_2$ to mean that $i(c_1) = i(c_2)$.

(c) To show that $A \subset B$, construct a function $f : A \to B$ such that $i_A = i_B \circ f$.

(d) To show that $A = B$, show that $A \subset B$ and $B \subset A$.

The standard laws of the algebra of sets follow from Definition 3. These are the commutative laws,

$$A \cup B = B \cup A \qquad A \cap B = B \cap A$$

the associative laws,

$$A \cup (B \cup C) = (A \cup B) \cup C \qquad A \cap (B \cap C) = (A \cap B) \cap C$$

the distributive laws,

$$A \cap (B \cup C) = (A \cap B) \cup (A \cap C) \qquad A \cup (B \cap C) = (A \cup B) \cap (A \cup C)$$

and the laws

$$A \cap B \subset A \qquad A \subset A \cup B$$

The void subset ϕ of a set S is defined as follows. To construct an element θ of ϕ, (i) construct an element u of S and (ii) prove that $0 = 1$. The inclusion map is defined by $i(\theta) \equiv u$. The definition of ϕ is negativistic, and we prefer to mention the void set as seldom as possible.

Definition 4   A family $(T,\lambda) \equiv \{\lambda(t)\}_{t \in T}$ of subsets of a set S is a rule λ which associates a subset $\lambda(t)$ of S to each element t of an *index set* T, equal subsets being associated to equal elements of T.

(An alternative approach to Definition 4 would be to define a family of subsets of S as a subset of the cartesian product $S \times T$.)

The set-theoretic operations can be generalized to arbitrary families (T,λ) of subsets of a set S. For example, the subset

$$\bigcup_{t \in T} \lambda(t) \equiv \{s \colon s \in \lambda(t) \text{ for some } t \text{ in } T\}$$

of S is called the *union* of the family (T,λ). More precisely, to construct an element u of $\bigcup_{t \in T} \lambda(t)$ we first construct an element t of T, and then construct an element v of $\lambda(t)$; then $i(u)$ is defined to be $i(v)$, and elements u_1 and u_2 are defined to be *equal* if $i(u_1) = i(u_2)$.

As another example, in case T is nonvoid the subset

$$\bigcap_{t \in T} \lambda(t) \equiv \{s \colon s \in S, s \text{ belongs to } \lambda(t) \text{ for all } t \text{ in } T\}$$

of S is called the *intersection* of the family (T,λ). More precisely, an element u of $\bigcap_{t \in T} \lambda(t)$ is a rule that associates an element u_t of $\lambda(t)$ to each element t of T, such that $i_{t_1}(a_{t_1}) = i_{t_2}(a_{t_2})$ whenever $t_1, t_2 \in T$. The image $i(u)$ of u is defined to be $i_{t_0}(a_{t_0})$, where t_0 is the distinguished element of the nonvoid set T. Elements u_1 and u_2 are *equal* if $i(u_1) = i(u_2)$.

In case (T,λ) has the property that $\lambda(t_1) = \lambda(t_2)$ if and only if $t_1 = t_2$, for all t_1, t_2 in T, then (T,λ) is called a *set of subsets* of S.

An important family of subsets of a set S is the family of *free* subsets. The free sets are indexed by the set $T \equiv F(S,\{0,1\})$ of functions from S to $\{0,1\}$. For each f in T the set $\lambda(f)$ consists of all s in S with $f(s) = 1$. (In other words, A is free if there exists a finite routine for deciding when an arbitrary element of S belongs to A.) The free sets $\lambda(f)$ and $\lambda(g)$ are equal if and only if $f = g$. Therefore the free sets are a set of subsets of S. For all f and g in T the following algebraic laws are valid:

$$\lambda(f) \cap \lambda(g) = \lambda(fg) \qquad \lambda(f) \cup \lambda(g) = \lambda(f + g - fg)$$

Each free set $\lambda(f)$ has a *complement* $-\lambda(f) \equiv \lambda(1 - f)$. The operation $\lambda(f) \rightarrow -\lambda(f)$ of complementation obeys the following algebraic laws:

$$-(-\lambda(f)) = \lambda(f)$$
$$-(\lambda(f) \cup \lambda(g)) = (-\lambda(f)) \cap (-\lambda(g))$$
$$-(\lambda(f) \cap \lambda(g)) = (-\lambda(f)) \cup (-\lambda(g))$$

A set S is called *discrete* if

$$D \equiv \{(s_1,s_2): s_1 \in S,\ s_2 \in S,\ s_1 = s_2\}$$

is a free subset of $S \times S$. This means there exists a finite routine for deciding when given elements of S are equal.

2. BOREL SETS

It is often necessary to consider families of sets that are closed under countable unions and countable intersections, and under a suitable operation of complementation. The operation of complementation entails difficulties. On the one hand, we do not wish to define complementation in terms of negation, but on the other hand, this seems to be the only general method available. The way out of this awkward position is to permit great flexibility in the formalism of complementation. A very flexible notion is the concept of a complemented set.

Definition 5 Let X be a set and F a nonvoid set of functions from X to $\mathbf{R}$. A *complemented set* in X (relative to F) consists of an ordered pair (A,B) of subsets of X such that for each x in A and y in B there exists f in F with $f(x) \neq f(y)$. The *union* $(A,B) \equiv \bigcup_{t \in T} (A_t,B_t)$ of a family $\{(A_t,B_t)\}_{t \in T}$ of complemented sets is defined by

$$A \equiv \bigcup_{t \in T} A_t \qquad B \equiv \bigcap_{t \in T} B_t$$

The *intersection* $(A,B) \equiv \bigcap_{t \in T} (A_t,B_t)$ is defined by

$$A \equiv \bigcap_{t \in T} A_t \qquad B \equiv \bigcup_{t \in T} B_t$$

The *complement* $-(A,B)$ of a complemented set (A,B) is defined by $-(A,B) \equiv (B,A)$. Complemented sets (A,B) and (C,D) are *equal* if $A = C$ and $B = D$.

The complement of a complemented set is clearly a complemented set. A union $(A,B) \equiv \bigcup_{t \in T} (A_t, B_t)$ of complemented sets is a complemented set. To see this, consider $x \in A$ and $y \in B$. Then $x \in A_t$ for some t. Since $y \in B_t$, there exists f in F with $f(x) \neq f(y)$. Therefore (A,B) is a complemented set. Similarly, an intersection of complemented sets is a complemented set.

We say that an element x of X *belongs* to the complemented set (A,B), and write $x \in (A,B)$, if $x \in A$.

A complemented set (A_1, B_1) is a *subset* of a complemented set (A_2, B_2), written $(A_1, B_1) \subset (A_2, B_2)$, if $A_1 \subset A_2$ and $B_2 \subset B_1$.

The following algebraic laws are valid:

$$- -(A,B) = (A,B)$$
$$- \bigcup_{t \in T} (A_t, B_t) = \bigcap_{t \in T} - (A_t, B_t)$$
and
$$- \bigcap_{t \in T} (A_t, B_t) = \bigcup_{t \in T} - (A_t, B_t)$$

In addition, all the usual finite algebraic laws which do not involve the operation of set complementation are satisfied. For instance, the distributive law

$$U \cup (V \cap W) = (U \cup V) \cap (U \cup W)$$

is valid for complemented sets U, V, and W. On the other hand, the law $U \cap (V \cup -V) = U$ is *not* valid. The infinite distributive law

$$(2.1) \qquad U \cup \bigcap_{t \in T} U_t = \bigcap_{t \in T} U \cup U_t$$

also fails. To see this, take $X = \mathbf{R}$ and let F be the set of all continuous functions from X to $\mathbf{R}$. Let $\{k_t\}$ be a sequence of integers, each of which is either 0 or 1, for which it is not known whether $k_t = 0$ for any t. Write

$$U_t \equiv \begin{cases} (\mathbf{R}, \phi) & \text{if } k_t = 1 \\ (\phi, \mathbf{R}) & \text{if } k_t = 0 \end{cases}$$

Write
$$U \equiv \bigcup_{t=1}^{\infty} - U_t$$

Then $U \cup U_t = (\mathbf{R}, \phi)$ for each t, and thus the right side of (2.1) equals $(\mathbf{R}, \phi)$. On the other hand, we have no way of constructing any element of $\mathbf{R}$ which belongs to $U \cup \bigcap_{t=1}^{\infty} U_t$. Hence (2.1) is *not* valid.

The invalid law (2.1) has the constructive substitute

$$(2.2) \qquad (U \cup -U) \cap (U \cup \bigcap_{t \in T} U_t) = (U \cup -U) \cap \bigcap_{t \in T} U \cup U_t$$

whose proof is left to the reader. The dual law,

$$(2.3) \qquad (U \cap -U) \cup (U \cap \bigcup_{t \in T} U_t) = (U \cap -U) \cup \bigcup_{t \in T} U \cap U_t$$

is also valid.

If to each complemented set (A,B) we associate the set A, we get a map j from the complemented sets in X into the subsets of X, which preserves all operations. Since F is nonvoid, (A,ϕ) is a complemented set for each subset A of X. Hence the map j is onto. The complemented set (X,ϕ) is denoted by X_0. Its complement (ϕ,X) is denoted by ϕ_0. Clearly $j(X_0) = X$ and $j(\phi_0) = \phi$.

Definition 6 Let $(\mathfrak{F},\varphi)$ be a family of complemented sets in a set X, relative to a set F of real-valued functions on X. A complemented set U in X will be called a *Borel set* generated by the family $(\mathfrak{F},\varphi)$ if it can be constructed inductively by means of the following principles of construction.

   (i)   Each complemented set $\varphi(t)$, with $t \in \mathfrak{F}$, is a Borel set.
   (ii) Each countable union of Borel sets already constructed is a Borel set.
   (iii) Each countable intersection of Borel sets already constructed is a Borel set.

The Borel sets generated by $(\mathfrak{F},\varphi)$ are closed under the operations of taking countable unions and countable intersections.

Corresponding to the inductive construction of the Borel sets, most of their properties are established by induction, as follows. To show that every Borel set has a given property P, first show that every set $\varphi(t)$, with $t \in \mathfrak{F}$, has property P, and then show that if $\{U_n\}$ is a sequence of complemented sets having property P, its union and intersection also have property P. As an example, it is easily seen that if $-\varphi(t)$ is a Borel set for all t in $\mathfrak{F}$, then $-U$ is a Borel set for all Borel sets U.

In most cases of interest the Borel sets are generated by the family F, as follows. Write $\mathfrak{F} \equiv F$. For each f in F, write

$$\varphi(f) \equiv (A_f, B_f)$$

where $A_f \equiv \{x \in X : f(x) > 0\}$ $B_f \equiv \{x \in X : f(x) \leq 0\}$

Then (F,φ) is a family of complemented sets, and the Borel sets generated by this family are called the *Borel sets generated by F*.

In case $\epsilon - f \in F$ for each f in F and each $\epsilon > 0$, the Borel sets generated by F are closed under complementation. This can be seen as follows: by the above remark it is enough to show that $-\varphi(f) \equiv (B_f, A_f)$ is a Borel set for each f in F. In fact

$$-\varphi(f) = \bigcap_{n=1}^{\infty} \varphi(n^{-1} - f)$$

Consider the special case $X = \mathbf{R}$, and let F be the set of continuous functions from $\mathbf{R}$ to $\mathbf{R}$. Then $(\mathbf{R}^{0+}, \mathbf{R}^{-})$ is a Borel set. Thus the effect of our definitions is to define $\mathbf{R}^{-}$ to be the complement of $\mathbf{R}^{0+}$. A different result would be obtained by basing the notion of complementation on negation (that is, by defining x to be in the complement of $\mathbf{R}^{0+}$ to mean that the assumption $x \in \mathbf{R}^{0+}$ leads to a contradiction), since we have no way to show that $x \in \mathbf{R}^{-}$ whenever the assumption $x \in \mathbf{R}^{0+}$ leads to a contradiction.

On the other hand, it is true that $x \in \mathbf{R}^{0+}$ if the assumption $x \in \mathbf{R}^{-}$ leads to a contradiction. This follows from Lemma 5 of Chap. 2.

Thus, if we were to base complementation on the notion of negation, the complement of $\mathbf{R}^{-}$ would be $\mathbf{R}^{0+}$, but the complement of $\mathbf{R}^{0+}$ would not be $\mathbf{R}^{-}$. Hence a set would not equal the complement of its complement.

3. NEIGHBORHOOD SPACES AND FUNCTION SPACES

Sometimes a set X comes with a family of subsets that can be used to define a notion of proximity; for instance, the real numbers come with the family of all open intervals. This observation leads classically to the idea of a topological space, but constructively it seems more natural to introduce the concept of a neighborhood space as follows.

Definition 7   A *neighborhood space* is a set X together with a family of subsets of X, called *neighborhoods*, such that if U_1 and U_2 are neighborhoods and $x \in U_1 \cap U_2$, then there exists a neighborhood U_3 with

$$x \in U_3 \subset U_1 \cap U_2$$

If x is an element of a neighborhood U, then U is called a *neighborhood of x*. A subset of X which is equal to the union of some family of neighborhoods is called an *open* set.

Thus a set V is open if and only if each x in V has a neighborhood U with $x \in U \subset V$. The intersection of finitely many open sets is open, and the union of an arbitrary family of open sets is open. The union of all neighborhoods which are included in a given subset Y of X is an open set which is called the *interior* of Y. It is the largest open subset of Y.

A subset Y of X is *closed* if each point x in X each of whose neighborhoods contains a point of Y is itself in Y. The intersection of any family of closed sets is a closed set. The *closure* of a subset Y of X consists of all x in X each of whose neighborhoods contains a point of Y. It is the smallest closed set containing Y. A subset Y of X is *dense* if its closure is X.

Associated to each neighborhood space X there is a subset Cont $(X,\mathbf{R})$ of $F(X,\mathbf{R})$, called the *weakly continuous* functions on X. A function $f: X \to \mathbf{R}$ belongs to Cont $(X,\mathbf{R})$ if and only if for each x in X and each $\epsilon > 0$ there exists a neighborhood U of x such that

$$|f(x) - f(y)| \leq \epsilon \qquad (y \in U)$$

The set Cont $(X,\mathbf{R})$ in practice turns out to be of minor interest, because it is not possible to get a good hold on its structure. To see this by an example, consider the real numbers $\mathbf{R}$ and let F denote the set of continuous functions from $\mathbf{R}$ to $\mathbf{R}$ as defined in Chap. 2. It is easy to see that every continuous function is weakly continuous. Therefore

$$F \subset \text{Cont } (\mathbf{R},\mathbf{R}) \subset F(\mathbf{R},\mathbf{R})$$

As previously remarked, Brouwer contends that all functions in $F(\mathbf{R},\mathbf{R})$ are continuous, so that these inclusions are actually equalities. He also contends that this assertion is a theorem. We support the first claim and reject the second: while reflection makes it extremely plausible that $F = F(\mathbf{R},\mathbf{R})$, to accept Brouwer's arguments as a proof would destroy the character of mathematics. It seems likely that we are in the tantalizing situation in which the equalities $F = \text{Cont } (\mathbf{R},\mathbf{R})$ and Cont $(\mathbf{R},\mathbf{R}) = F(\mathbf{R},\mathbf{R})$ will never be proved and never be counterexampled.

This situation is typical. For most sets X we shall be interested in a certain well-structured subset of $F(X,\mathbf{R})$ (rather than in $F(X,\mathbf{R})$ itself), which it is not possible to realize as the set of weakly continuous functions relative to any neighborhood structure on X. Therefore it makes sense to focus attention on subsets of $F(X,\mathbf{R})$ instead of on neighborhood structures. For convenience these sets are required to satisfy certain minimal restrictions.

Definition 8 A *function space* X is a set X together with a subset F of $F(X,\mathbf{R})$, called the set of *continuous functions* or the *topology*, satisfying the following conditions.

(a) F contains the constant functions

(b) Sums and products of elements of F are in F

(c) The composition $h \circ f$ of an element f of F and a continuous function $h \colon \mathbf{R} \to \mathbf{R}$ is in F

(d) Uniform limits of elements of F are in F; that is, if for each $\epsilon > 0$ there exists g in F, with $|g(x) - f(x)| \leq \epsilon$ for all x in X, then $f \in F$

The set $F(X,\mathbf{R})$ is always a topology on X. In particular, $F(\mathbf{R},\mathbf{R})$ is a topology on $\mathbf{R}$. A more interesting topology on $\mathbf{R}$ is the set of continuous functions.

Let F be a topology on a set X. Since

$$\max\ \{f,g\}\ = f + \max\ \{g - f,\, 0\}\ = f + \tfrac{1}{2}(g - f + |g - f|)$$

$\max\ \{f,g\} \in F$ for all functions f and g in F. Similarly, $\min\ \{f,g\} \in F$ whenever $f,\, g \in F$. Since

$$fg = \tfrac{1}{2}((f + g)^2 - f^2 - g^2)$$

the requirement that products of elements of F belong to F is actually superfluous.

A topology F for a set X introduces a notion of proximity into X. As remarked above, a notion of proximity is classically introduced into a set X not by giving a family of functions but by giving a family of subsets, either open sets or neighborhoods. Classically this is equivalent to giving a family of functions from X to the set $\{0,1\}$, while constructively there is a vast difference, since only the all-too-rare free subsets of X correspond to such functions. This is because functions are sharply defined, whereas most sets are fuzzy around the edges. This fuzziness of sets is another reason to focus attention on function spaces instead of neighborhood spaces. The truth of the matter is that neither function spaces nor neighborhood spaces are as important as certain related structures—metric spaces and normed linear spaces—which will be studied in subsequent chapters.

One can pass from a neighborhood structure on a set X to a topology, or go the other way.

Proposition 1 *The set* Cont $(X,\, \mathbf{R})$ *of weakly continuous functions on a neighborhood space X is a topology for X.*

Proof Condition (*a*) of Definition 8 is obviously satisfied. The proofs of (*b*), (*c*), and (*d*) are similar to proofs of some of the propositions of Chap. 2.

Conversely, a topology F on a set X induces a neighborhood structure on X, obtained by taking neighborhoods to be sets of the form

$$f^{-1}(0, \infty) \equiv \{x \in X : f(x) > 0\}$$

where f is any element of F. To see this, consider elements f, g in F and define $h \equiv \min \{f,g\}$. Then

$$h^{-1}(0, \infty) = f^{-1}(0, \infty) \cap g^{-1}(0, \infty)$$

and thus the intersection of neighborhoods is a neighborhood, and X is therefore a neighborhood space.

Any set X can be topologized—take $F \equiv F(X,\mathbf{R})$. This is the *finest* topology, and the most natural. Its lack of structure keeps it from being important.

At the other extreme, X can be topologized by letting F consist of the constant functions and nothing else. This topology is too small to be useful.

Any subset F_0 of $F(X,\mathbf{R})$ can, of course, be used to generate a topology F for X, defined to be the smallest subset of $F(X,\mathbf{R})$ containing F_0 and satisfying (*a*) to (*d*) of Definition 8. The set F is just the set of all f in $F(X,\mathbf{R})$ obtained inductively from the following principles of construction.

(*a*) F_0 and the constant functions are in F

(*b*) Sums and products of elements of F are in F

(*c*) $h \circ f$ is in F if $f \in F$ and $h : \mathbf{R} \to \mathbf{R}$ is continuous

(*d*) A uniform limit of elements of F is in F

PROBLEMS

1. Prove the algebraic laws stated following Definition 3.

2. A family $(T,\lambda) \equiv \{\lambda(t)\}_{t \in T}$ of sets is a rule which assigns to each t in a discrete set T a set $\lambda(t)$. (The sets $\lambda(t)$ are not necessarily subsets of some fixed set S.) Define the *exterior union* (or *disjoint union*) of such a family. In case the sets $\lambda(t)$ are all subsets of a fixed set S, relate the exterior union to the *interior union* $\bigcup_{t \in T} \lambda(t)$ defined previously.

3. Show that the family of free subsets of a set X need not be closed with respect to countable unions.

4. Let S be a discrete set, and f an element of $F(S \times S, \{0,1\})$ such that elements s_1 and s_2 of S are equal if and only if $f(s_1,s_2) = 1$. Write $s_1 \neq s_2$ if $f(s_1,s_2) = 0$. Show that this is an inequality relation on S (in the sense that at most one of the relations $s_1 = s_2$ and $s_1 \neq s_2$ can hold for given s_1 and s_2).

5. Let S be a discrete set, and (T,λ) the family of free subsets of S. Show that if $a, b \in S$ and $a \neq b$, there exists t in T with $a \in \lambda(t)$ and $b \in -\lambda(t)$.

6. Call elements a and b of a set A *weakly unequal* if there exists a function $f: \{a,b\} \to \mathbf{R}$ with $f(a) \neq f(b)$. Show that a and b are weakly unequal if and only if $a = b$ implies $0 = 1$.

7. Show that if the Borel subset A of X is generated by the set $F \subset F(X,\mathbf{R})$, then A is generated by some countable subset of F.

8. Let A and B be arbitrary sets, and $\neq$ an inequality relation for B. Show how to define an inequality relation for $F(A,B)$.

9. Let λ be a function from a set A into $F(A,\{0,1\})$. Show that there exists f in $F(A,\{0,1\})$ with $f \neq \lambda(a)$ for all a in A. (This is a constructive version of Cantor's result that a set is always smaller than the set of its subsets.)

10. Give an example of a complemented subset of $\mathbf{R}$, relative to the family F of all continuous functions from $\mathbf{R}$ to $\mathbf{R}$, which we shall probably never be able to express as a Borel set generated by F.

11. Prove (2.2) and (2.3).

12. A function space (X,F) is *connected* if for each f in F the closure of $f(X)$ is a convex subset of $\mathbf{R}$. Show that a product $X \equiv \prod_{n=1}^{\infty} X_n$ of connected spaces (X_n,F_n) is connected, where the topology for X is the topology generated by the functions $\{x_n\} \to f_i(x_i)$ ($1 \leq i < \infty, f_i \in F_i$).

13. Prove that X is connected if and only if for arbitrary x, y in X, arbitrary bounded elements $f_1, \ldots, f_n$ of F, and an arbitrary $\epsilon > 0$ there exists $x_1, \ldots, x_N$ in X, with $x_1 = x$ and $x_N = y$, such that

$$|f_i(x_j) - f_i(x_{j+1})| \leq \epsilon \qquad (1 \leq i \leq n, 1 \leq j \leq N - 1)$$

NOTES

The notion of equality, contrary to classical usage, is a convention. We define what it means for elements of a given set to be equal, but the notion of equality of elements of different sets is not defined. The only way in which elements of different sets A and B can be regarded as equal is by realizing A and B as subsets of a third set C. For this reason we define the operations of union and intersection only for sets which are given as subsets of a given set. (See, however, Prob. 2.)

The concept of a complemented set suffers from what appears to be a glaring deficiency. If X is a set with many elements, it seems unnatural to call (ϕ,ϕ) a complemented set in X, especially since (ϕ,X) is also a complemented set. On the other hand, there is nothing to force us to consider complemented sets like (ϕ,ϕ). In the applications there will always be some condition that keeps the complemented sets (A,B) with which we are concerned from having the property that both A and B are ridiculously small.

A Borel set generated by a given family F of continuous functions is a much more palpable object than an arbitrary complemented set. It would be possible to work with Borel sets exclusively, but to consider the more general notion of a complemented set adds additional insight.

Brouwer's contention that all elements of $F(\mathbf{R},\mathbf{R})$ are continuous seems to contradict claims of certain recursive function theorists, who give examples of elements of $F(\mathbf{R},\mathbf{R})$ that are not continuous. In both instances, the claims are based on extramathematical considerations. Brouwer analyzes all possible techniques for constructing elements f of $F(\mathbf{R},\mathbf{R})$, and comes to the conclusion that all such f are continuous. The recursive function theorists analyze the possibilities for constructing real numbers, and come to the conclusion that they all possess a certain property (i.e., they are recursive). In addition, they show how to construct a discontinuous function on the set of recursive real numbers. These two positions are, in fact, compatible. They do not contradict each other, because it is possible to believe both (a) that all constructive real numbers are recursive and (b) that without making use of some unprovable hypothesis (such as the hypothesis that all constructive real numbers are recursive) the only elements of $F(\mathbf{R},\mathbf{R})$ that can be constructed are continuous. Extramathematical considerations of both types (especially the first) are useful in indicating that we should not try to do certain things constructively, but they have no place in the actual development of constructive mathematics.

Definition 8 should not be taken too seriously. The purpose is merely to list a minimal number of properties that the set of all continuous functions should be expected to have. Other properties could be added. To find a complete list seems to be a nontrivial and interesting problem.

CHAPTER *4*

METRIC SPACES

The concept of a metric is defined, some examples are studied, and various techniques for constructing metrics are developed. The neighborhood structure of a metric space is defined, and the notions of weakly continuous and uniformly continuous functions are introduced. Completeness is defined in Sec. 3, and the construction of the completion is carried through. Following Brouwer, we define a compact space to be a metric space that is complete and totally bounded. Compact and locally compact spaces are studied in Secs. 4 and 5. Constructivizations of various classical results, such as Ascoli's theorem, the Stone-Weierstrass theorem, and the Tietze extension theorem, are given. The concept of a located set, due to Brouwer, plays an important role. Crucial for later developments is Theorem 8, a partial substitute for the classical result that a closed subset of a compact space is compact.

75

The sets that occur in analysis are constructed from the real numbers by certain standard methods, for instance by the formation of subsets or cartesian products of sets already defined. They often carry a structure, such as a function space structure or a neighborhood structure, built up at the same time the sets themselves are constructed. Of special importance is the structure given by a metric, where a metric on a set X is a function on $X \times X$ satisfying certain properties that we intuitively associate with distance. A metric is yet another way of introducing a notion of proximity into X.

1. FUNDAMENTAL DEFINITIONS AND CONSTRUCTIONS

Guided by our knowledge of the real numbers and our intuition, we introduce the notion of distance as follows.

Definition 1 A *metric* on a set X is a function $\rho: X \times X \to \mathbf{R}^{0+}$ such that

    (*a*) $\rho(x,y) = 0$ if and only if $x = y$

    (*b*) $\rho(x,y) = \rho(y,x)$

    (*c*) $\rho(x,y) \le \rho(x,z) + \rho(z,y)$

Elements x and y of X are *unequal*, $x \ne y$, if and only if $\rho(x,y) > 0$.

A set X which is endowed with a metric is called a *metric space*. Inequality (*c*) is called the *triangle inequality*.

It often happens that in place of (*a*) the weaker condition

$$(a') \quad \rho(x,y) = 0 \quad \text{if } x = y$$

is satisfied. In this case ρ is called a *pseudometric* on X, and X is called a *pseudometric space*. Associated to each pseudometric space (X,ρ) is a metric space (X_0,ρ) obtained by taking X_0 to be the set X with the equality relation $x = y$ modified to mean that $\rho(x,y) = 0$. The theory of pseudometric spaces therefore is essentially no different from the theory of metric spaces.

The most important example of a metric space is the set of real numbers, metrized by defining

$$\rho(x,y) \equiv |x - y|$$

In order to define metrics for other sets, it is best to establish some general methods of metrization.

Definition 2 Let $f: Y \to X$ be a function from a set Y to a pseudometric space X. The *induced pseudometric* ρ^* on Y is defined by

$$(1.1) \qquad \rho^*(y_1,y_2) \equiv \rho(f(y_1),f(y_2)) \qquad (y_1, y_2 \in Y)$$

For ρ^* to be a metric it is sufficient that ρ be a metric and that $y_1 = y_2$ whenever $f(y_1) = f(y_2)$.

As an example of the use of Definition 2, the inclusion map $i: Y \to X$ from a subset Y of a metric space X into X induces a metric on Y. In particular the above metric on $\mathbf{R}$ induces metrics on Q, Z, and the various intervals.

As another example, if $f: X \to \mathbf{R}$ is any real-valued function on a set X, the pseudometric ρ^* is given by

$$\rho^*(x,y) = |f(x) - f(y)|$$

Definition 3 Let $(X_1,\rho_1), \ldots, (X_n,\rho_n)$ be metric spaces. The *product metric* ρ on the cartesian product space $X \equiv X_1 \times \cdots \times X_n$ is given by

$$(1.2) \qquad \rho(x,y) \equiv \sum_{i=1}^{n} \rho_i(x_i,y_i)$$

for all $x \equiv (x_1, \ldots, x_n)$ and $y \equiv (y_1, \ldots, y_n)$ in X.

There are other ways of metrizing a finite product of metric spaces. In fact $\mathbf{R}^n$ is usually given the metric d defined by

$$d(x,y) \equiv \Big(\sum_{i=1}^{n} (x_i - y_i)^2 \Big)^{1/2}$$

to correspond to the geometry of $\mathbf{R}^n$ as we know it from experience. This metric is so important that we digress at this point to give the proof that it actually is a metric. To this end we establish two basic inequalities. The first is known as the *Cauchy-Schwarz inequality*.

Proposition 1 *Let* $x_1, \ldots, x_n, y_1, \ldots, y_n$ *be real numbers. Then*

$$(1.3) \qquad \Big| \sum_{i=1}^{n} x_i y_i \Big| \leq \Big(\sum_{i=1}^{n} x_i^2 \Big)^{1/2} \Big(\sum_{i=1}^{n} y_i^2 \Big)^{1/2}$$

Proof We compute

$$\sum_{i=1}^{n} x_i{}^2 \sum_{i=1}^{n} y_i{}^2 - \Big(\sum_{i=1}^{n} x_i y_i\Big)^2$$

$$= \tfrac{1}{2} \sum_{i=1}^{n} x_i{}^2 \sum_{j=1}^{n} y_j{}^2 + \tfrac{1}{2} \sum_{j=1}^{n} x_j{}^2 \sum_{i=1}^{n} y_i{}^2 - \sum_{i=1}^{n} x_i y_i \sum_{j=1}^{n} x_j y_j$$

$$= \sum_{i,j=1}^{n} \tfrac{1}{2}(x_i{}^2 y_j{}^2 + x_j{}^2 y_i{}^2 - 2 x_i y_j x_j y_i)$$

$$= \sum_{i,j=1}^{n} \tfrac{1}{2}(x_i y_j - x_j y_i)^2 \geq 0$$

This is equivalent to (1.3).

The next inequality is called *Minkowski's inequality*.

Proposition 2 *Let* $x_1, \ldots, x_n, y_1, \ldots, y_n$ *be real numbers. Then*

$$(1.4) \qquad \Big(\sum_{i=1}^{n} (x_i - y_i)^2\Big)^{1/2} \leq \Big(\sum_{i=1}^{n} x_i{}^2\Big)^{1/2} + \Big(\sum_{i=1}^{n} y_i{}^2\Big)^{1/2}$$

Proof We compute, using Proposition 1,

$$\sum_{i=1}^{n} (x_i - y_i)^2 = \sum_{i=1}^{n} x_i{}^2 + \sum_{i=1}^{n} y_i{}^2 - 2\sum_{i=1}^{n} x_i y_i$$

$$\leq \sum_{i=1}^{n} x_i{}^2 + \sum_{i=1}^{n} y_i{}^2 + 2\Big(\sum_{i=1}^{n} x_i{}^2\Big)^{1/2}\Big(\sum_{i=1}^{n} y_i{}^2\Big)^{1/2}$$

$$= \Big(\big(\sum_{i=1}^{n} x_i{}^2\big)^{1/2} + \big(\sum_{i=1}^{n} y_i{}^2\big)^{1/2}\Big)^2$$

This is equivalent to (1.4).

Corollary *The function d is a metric on* $\mathbf{R}^n$.

Proof Clearly d satisfies conditions (*a*) and (*b*) of Definition 1. By Minkowski's inequality,

$$d(x,z) = \Big\{\sum_{i=1}^{n} ((x_i - y_i) - (y_i - z_i))^2\Big\}^{1/2}$$

$$\leq \Big(\sum_{i=1}^{n} (x_i - y_i)^2\Big)^{1/2} + \Big(\sum_{i=1}^{n} (y_i - z_i)^2\Big)^{1/2}$$

$$= d(x,y) + d(y,z)$$

for all $x \equiv (x_1, \ldots, x_n)$, $y \equiv (y_1, \ldots, y_n)$ and $z \equiv (z_1, \ldots, z_n)$ in $\mathbf{R}^n$. Therefore d also satisfies (c).

Related to the metric d on $\mathbf{R}^n$ is a certain metric on the set F of continuous functions on a compact interval $I \equiv [a,b]$. For arbitrary elements f, g of F we define

$$d(f,g) \equiv (\int |f(x) - g(x)|^2 \, dx)^{1/2}$$

Minkowski's inequality for integrals,

$$d(f,g) \leq d(f,h) + d(g,h) \qquad (f, g, h \in F)$$

follows from (1.4) by approximation.

There is also a *Cauchy-Schwarz inequality for integrals,*

$$|\int f(x)g(x) \, dx| \leq (\int f(x)^2 \, dx \int g(x)^2 \, dx)^{1/2}$$

which follows from (1.3) by approximation.

The Cauchy-Schwarz and Minkowski inequalities admit many other generalizations, as will be seen later.

A metric space (X,ρ) is called *bounded* if there exists a constant $C > 0$, called a *bound* for (X,ρ), such that $\rho(x,y) \leq C$ for all x and y in X. A subset Y of a nonvoid metric space X is *bounded* if, for all (equivalently, some) x in X, the set $Y \cup \{x\}$ with the induced metric ρ^* is a bounded metric space. (Exercise: a subset Y of $\mathbf{R}$ can be bounded as a metric space but *not* bounded as a subset of $\mathbf{R}$.)

A countable product of metric spaces can be metrized in case each factor is bounded.

Definition 4 Let $\{(X_n,\rho_n)\}$ be a sequence of metric spaces, each bounded by 1. The *product metric* ρ on $X \equiv \prod_{n=1}^{\infty} X_n$ is defined by

$$\rho(\{x_n\},\{y_n\}) \equiv \sum_{n=1}^{\infty} 2^{-n}\rho(x_n,y_n) \qquad (\{x_n\}, \{y_n\} \in X)$$

Definition 4 is useful even when the spaces (X_n,ρ_n) are not necessarily bounded, because there is a standard method for transforming an arbitrary metric into a metric bounded by 1.

Proposition 3 *Let (X,ρ) be a metric space. Let $h\colon \mathbf{R}^{0+} \to \mathbf{R}^{0+}$ satisfy the conditions*

    (i) $h(u) = 0$ *if and only if* $u = 0$

    (ii) $h(u + v) \leq h(u) + h(v)$ $(u,\, v \in \mathbf{R}^{0+})$

Then $d \equiv h \circ \rho$ is a metric on X.

Proof Condition (a) of Definition 1 is satisfied because of (i). Condition (b) is obviously satisfied. To check (c) note that

$$
\begin{aligned}
d(x,z) = h(\rho(x,z)) &\leq h(\rho(x,y) + \rho(y,z)) \\
&\leq h(\rho(x,y)) + h(\rho(y,z)) = d(x,y) + d(y,z)
\end{aligned}
$$

for all x, y, z in X.

Corollary *If ρ is any metric on a set X, then the function ρ' defined by*

$$
\rho'(x,y) \equiv \min\,\{\rho(x,y),1\} \qquad (x,\, y \in X)
$$

is a bounded metric on X.

Proof The proof is obvious.

This corollary leads us to define the *product* of an arbitrary sequence $\{(X_n,\rho_n)\}$ of metric spaces to be the product of the associated sequence $\{(X_n,\rho_n')\}$.

The definition of a continuous function on a compact interval generalizes to arbitrary metric spaces, leading to the important idea of a uniformly continuous function.

Definition 5 A function $f\colon X_1 \to X_2$ from a metric space (X_1,ρ_1) to a metric space (X_2,ρ_2) is *uniformly continuous* if there exists $\omega\colon \mathbf{R}^+ \to \mathbf{R}^+$ such that

$$
(1.5) \qquad\qquad\qquad \rho_2(f(x),f(y)) \leq \epsilon
$$

whenever x, $y \in X_1$ and $\rho_1(x,y) \leq \omega(\epsilon)$. The function ω is called the *modulus of continuity* of f on X_1.

As an example, for each point x_0 in an arbitrary metric space X the function $x \to \rho(x,x_0)$ is uniformly continuous on X with modulus of continuity $\epsilon \to \epsilon$. To see this, consider points x and y in X. Then $\rho(x,x_0) \leq \rho(x,y) + \rho(y,x_0)$, or $\rho(x,x_0) - \rho(y,x_0) < \rho(x,y)$. Similarly, $\rho(y,x_0) - \rho(x,x_0) \leq \rho(x,y)$. Therefore

$$(1.6) \qquad |\rho(x,x_0) - \rho(y,x_0)| \leq \rho(x,y)$$

It follows that $|\rho(x,x_0) - \rho(y,x_0)| \leq \epsilon$ whenever $\rho(x,y) \leq \epsilon$.

As another example, for each $x \equiv \{x_n\}$ in the product (X,ρ) of the metric spaces $\{(X_n,\rho_n)\}_{n=1}^{\infty}$ (all bounded by 1) and each n in Z^+, write $\pi_n(x) \equiv x_n$. Then the *projection* $\pi_n\colon X \to X_n$ of X onto X_n is uniformly continuous with modulus of continuity $\epsilon \to 2^{-n}\epsilon$.

Definition 6 A sequence $\{f_n\}$ of functions from a set X_1 to a metric space X_2 *converges uniformly* to a function $f\colon X_1 \to X_2$ if for each $\epsilon > 0$ there exists N_ϵ in Z^+ such that

$$(1.7) \qquad \rho(f_n(x),f(x)) \leq \epsilon \qquad (n \geq N_\epsilon,\, x \in X_1)$$

The proofs of the following statements, which are similar to certain proofs of Chap. 2, are left to the reader.

Proposition 4 *The sum $f + g$ and product fg of uniformly continuous functions $f\colon X \to \mathbf{R}$ and $g\colon X \to \mathbf{R}$ are uniformly continuous, and f^{-1} is uniformly continuous if $|f(x)| \geq c$ for all x in X and some $c > 0$. The composition $f_2 \circ f_1$ of uniformly continuous functions $f_1\colon X_1 \to X_2$, $f_2\colon X_2 \to X_3$ is uniformly continuous. The limit $f\colon X_1 \to X_2$ of a uniformly convergent sequence $\{f_n\}$ of uniformly continuous functions from a metric space X_1 to a metric space X_2 is uniformly continuous.*

A uniformly continuous function f from a metric space X_1 to a metric space X_2 which has a uniformly continuous inverse is called a *metric equivalence*, and X_1 and X_2 are said to be *equivalent* metric spaces.

Two metrics ρ_1 and ρ_2 on the same set X for which the identity function from X to itself is a metric equivalence of (X,ρ_1) with (X,ρ_2) are called *equivalent metrics* on X. The product metric ρ on $\mathbf{R}^n$ is equivalent to the metric d. An arbitrary metric ρ is equivalent to the bounded metric ρ' of the corollary to Proposition 3. If (X_n,ρ_n) and $(Y_n,\tilde{\rho}_n)$ are equivalent metric spaces $(1 \leq n < \infty)$, then the products of the sequences $\{(X_n,\rho_n)\}$ and $\{(Y_n,\tilde{\rho}_n)\}$ are equivalent.

A notion of convergence can be defined for an arbitrary metric space.

Definition 7 A sequence $\{x_n\}$ of elements of a metric space X *converges* to an element x of X, written either

$$\lim_{n \to \infty} x_n = x$$

or
$$x_n \to x \qquad \text{as } n \to \infty$$
if
$$\lim_{n \to \infty} \rho(x_n,x) = 0$$

In case $X = \mathbf{R}$, this coincides with the definition of convergence already given.

2. ASSOCIATED STRUCTURES

Every metric space has a neighborhood structure and a topology.

Definition 8   The *open sphere* of radius $r > 0$ about a point x in a metric space X is the subset

$$S(x,r) \equiv \{y \in X : \rho(x,y) < r\}$$

of X. The neighborhood structure of X is defined by taking the neighborhoods to be the open spheres.

To see that the open spheres give a neighborhood structure, note first that if $y \in S(x,r)$ and $0 < t \leq r - \rho(x,y)$ then $S(y,t) \subset S(x,r)$, because if $z \in S(y,t)$, then

$$\rho(z,x) \leq \rho(z,y) + \rho(x,y) < t + \rho(x,y) \leq r$$

It follows that if $S(x_0,r_0)$ and $S(x_1,r_1)$ are open spheres containing a point x of X, then

$$x \in S(x,t) \subset S(x_0,r_0) \cap S(x_1,r_1)$$

where $t \equiv \min \{r_0 - \rho(x,x_0), r_1 - \rho(x,x_1)\}$. Therefore the open spheres give a neighborhood structure.

The notions of open sets, closed sets, and dense sets, and of interiors and closures when applied to subsets of a metric space X refer to this neighborhood structure. A subset Y of X is closed if and only if it contains all points that are limits of sequences of points of Y.

The *closed sphere* of radius $r > 0$ about a point x in a metric space X is the set

$$Sc(x,r) \equiv \{y \in X : \rho(x,y) \leq r\}$$

It is a closed subset of X.

A subset A of X is *located* if the *distance*

$$\rho(x,A) \equiv \inf \{\rho(x,y) : y \in A\}$$

from x to A exists, for every x in X. If A is located, then the closure $\bar{A}$ of A is located, and $\rho(x,\bar{A}) = \rho(x,A)$ for all x in X.

If A is located, the distance function $x \to \rho(x,A)$ satisfies the inequality

$$(2.1) \qquad \rho(x,A) \leq \rho(x,y) + \rho(y,A) \qquad (x, y \in X)$$

because (2.1) is equivalent to the assertion that $\rho(x,A) \leq \rho(x,y) + \rho(y,z)$ for all z in A, and this is true because $\rho(x,A) \leq \rho(x,z) \leq \rho(x,y) + \rho(y,z)$.

The *metric complement* $-A$ of a located subset A of a metric space X is the set

$$-A \equiv \{x : x \in X, \rho(x,A) > 0\}$$

Because of (2.1), $-A$ is an open set.

The topology associated with the neighborhood structure of a metric space X is called the *weak topology*, and continuous functions in the weak topology are *weakly continuous*. A weakly continuous function $f \colon X \to \mathbf{R}$ has the property that $f(x_n) \to f(x)$ as $n \to \infty$ whenever $x_n \to x$ as $n \to \infty$.

The weak topology is not of much use. More useful is the uniform topology, defined as follows.

Definition 9 The *uniform topology* F for a metric space X consists of all uniformly continuous functions $f \colon X \to \mathbf{R}$.

Since every uniformly continuous function is weakly continuous, the weak topology is larger than the uniform topology.

Other topologies for a metric space X are obtained by considering the set of all $f \colon X \to \mathbf{R}$ which are uniformly continuous on each element of a given family of subsets of X, as was done in defining the continuous functions on an arbitrary real interval I.

3. COMPLETENESS

The construction which led from Q to $\mathbf{R}$ leads from a general metric space X to a metric space $\tilde{X}$, the *completion* of X.

Definition 10 A sequence $\{x_n\}$ of elements of a metric space X is *regular* if

$$(3.1) \qquad \rho(x_m,x_n) \leq m^{-1} + n^{-1} \qquad (m, n \in Z^+)$$

The set $\tilde{X}$ of all regular sequences of elements of X is called the *comple-*

tion of X. Elements $\{x_n\}$ and $\{y_n\}$ of $\tilde{X}$ are *equal* if

$$(3.2) \qquad \rho(x_n, y_n) \leq 2n^{-1} \qquad (n \in Z^+)$$

The *distance* $\rho(x, y)$ between elements $x \equiv \{x_n\}$ and $y \equiv \{y_n\}$ of $\tilde{X}$ is

$$(3.3) \qquad \rho(x, y) \equiv \lim_{n \to \infty} \rho(x_n, y_n)$$

The limit (3.3) exists because

$$\begin{aligned}
\rho(x_n, y_n) - \rho(x_m, y_m) &\leq \rho(x_n, x_m) + \rho(x_m, y_m) + \rho(y_m, y_n) - \rho(x_m, y_m) \\
&= \rho(x_n, x_m) + \rho(y_m, y_n) \\
&\leq 2(m^{-1} + n^{-1}) \qquad (m, n \in Z^+)
\end{aligned}$$

The operation $\rho \colon \tilde{X} \times \tilde{X} \to \mathbf{R}$ is a function because if $x \equiv \{x_n\}$, $x' \equiv \{x_n'\}$, $y \equiv \{y_n\}$, $y' \equiv \{y_n'\}$ are elements of $\tilde{X}$ with $x = x'$ and $y = y'$, then

$$\begin{aligned}
\rho(x, y) &= \lim_{n \to \infty} \rho(x_n, y_n) \\
&\leq \lim_{n \to \infty} \{\rho(x_n, x_n') + \rho(x_n', y_n') + \rho(y_n', y_n)\} \\
&= \rho(x', y')
\end{aligned}$$

Similarly $\rho(x', y') \leq \rho(x, y)$. Therefore $\rho(x, y) = \rho(x', y')$.

Theorem 1 *The function ρ is a metric for $\tilde{X}$.*

Proof Condition (b) of Definition 1 is clearly satisfied. To check (c), consider points $x \equiv \{x_n\}$, $y \equiv \{y_n\}$, and $z \equiv \{z_n\}$ in $\tilde{X}$. Then

$$\begin{aligned}
\rho(x, y) &= \lim_{n \to \infty} \rho(x_n, y_n) \\
&\leq \lim_{n \to \infty} \rho(x_n, z_n) + \lim_{n \to \infty} \rho(z_n, y_n) \\
&= \rho(x, z) + \rho(z, y)
\end{aligned}$$

As to condition (a), it is obvious that $\rho(x, y) = 0$ if $x = y$. Conversely, if $\rho(x, y) = 0$, then for all n in Z^+,

$$\begin{aligned}
\rho(x_n, y_n) &\leq \rho(x_n, x_m) + \rho(x_m, y_m) + \rho(y_m, y_n) \\
&< 2(n^{-1} + m^{-1}) + \rho(x_m, y_m) \to 2n^{-1} \qquad \text{as } m \to \infty
\end{aligned}$$

Therefore $x = y$.

The *inclusion map* $i: X \to \tilde{X}$ is defined for each x in X by taking $i(x)$ to be the regular sequence each of whose terms is x. The map i realizes X as a subset of $\tilde{X}$, with preservation of metric:

$$\rho(x,y) = \rho(i(x),i(y)) \qquad (x, y \in X)$$

The subset X is dense in $\tilde{X}$.

Definition 11 A *Cauchy sequence* $\{x_n\}$ of elements of a metric space X is a sequence such that for each $\epsilon > 0$ there exists a positive integer N_ϵ with

$$\rho(x_m,x_n) \leq \epsilon \qquad (m, n \geq N_\epsilon)$$

The metric space X is called *complete* if every Cauchy sequence converges.

It is obvious that a convergent sequence in an arbitrary metric space is a Cauchy sequence.

Theorem 2 *The completion $\tilde{X}$ of an arbitrary metric space X is complete.*

Proof The proof is the same as the proof of Theorem 2 of Chap. 2. It is therefore omitted.

Proposition 5 *A metric space X is complete if and only if $X = \tilde{X}$.*

Proof By Theorem 2, if $X = \tilde{X}$, then X is complete. Assume conversely that X is complete. Now, X is a subset of $\tilde{X}$, or $X \subset \tilde{X}$. To show that $\tilde{X} \subset X$, consider an arbitrary element $x \equiv \{x_n\}$ of $\tilde{X}$. Then $\{x_n\}$ is a Cauchy sequence of elements of X. It therefore converges to an element x_0 of X. For each n in Z^+ we have

$$\rho(x_0,x_n) = \lim_{m \to \infty} \rho(x_m,x_n) \leq \lim_{m \to \infty} (n^{-1} + m^{-1}) \leq n^{-1}$$

Therefore $x = i(x_0)$. It follows that $\tilde{X} \subset X$.

A closed subset Y of a complete metric space X is complete, since any Cauchy sequence $\{y_n\}$ of elements of Y converges to a point y_0 in X, which (because Y is closed) is in Y. Conversely, a complete subset Y of an arbitrary metric space X is closed, since if the point x_0 in X is in the closure of Y, then $x_0 = \lim_{n \to \infty} y_n$ for some sequence $\{y_n\}$ of points in Y. Since Y is complete, the Cauchy sequence $\{y_n\}$ converges to a point y_0 in Y. Then $x_0 = y_0$ is in Y.

A complete metric space stays complete under an equivalent metric. Therefore the following result has intrinsic significance.

Theorem 3 *The product (X,ρ) of a sequence $\{(X_n,\rho_n)\}$ of complete metric spaces is complete.*

Proof By the above remark, there is no loss of generality in assuming that (X_n,ρ_n) is bounded by 1 for all n. Let $\{x^k\}$ be a Cauchy sequence of elements of X. For each k, x^k is a sequence $x^k \equiv \{x_n{}^k\}_{n=1}^{\infty}$, with $x_n{}^k \in X_n$. For each n the sequence $\{x_n{}^k\}_{k=1}^{\infty}$ is a Cauchy sequence, because

$$\rho_n(x_n{}^k,x_n{}^j) \leq 2^n\rho(x^k,x^j) \qquad (j, k \in Z^+)$$

whose limit we denote by $x_n{}^0 \in X_n$. The sequence $\{x^k\}$ converges to the point $x^0 \equiv \{x_n{}^0\}$ of X, since for all N in Z^+ we have

$$\begin{aligned}
\rho(x^k,x^0) &\equiv \sum_{n=1}^{\infty} 2^{-n}\rho_n(x_n{}^k,x_n{}^0) \\
&\leq \sum_{n=1}^{N} 2^{-n}\rho_n(x_n{}^k,x_n{}^0) + 2^{-N} \\
&\leq 2^{-N+1}
\end{aligned}$$

if k is sufficiently large.

The next lemma is useful in constructing uniformly continuous functions with values in a complete metric space.

Lemma 1 *Let Y be a dense subset of a metric space X and $f\colon Y \to Z$ a uniformly continuous function from Y to a complete metric space Z with modulus of continuity ω. Then there exists a uniformly continuous function $g\colon X \to Z$ with modulus of continuity $\frac{1}{2}\omega$ such that $f(y) = g(y)$ for all y in Y.*

Proof Since Y is dense in X, for each x in X there exists a sequence $\{y_n\}$ of points of Y that converges to x. If ϵ is any positive number, then $\rho(f(y_m),f(y_n)) \leq \epsilon$ whenever $\rho(y_m,y_n) \leq \omega(\epsilon)$. Therefore $\{f(y_n)\}$ is a Cauchy sequence of elements of Z, whose limit we denote by $g(x)$. Clearly $g\colon X \to Z$ is a function which equals f on Y. To show that $\frac{1}{2}\omega$ is a modulus of continuity for g, consider $\epsilon > 0$ and points x^1 and x^2 in X with $\rho(x^1,x^2) \leq \frac{1}{2}\omega(\epsilon)$. Let $\{y_n{}^1\}$ and $\{y_n{}^2\}$ be sequences of points of Y converging respectively to x^1 and x^2. Then $\rho(y_n{}^1,y_n{}^2) \leq \omega(\epsilon)$ for all sufficiently large n, and thus $\rho(f(y_n{}^1),f(y_n{}^2)) \leq \epsilon$. Therefore

$$\rho(f(x^1),f(x^2)) = \lim_{n \to \infty} \rho(f(y_n{}^1),f(y_n{}^2)) \leq \epsilon$$

The next result is called the *Baire category theorem.*

Theorem 4 *Let $\{U_n\}$ be a sequence of dense open sets in a complete metric space X. Then the intersection*

$$U \equiv \bigcap_{n=1}^{\infty} U_n$$

is also dense in X.

Proof Let $S(x_0,r_0)$ be any open sphere in X. Since U_1 is dense, there exists a closed sphere $Sc(x_1,r_1)$ with $r_1 \leq 1$, such that

$$Sc(x_1,r_1) \subset S(x_0,r_0) \cap U_1$$

Continuing by induction, we construct a sequence $\{Sc(x_n,r_n)\}$ of closed spheres such that $r_n \leq n^{-1}$ and

$$Sc(x_n,r_n) \subset S(x_{n-1},r_{n-1}) \cap U_n$$

for all n. Then $x_m \in S(x_n,r_n)$ whenever $m \geq n$, so that $\rho(x_m,x_n) \leq r_n \leq n^{-1}$. Therefore $\{x_n\}$ is a Cauchy sequence. Since all except finitely many terms of this sequence lie in any given closed sphere $Sc(x_n,r_n)$, the limit x lies in each of these spheres. Therefore

$$x \in S(x_0,r_0) \cap \bigcap_{n=1}^{\infty} U_n$$

It follows that U is dense in X.

Theorem 4 is one of the most useful versions of Cantor's diagonal technique (which was used in the proof of Theorem 1 of Chap. 2). The reader should show that the latter theorem is a corollary of Theorem 4.

A located subset Y of a metric space X is *nowhere dense* in X if the metric complement $-Y$ of Y is dense. With this definition we reformulate Theorem 4.

Corollary *Let* $\{Y_n\}$ *be a sequence of nowhere-dense subsets of a metric space* X. *Then every open sphere* $S(x,r)$ *in* X *contains a point* y *whose distance to each* Y_n *is positive.*

Proof By Theorem 4 there exists a point y in $S(x,r) \cap \bigcap_{n=1}^{\infty} - Y_n$. The distance of y to Y_n is positive because $y \in -Y_n$.

A more significant concept for metric spaces than boundedness is the concept of total boundedness, defined as follows.

Definition 12 A metric space X is *totally bounded* if for each $\epsilon > 0$ there exists a finite subset $\{x_1, \ldots, x_n\}$ (where n is a positive integer depending on ϵ) of X, called an ϵ *approximation to* X, such that for each x in X at least one of the numbers $\rho(x,x_1), \ldots, \rho(x,x_n)$ is less than ϵ.

The reader can show as an exercise that X is totally bounded if for each $\epsilon > 0$ there exists a subfinite ϵ approximation $\{x_1, \ldots, x_n\}$ to X.

The property of total boundedness is preserved under passage to an equivalent metric. The product of a sequence of totally bounded metric spaces is totally bounded.

Call a metric space X *separable* if it has a countable dense subset. The product of a sequence of separable metric spaces is separable. Every totally bounded metric space is separable.

4. COMPACTNESS

Of special importance are those metric spaces which are both complete and totally bounded.

Definition 13 A *compact* metric space, or simply a *compact space*, is a metric space which is complete and totally bounded.

The compact intervals of real numbers are compact in the sense of Definition 13.

Total boundedness is the more important of the two properties which occur in the definition of compactness, since a metric space which is totally bounded but not necessarily complete can always be compactified by passing to its completion.

The following propositions show that compactness is preserved by many of the standard operations on metric spaces.

Proposition 6 *The product of a sequence of compact spaces is compact.*

Proof   This is an immediate consequence of the earlier remarks that such a product is (i) complete and (ii) totally bounded.

Proposition 7   *A closed located subset Y of a compact space X is compact.*

Proof   The subset Y is complete by the remark following Proposition 5. To show that Y is totally bounded, consider $\epsilon > 0$ and let $\{x_1, \ldots, x_n\}$ be a $\frac{1}{3}\epsilon$ approximation to X. For each i choose y_i in Y with $\rho(x_i, y_i) < \rho(x_i, Y) + \frac{1}{3}\epsilon$.

Consider an arbitrary y in Y. Then $\rho(y, x_i) < \frac{1}{3}\epsilon$ for some i, and thus

$$\rho(x_i, Y) \leq \rho(y, x_i) < \frac{1}{3}\epsilon$$

This gives

$$\rho(y, y_i) \leq \rho(y, x_i) + \rho(x_i, y_i) < \tfrac{1}{3}\epsilon + \tfrac{1}{3}\epsilon + \tfrac{1}{3}\epsilon = \epsilon$$

Thus the subfinite set $\{y_i : 1 \leq i \leq n\}$ is an ϵ approximation to Y. Since ϵ is any positive number, it follows that Y is totally bounded.

Proposition 8   *The image $f(X)$ of a totally bounded metric space X under a uniformly continuous function $f\colon X \to Y$ is totally bounded.*

Proof   Let ω be a modulus of continuity for f. For each $\epsilon > 0$ let $\{x_1, \ldots, x_n\}$ be an $\omega(\frac{1}{2}\epsilon)$ approximation to X. Then if $f(x)$ is any point in $f(X)$, we have

$$\rho(x, x_i) < \omega(\tfrac{1}{2}\epsilon)$$

for some value of i. Therefore $\rho(f(x), f(x_i)) \leq \frac{1}{2}\epsilon < \epsilon$. It follows that $\{f(x_1), \ldots, f(x_n)\}$ is an ϵ approximation to $f(X)$.

Corollary   *Let $f\colon X \to \mathbf{R}$ be a uniformly continuous function on a totally bounded metric space X. Then the least upper bound and greatest lower bound of the subset $f(X)$ of $\mathbf{R}$, called respectively the* supremum *or* sup *and the* infimum *or* inf *of f on X, and written* sup $\{f(x) : x \in X\}$ *and* inf $\{f(x) : x \in X\}$, *exist.*

Proof   By Proposition 8, we see that $f(X)$ is totally bounded. By Theorem 3 of Chap. 2, the l.u.b. and g.l.b. of $f(X)$ exist.

Proposition 9   *A compact subset Y of a metric space X is closed and located.*

Proof Since Y is compact, it is closed. If x is any point of X, then $y \to \rho(x,y)$ is a uniformly continuous function on Y, as we have seen. By the corollary to Proposition 8, its infimum $\rho(x,Y)$ exists. Thus Y is located.

Proposition 10 *A subset X of $\mathbf{R}^n$ is compact if and only if it is closed, located, and bounded.*

Proof Proposition 9, together with the fact that a totally bounded space is bounded, implies that the conditions are necessary. Assume conversely that X is closed, located, and bounded. Because X is bounded, there exists $c > 0$ such that $X \subset [-c,c]^n$. The product $[-c,c]^n$ is compact, by Proposition 6. As a closed located subset of a compact space, X is compact, by Proposition 7.

A compact space X will be given the *uniform topology*, so that a continuous function $f: X \to \mathbf{R}$ is one which is uniformly continuous. More generally, we define a *continuous function* from X to an arbitrary metric space Y to be one which is uniformly continuous.

Some of the most important metric spaces of analysis are spaces of functions. Here is a basic example.

Definition 14 For each compact space X and each metric space Y, we shall use $C(X,Y)$ to denote the set of all (uniformly) continuous functions from X to Y. For $C(X,\mathbf{R})$ we write simply $C(X)$. The metric ρ on $C(X,Y)$ is defined by

$$(4.1) \quad \rho(f,g) \equiv \sup \{\rho(f(x),g(x)) : x \in X\} \qquad (f, g \in C(X,Y))$$

In case $Y = \mathbf{R}$, the metric ρ is related to the *norm*

$$(4.2) \qquad \|f\| \equiv \sup \{|f(x)| : x \in X\}$$

on $C(X)$ by the equality

$$\rho(f,g) \equiv \|f - g\|$$

A subset S of $C(X,Y)$ is *equicontinuous* if the functions in S have a common modulus of continuity ω. This concept is important because it affords a means of proving that certain subsets of $C(X,Y)$ are totally bounded. The result in question is called *Ascoli's theorem* or *Arzela's theorem*.

Theorem 5 *Let S be an equicontinuous subset of $C(X,Y)$ such that for each $\epsilon > 0$ there exists an ϵ approximation $\{x_1, \ldots, x_n\}$ to X for which the subset*

$$A \equiv \{(f(x_1), \ldots, f(x_n)) : f \in S\}$$

of Y^n is totally bounded. Then S is totally bounded.

Proof Let ω be a common modulus of continuity for the functions in S. Consider $\epsilon > 0$, and let $x_1, \ldots, x_n$ be an $\omega(\epsilon/3)$ approximation to X, such that the set A is totally bounded. Let $f_1, \ldots, f_m$ be elements of S such that the points

$$u_i \equiv (f_i(x_1), \ldots, f_i(x_n)) \qquad (1 \leq i \leq m)$$

approximate A to within $\epsilon/4$. Then for an arbitrary f in S,

$$\sum_{j=1}^{n} \rho(f_i(x_j), f(x_j)) < \frac{\epsilon}{4}$$

for some value of i, which we fix. For an arbitrary x in X there exists x_j with $\rho(x, x_j) < \omega(\epsilon/3)$. Then

$$\rho(f_i(x), f(x)) \leq \rho(f_i(x), f_i(x_j)) + \rho(f_i(x_j), f(x_j)) + \rho(f(x_j), f(x))$$
$$\leq \frac{\epsilon}{3} + \frac{\epsilon}{4} + \frac{\epsilon}{3}$$

Since x is an arbitrary point of X, it follows that $\rho(f_i, f) < \epsilon$. Thus $\{f_1, \ldots, f_m\}$ is an ϵ approximation to S. Since ϵ is arbitrary, S is totally bounded.

To get a good application of Theorem 5, we need three lemmas.

Lemma 2 *A finite family $[a_1, b_1], \ldots, [a_n, b_n]$ of compact intervals has a common point x if and only if*

$$(4.3) \qquad\qquad a_i \leq b_j \qquad (1 \leq i, j \leq n)$$

Proof The condition is clearly necessary, since if x exists, then $a_i \leq x \leq b_j$ for all i and j. Assume conversely that (4.3) holds. Then

$$\min \{b_1, \ldots, b_n\} - \max \{a_1, \ldots, a_n\} \geq 0$$

Therefore the point

$$x \equiv \min \{b_1, \ldots, b_n\}$$

belongs to each of the given intervals.

Lemma 3 *Let $c_1, \ldots, c_n$ be real numbers, and $r_1, \ldots, r_n$ be non-negative real numbers. Then there exists a real number x with*

(4.4) $|x - c_i| \leq r_i \qquad (1 \leq i \leq n)$

if and only if

(4.5) $|c_i - c_j| \leq r_i + r_j \qquad (1 \leq i, j \leq n)$

Proof The condition is obviously necessary. Assume conversely that (4.5) is satisfied. Then $c_i - r_i \leq c_j + r_j$ for all i and j. The intervals $[c_i - r_i, c_i + r_i]$ therefore have a common point x, by Lemma 2. This is equivalent to (4.4).

Lemma 4 *Let K and r_{ij} ($1 \leq i, j \leq n$) be positive real numbers. Then the set Y of points $x \equiv (x_1, \ldots, x_n)$ in $\mathbf{R}^n$ such that*

(4.6) $|x_i| \leq K, |x_i - x_j| \leq r_{ij} \qquad (1 \leq i, j \leq n)$

is totally bounded.

Proof For concreteness, let $\mathbf{R}^n$ have the metric ρ. For each $x \equiv (x_1, \ldots, x_n)$ in $\mathbf{R}^n$, write

$$c_x \equiv \max \{|x_i - x_j| r_{ij}^{-1} : 1 \leq i, j \leq n\}$$
and $$f(x) \equiv \max \{1, c_x\}^{-1} x$$

Then f is a continuous function from $[-K, K]^n$ into Y, which is onto because $f(x) = x$ for all x in Y. Since $[-K, K]^n$ is totally bounded, so is Y.

Theorem 6 *Let $\lambda : \mathbf{R}^{0+} \to \mathbf{R}^{0+}$ be a continuous function such that $\lambda(0) = 0$,*

$$\lambda(x) < \lambda(y) \qquad (x < y, \text{ and } x, y \in \mathbf{R}^{0+})$$
and $$\lambda(x + y) \leq \lambda(x) + \lambda(y) \qquad (x, y \in \mathbf{R}^{0+})$$

Let K be a positive constant. Let S consist of all continuous functions f

from a compact space X to $\mathbf{R}$ such that $\|f\| \leq K$ and

$$|f(x) - f(y)| \leq \lambda(\rho(x,y)) \qquad (x, y \in X)$$

Then S is compact.

Proof Since S is obviously closed in $C(X)$, to show that S is compact it is enough to show that it is totally bounded. Since S is equicontinuous, by Theorem 5 it is enough to show that for every finite sequence $x_1, \ldots,$ x_n of unequal points of X the subset

$$A \equiv \{(f(x_1), \ldots, f(x_n)) : f \in S\}$$

of $\mathbf{R}^n$ is totally bounded. The set

$$B \equiv \{u \equiv (u_1, \ldots, u_n) \in \mathbf{R}^n : |u_i| \leq K, |u_i - u_j| \\ \leq \lambda(\rho(x_i,x_j)), 1 \leq i, j \leq n\}$$

is totally bounded, by Lemma 4. Therefore it is enough to show that $A = B$ or (since obviously $A \subset B$) that $B \subset A$. Since X is totally bounded, $x_1, \ldots, x_n$ can be extended to a sequence $\{x_k\}_{k=1}^{\infty}$ of elements of X which is dense in X. Consider an element u of B. We continue the finite sequence $u_1, \ldots, u_n$ to an infinite sequence $\{u_k\}$ such that

$$(4.7) \qquad |u_i| \leq K, |u_i - u_j| \leq \lambda(\rho(x_i,x_j)) \qquad (i, j \in Z^+)$$

This is done inductively. Assume $u_1, \ldots, u_m$ have already been constructed to satisfy (4.7). Then, by Lemma 3, u_{m+1} can be constructed to satisfy (4.7) because

$$|u_i - u_j| \leq \lambda(\rho(x_i,x_j)) \leq \lambda[\rho(x_i,x_{m+1}) + \rho(x_{m+1},x_j)] \\ \leq \lambda[\rho(x_i,x_{m+1}) + \lambda(\rho(x_j,x_{m+1}))] \qquad (1 \leq i, j \leq m)$$

by the hypotheses on the function λ. Thus the sequence $\{u_k\}$ is constructed inductively. Let Y be the dense subset $\{x_1, x_2, \ldots\}$ of X. Define $h: Y \to \mathbf{R}$ by $h(x_i) = u_i$. By (4.7), we see that h is uniformly continuous. It follows from Lemma 1 that h has a uniformly continuous extension f to X. By (4.7) and the fact that Y is dense, we see that $f \in S$. Since

$$u = \{f(x_1), \ldots, f(x_n)\}$$

$u \in A$. Thus $B \subset A$, as was to be proved.

Corollary *Let X be compact, and let K, c, and α be positive numbers, with $\alpha \leq 1$. Let S be the set of all f in $C(X)$ with $\|f\| \leq K$ satisfying the Lipschitz condition*

$$|f(x) - f(y)| \leq c\rho(x,y)^\alpha \qquad (x,\, y \in X)$$

Then S is compact.

Proof   The result follows from Theorem 6, since the function $\lambda \colon \mathbf{R}^{0+} \to \mathbf{R}^{0+}$ defined by $\lambda(t) \equiv ct^\alpha$ satisfies the hypotheses.

Let X be an arbitrary metric space. For each located subset B of X the function $x \to \rho(x,B)$ is uniformly continuous, by (2.1). Therefore the supremum

$$m(A,B) \equiv \sup\,\{\rho(x,B) \colon x \in A\}$$

exists for each compact set $A \subset X$. The function ρ defined by

$$\rho(A,B) \equiv \max\,\{m(A,B), m(B,A)\}$$

is well defined on $\zeta \times \zeta$, where ζ is the family of all compact subsets of X. We shall show that ρ is a metric on ζ. Clearly $\rho(A,B) = \rho(B,A)$, and $\rho(A,A) = 0$. Assume that $\rho(A,B) = 0$. Then $m(A,B) = 0$, so $\rho(x,B) = 0$ for all x in A. Since B is closed, it follows that $x \in B$ whenever $x \in A$, or $A \subset B$. Similarly, $B \subset A$. Thus $A = B$ whenever $\rho(A,B) = 0$. It remains to check that ρ satisfies the triangle inequality. To this end consider arbitrary compact subsets A, B, and C of X and an arbitrary x in X. Then, by (2.1),

$$\rho(x,C) \leq \inf\,\{\rho(x,y) + \rho(y,C) \colon y \in B\}$$
$$\leq \inf\,\{\rho(x,y) \colon y \in B\} + m(B,C) \leq \rho(x,B) + \rho(B,C)$$

Therefore

$$m(A,C) \leq m(A,B) + \rho(B,C) \leq \rho(A,B) + \rho(B,C)$$

Similarly $m(C,A) \leq \rho(A,B) + \rho(B,C)$. Therefore $\rho(A,C) \leq \rho(A,B) + \rho(B,C)$, and thus ρ satisfies the triangle inequality.

Our next theorem will be the *Stone-Weierstrass theorem*, the basic result in the theory of approximation by real-valued functions on compact spaces. Some groundwork is necessary.

Definition 15 A *polynomial* (of degree N in n variables) is a function $p: \mathbf{R}^n \to \mathbf{R}$ of the form

$$p(x_1, \ldots, x_n) = \sum_{0 \le i_1 + \cdots + i_n \le N} a_{i_1 \ldots i_n} x_1^{i_1} \cdots x_n^{i_n}$$

If $p(0) = 0$, the polynomial p is *strict*.

Lemma 5 *For each $\epsilon > 0$ there exists a strict polynomial $p: \mathbf{R} \to \mathbf{R}$ such that*

$$\big| \, |x| - p(x) \big| \le \epsilon \qquad (|x| \le 1)$$

Proof The nth derivative of the function

$$t \to (1 - t)^{1/2} \equiv \exp \left(\tfrac{1}{2} \ln (1 - t) \right)$$

on the interval $(-1,1)$ is

$$(-1)^n \frac{1}{2} \cdot \left(-\frac{1}{2} \right) \cdots \frac{3 - 2n}{2} (1 - t)^{1/2 - n}$$
$$= -2^{-n} 1 \cdot 3 \cdot 5 \cdots (2n - 3)(1 - t)^{1/2 - n}$$

This function therefore has the Taylor's series

$$(4.8) \qquad 1 - \sum_{n=1}^{\infty} (2^n n!)^{-1} 1 \cdot 3 \cdots (2n - 3) t^n$$

converging, by Theorem 8 of Chap. 2, to $(1 - t)^{1/2}$ on the interval $(-1,1)$. Since $(1 - t)^{1/2} \to 0$ as $t \to 1$, the series $\sum_{n=1}^{\infty} (2^n n!)^{-1} 1 \cdot 3 \cdots (2n - 3)$ converges to 1. By the comparison test the series (4.8) converges on the interval $[-1,1]$. Since (4.8) converges to $(1 - t)^{1/2}$ on $(-1,1)$, it therefore converges to $(1 - t)^{1/2}$ on $[-1,1]$. We can therefore choose the positive integer N so large that

$$\big| (1 - t)^{1/2} - g(t) \big| \le \frac{\epsilon}{2} \qquad (t \in [0,1])$$

where $$g(t) \equiv 1 - \sum_{n=1}^{N} (2^n n!)^{-1} 1 \cdot 3 \cdots (2n - 3) t^n$$

Then $\quad \big| \, |x| - g(1 - x^2) \big| = \big| (1 - (1 - x^2))^{1/2} - g(1 - x^2) \big| \le \frac{\epsilon}{2}$
$$(|x| \le 1)$$

The polynomial p defined by

$$p(x) \equiv g(1 - x^2) - g(1)$$

therefore satisfies our requirements.

Definition 16   Let G be a family of real-valued functions on a set X. Then $\mathfrak{A}(G)$ is the family of all real-valued functions f on X of the form $f \equiv p \circ g$, where $p \colon \mathbf{R}^n \to \mathbf{R}$ is a strict polynomial and where $g \colon X \to \mathbf{R}^n$ has the form

$$g(x) \equiv (g_1(x), \ \ldots \ , g_n(x))$$

for some $g_1, \ \ldots \ , g_n$ in G.

$(\mathfrak{A}(G)$ is the smallest family of real-valued functions on X containing G and closed with respect to the operations of addition, multiplication, and multiplication by real numbers.)

Lemma 6   *Let G be any family of continuous functions on a compact space X. Let f and g be any functions belonging to the closure H of $\mathfrak{A}(G)$ in $C(X)$. Then $|f|$, $\max \{f,g\}$, and $\min \{f,g\}$ also belong to H.*

Proof   Clearly $H = \mathfrak{A}(H)$. Choose $c > 0$ so small that $\|cf\| \leq 1$. For the polynomial function p of Lemma 5 we have

$$\big|\, |cf(x)| - p(cf(x)) \big| \leq \epsilon \qquad (x \in X)$$

Since $p \circ (cf) \in H$, and since ϵ is arbitrary, it follows that $|cf| \in H$. Therefore $|f| \in H$, and therefore

$$\max \{f,g\} = \tfrac{1}{2}(f + g + |f - g|) \in H$$

Similarly, $\min \{f,g\} \in H$.

Definition 17   Let X be a compact space. A family $G \subset C(X)$ is *separating* if there exists a function $\delta \colon \mathbf{R}^+ \to \mathbf{R}^+$ such that whenever $\epsilon > 0$, $x, y \in X$, and $\rho(x,y) \geq \epsilon$ there exists g in G satisfying

$$|g(z)| \leq \epsilon \qquad (z \in X, \ \rho(x,z) \leq \delta(\epsilon))$$

and

$$(4.9) \qquad |g(z) - 1| \leq \epsilon \qquad (z \in X, \ \rho(y,z) \leq \delta(\epsilon))$$

and, whenever $\epsilon > 0$ and $y \in X$, there exists g in G satisfying (4.9).

Lemma 7 *Let G be a separating family of continuous functions on a compact space X, and let H be the closure of $\mathfrak{A}(G)$ in $C(X)$. Let h be a function in H with $c \equiv \inf \{|h(x)| : x \in X\} > 0$. Then $h^{-1} \in H$.*

Proof Write $\lambda \equiv \|h\|^{-2}h^2$. Then $c^2\|h\|^{-2} \leq \lambda \leq 1$, and thus $\|1 - \lambda\| \leq 1 - c^2\|h\|^{-2}$. Hence the series

$$\sum_{n=0}^{\infty} \|h\|^{-2}h(1 - \lambda)^n$$

converges uniformly on X to the function

$$\|h\|^{-2}h(1 - (1 - \lambda))^{-1} = h^{-1}$$

Since each term of this series belongs to H, it follows that h^{-1} belongs to H.

We come now to the Stone-Weierstrass theorem.

Theorem 7 *Let G be a separating family of continuous functions on a compact space X. Then $\mathfrak{A}(G)$ is dense in $C(X)$.*

Proof Let H be the closure of $\mathfrak{A}(G)$ in $C(X)$. Clearly $H = \mathfrak{A}(H)$. We must show that $H = C(X)$. First we show that $1 \in H$. To this end, let $x_1, \ldots, x_n$ be a $\delta(\frac{1}{2})$ approximation to X, where δ is the function of Definition 17. Thus for each i $(1 \leq i \leq n)$ there exists g_i in G with $g_i(z) \geq \frac{1}{2}$ whenever $\rho(x_i,z) \leq \delta(\frac{1}{2})$. By Lemma 6, we have $h \equiv \max \{g_1, \ldots, g_n\} \in H$. Now, for each z in X we have $\rho(x_i,z) \leq \delta(\frac{1}{2})$, and therefore $g_i(z) \geq \frac{1}{2}$, for at least one value of i. It follows that $h \geq \frac{1}{2}$. By Lemma 7, we have $h^{-1} \in H$. Thus $hh^{-1} = 1 \in H$.

Now, let y be any point in X, and r any real number, with $0 < r \leq \frac{1}{4}$. Let $\{x_1, \ldots, x_n\}$ be a $c \equiv \min \{r,\delta(r)\}$ approximation to X. The set $\{1, \ldots, n\}$ is the union of finite sets S and T, with $\rho(x_i,y) > r$ for all i in S and $\rho(x_i,y) < 2r$ for all i in T. By Definition 17, for each i in S there exists g_i in G such that $g_i(z) \leq r \leq \frac{1}{4}$ whenever $\rho(x_i,z) \leq \delta(r)$, and $g_i(z) \geq 1 - r \geq \frac{3}{4}$ whenever $\rho(y,z) \leq \delta(r)$. Write

$$h_i \equiv \min \{1, \max \{0, 2g_i - \tfrac{1}{2}\}\} \qquad (i \in S)$$

Then $h_i \in H$, $0 \leq h_i \leq 1$, $h_i(z) = 0$ whenever $\rho(x_i,z) \leq \delta(r)$, and $h_i(z) = 1$ whenever $\rho(y,z) \leq \delta(r)$. Write

$$\lambda \equiv \prod_{i \in S} h_i$$

Then $\lambda \in H$, $0 \leq \lambda \leq 1$, and $\lambda(z) = 1$ whenever $\rho(y,z) \leq \delta(r)$. Consider z in X with $\rho(y,z) \geq 3r$, and choose i $(1 \leq i \leq n)$ with $\rho(x_i,z) \leq c \leq r$.

Then
$$\rho(x_i,y) \geq \rho(y,z) - \rho(x_i,z) \geq 3r - r = 2r$$

Hence $i \in S$. Since $\rho(x_i,z) \leq c \leq \delta(r)$, we have

$$\lambda(z) = h_i(z) = 0$$

We have thus proved that for each y in X and each r with $0 < r \leq \frac{1}{4}$ there exists λ in H with $0 \leq \lambda \leq 1$, $\lambda(z) = 1$ whenever $\rho(y,z) \leq \delta(r)$, and $\lambda(z) = 0$ whenever $\rho(y,z) \geq 3r$.

We are now prepared to show that every f in $C(X)$ is in H. Let ω be a modulus of continuity for f. Let ϵ be any positive constant, and write $r \equiv \min \{\frac{1}{4}, \frac{1}{4}\omega(\epsilon)\}$. Let $y_1, \ldots, y_m$ be a $\delta(r)$ approximation to X. By the above, for $1 \leq j \leq m$ there exists λ_j in H with $0 \leq \lambda_j \leq 1$, $\lambda_j(z) = 1$ whenever $\rho(y_j,z) \leq \delta(r)$, and $\lambda_j(z) = 0$ whenever $\rho(y_j,z) \geq 3r$. Thus $\lambda \equiv \sum_{j=1}^{m} \lambda_j$ is in H and $\lambda \geq 1$, since for each z in H we have $\rho(y_j,z) \leq \delta(r)$ for some j, and therefore $\lambda(z) \geq \lambda_j(z) \geq 1$. By Lemma 7, we have $\lambda^{-1} \in H$. The function

$$g \equiv \sum_{j=1}^{m} f(y_j)\lambda^{-1}\lambda_j$$

therefore belongs to H. Let z be any point of X. Since $\lambda_j(z)$ vanishes whenever $\rho(y_j,z) \geq 3r$, and $|f(z) - f(y_j)| \leq \epsilon$ whenever $\rho(y_j,z) \leq \omega(\epsilon)$, and $3r < \omega(\epsilon)$, we have

$$\begin{aligned}
|f(z) - g(z)| &= \Big| \sum_{j=1}^{m} (f(z) - f(y_j))\lambda^{-1}(z)\lambda_j(z) \Big| \\
&\leq \sum_{j=1}^{m} |f(z) - f(y_j)|\lambda^{-1}(z)\lambda_j(z) \\
&\leq \sum_{j=1}^{m} \epsilon\lambda^{-1}(z)\lambda_j(z) = \epsilon
\end{aligned}$$

Since ϵ is an arbitrary positive constant, it follows that $f \in H$, as was to be proved.

Corollary 1 *Let $\{(X_n,\rho_n)\}_{n=1}^{\infty}$ be compact spaces, with product (X,ρ). Let G be the set of all functions on X of the form*

$$g \equiv f_n \circ \pi_n$$

where $\pi_n\colon X \to X_n$ is the projection of X onto X_n and $f_n\colon X_n \to \mathbf{R}$ is continuous. Then the set H of all h in $C(X)$ of the form $h = h_1 + \cdots + h_n$, where each h_i is a finite product of functions in G, is dense in $C(X)$.

Proof Clearly $H = \mathfrak{A}(G)$. To finish the proof it is enough, by Theorem 7, to show that G is separating. To this end consider $\epsilon > 0$ and let $x \equiv \{x_n\}$ and $y \equiv \{y_n\}$ be points of X with $\rho(x,y) \geq \epsilon$. Choose N in Z^+ with $\sum_{n=N+1}^{\infty} 2^{-n} < \epsilon/2$. By Definition 4, we have

$$\sum_{n=1}^{N} 2^{-n}\rho_n(x_n,y_n) > \rho(x,y) - \frac{\epsilon}{2} \geq \frac{\epsilon}{2}$$

Therefore $\rho_n(x_n,y_n) > \epsilon/2$ for some n with $n \leq N$. Write $g \equiv f_n \circ \pi_n$, where $f_n\colon X_n \to \mathbf{R}$ is the function

$$f_n(z_n) \equiv (\rho_n(x_n,y_n))^{-1}\rho_n(x_n,z_n) \qquad (z_n \in X_n)$$

If $z \equiv \{z_n\} \in X$ and $\rho(x,z) \leq \epsilon^2 2^{-N-1}$, then

$$\begin{aligned}
|g(z)| = |f_n(z_n)| &\leq 2\epsilon^{-1}\rho_n(x_n,z_n) \leq 2\epsilon^{-1}2^n\rho(x,z) \\
&\leq 2\epsilon^{-1}2^N\rho(x,z) \leq \epsilon
\end{aligned}$$

Similarly if $\rho(y,z) \leq \epsilon^2 2^{-N-1}$, then

$$\begin{aligned}
|g(z) - 1| = |f_n(z_n) - f_n(y_n)| &\leq 2\epsilon^{-1}|\rho_n(x_n,z_n) - \rho_n(x_n,y_n)| \\
&\leq 2\epsilon^{-1}\rho_n(y_n,z_n) \leq 2\epsilon^{-1}2^N\rho(y,z) \leq \epsilon
\end{aligned}$$

Therefore G is separating.

Corollary 2 *Let X be a compact space, and let G consist of all functions $x \to \rho(x,x_0)$, with $x_0 \in X$. Then $\mathfrak{A}(G)$ is dense in $C(X)$.*

Proof Consider $\epsilon > 0$ and x, y in X with $\rho(x,y) \geq \epsilon$. Define the function g in $\mathfrak{A}(G)$ by

$$g(z) \equiv \rho(x,y)^{-1}\rho(x,z)$$

Then if $\rho(x,z) \leq \epsilon^2$

$$|g(z)| \leq \epsilon^{-1}\rho(x,z) \leq \epsilon$$

and if $\rho(y,z) \leq \epsilon^2$,

$$|g(z) - 1| = \rho(x,y)^{-1}|\rho(x,z) - \rho(x,y)| \leq \epsilon^{-1}\rho(y,z) \leq \epsilon$$

Therefore G is separating.

Corollary 3   *Every continuous function f on a compact set $X \subset \mathbf{R}^n$ can be arbitrarily closely approximated on X by polynomial functions $p\colon \mathbf{R}^n \to \mathbf{R}$.*

Proof   First consider the case in which $n = 1$. By Lemma 5, the function $x \to |x - x_0|$ can be arbitrarily closely approximated on X by polynomials. The theorem then follows from Corollary 2.

Next consider the case $X \equiv [a,b]^n$ for some compact interval $[a,b]$. The result then follows from Corollary 1.

Finally, consider the general case. Since X is bounded, there exists a compact interval $[a,b]$ with $X \subset [a,b]^n$. By the case already considered, each of the functions $x \to \rho(x,x_0)$ $(x_0 \in [a,b]^n)$ can be arbitrarily closely approximated by polynomials on $[a,b]^n$. The result then follows by Corollary 2.

The case $n = 1$ and $X \equiv [-1,1]$ of Corollary 3 is the famous *Weierstrass approximation theorem*.

We turn now to the study of sufficient conditions for a subset of a compact space to be compact, first proving that every compact space is a union of finitely many compact sets of arbitrarily small diameters.

Proposition 11   *For every compact space X and every $\epsilon > 0$ there exist finitely many compact subsets $X_1, \ldots, X_n$ of X of diameters at most ϵ whose union is X.*

Proof   By the total boundedness of X, there exist subsets $X_1{}^1, \ldots, X_n{}^1$ of X, each consisting of one point, such that for each x in X at least one of the numbers $\rho(x,X_j{}^1)$ $(1 \leq j \leq n)$ is less than $3^{-2}\epsilon$.

We define inductively sequences $\{X_1{}^i\}_{i=1}^\infty, \ldots, \{X_n{}^i\}_{i=1}^\infty$ of subfinite subsets of X such that

    (a) $X_j{}^1 \subset X_j{}^2 \subset \cdots$ $(1 \leq j \leq n)$

    (b) $\rho(x,X_j{}^i) < 3^{-i}\epsilon$ $(i \in Z^+, 1 \leq j \leq n, x \in X_j{}^{i+1})$

    (c) $\rho(x,X_j{}^{i+1}) < 3^{-i-2}\epsilon$ $(i \in Z^+, 1 \leq j \leq n, \rho(x,X_j{}^i) < 3^{-i-1}\epsilon)$

The sets $X_j{}^1$ $(1 \leq j \leq n)$ have already been defined. Assume therefore that $X_1{}^\imath, \ldots, X_n{}^\imath$ have been defined, and satisfy the relevant conditions. Let $\{x_1, \ldots, x_N\}$ be a $3^{-\imath-2}\epsilon$ approximation to X. Partition the set $\{(m,j): 1 \le m \le N,\ 1 \le j \le n\}$ into two sets S and T such that

$$\rho(x_m, X_j{}^\imath) < 3^{-\imath}\epsilon \qquad ((m,j) \in S)$$

and

$$\rho(x_m, X_j{}^\imath) > \tfrac{1}{2} 3^{-\imath}\epsilon \qquad ((m,j) \in T)$$

Enlarge each of the sets $X_j{}^\imath$ by adjoining all those points x_m for which $(m,j) \in S$. The resulting sets $X_j{}^{\imath+1}$ clearly satisfy (a) and (b). To check (c) consider x in X with $\rho(x, X_j{}^\imath) < 3^{-\imath-1}\epsilon$, and choose x_m with $\rho(x,x_m) < 3^{-\imath-2}\epsilon$. Then

$$\begin{aligned} \rho(x_m, X_j{}^\imath) &\le \rho(x_m,x) + \rho(x, X_j{}^\imath) \\ &< 3^{-\imath-2}\epsilon + 3^{-\imath-1}\epsilon < \tfrac{1}{2} 3^{-\imath}\epsilon \end{aligned}$$

Therefore $(m,j) \in S$, and thus $x_m \in X_j{}^{\imath+1}$. It follows that $\rho(x, X_j{}^{\imath+1}) < 3^{-\imath-2}\epsilon$. This completes the construction of the sets $X_j{}^\imath$.

For $1 \le j \le n$ write $Y_j \equiv \bigcup_{\imath=1}^{\infty} X_j{}^\imath$, and let X_j be the closure of Y_j. By (b), the distance of any point in Y_j to $X_j{}^\imath$ is at most

$$d_\imath \equiv \sum_{k=i}^{\infty} 3^{-k}\epsilon = \tfrac{1}{2} 3^{-\imath+1}\epsilon$$

Therefore Y_j is totally bounded, and its closure X_j is compact. The diameter of X_j is at most $2d_1 = \epsilon$.

To show that $X = \bigcup_{j=1}^{n} X_j$, consider any x in X. Choose j ($1 \le j \le n$) with $\rho(x, X_j{}^1) < 3^{-2}\epsilon$. It follows from (c) by induction on i that $\rho(x, X_j{}^\imath) < 3^{-i-1}\epsilon$ for all i. Therefore $x \in X_j$.

Let $P(\alpha)$ be a property of a real number α and $\{\alpha_n\}$ a sequence of real numbers such that $P(\alpha)$ is valid for each real number α with $\alpha \ne \alpha_n$ for all n. Then $P(\alpha)$ is said to hold for *all except countably many* real numbers, the sequence $\{\alpha_n\}$ is called the *excluded sequence*, and a real number α with $\alpha \ne \alpha_n$ for all n is called *admissible*.

Theorem 8 *Let $f\colon X \to \mathbf{R}$ be a continuous function on a compact space X. Then for all except countably many real numbers $\alpha > a \equiv \inf \{f(x) : x \in X\}$ the set*

$$X_\alpha \equiv \{x \in X : f(x) \le \alpha\}$$

is compact. Moreover, if α is admissible and ϵ is any positive constant, there exists $\delta > 0$ such that X_t is compact for all admissible t with $|\alpha - t| < \delta$ and satisfies the inequality

$$\rho(X_t, X_\alpha) \leq \epsilon$$

where ρ is the metric on the family of compact subsets of X.

Proof For each k in Z^+ represent X as a finite union $X = \bigcup_{j=1}^{N(k)} X_j^k$ of compact sets of diameters less than k^{-1}. Define the sequence $\{\alpha_n\}$ to be an ordering of the numbers

$$c_{jk} \equiv \inf \{f(x) : x \in X_j^k\} \qquad (k \in Z^+, 1 \leq j \leq N(k))$$

Let α be admissible, with $\alpha > a$. For each j and k with $c_{jk} < \alpha$ choose x_j^k in X_j^k with $f(x_j^k) < \alpha$. Consider any point x in X_α and any positive integer k. Choose j ($1 \leq j \leq N(k)$) with $x \in X_j^k$. Then $c_{jk} \leq f(x) \leq \alpha$. Therefore $c_{jk} < \alpha$ and $\rho(x, x_j^k) < k^{-1}$. The elements x_j^k for which $c_{jk} < \alpha$ are thus a k^{-1} approximation to X_α. Therefore X_α is totally bounded. Since X_α is also closed, it is compact. If ϵ is any positive constant, choose k in Z^+ with $k^{-1} < \epsilon$. Write

$$\delta \equiv \min \{|\alpha - c_{jk}| : 1 \leq j \leq N(k)\}$$

Consider any admissible t with $|\alpha - t| < \delta$. Then, for each j, $c_{jk} \leq t$ if and only if $c_{jk} \leq \alpha$. Let x be any point of X_t, so that $f(x) \leq t$. Now, $x \in X_j^k$ for some j, and $c_{jk} \leq f(x) \leq t$. Therefore $c_{jk} \leq \alpha$, hence $c_{jk} < \alpha$. Hence $x_j^k \in X_\alpha$. It follows that

$$\rho(x, X_\alpha) \leq \rho(x, x_j^k) \leq \operatorname{diam} X_j^k \leq k^{-1} < \epsilon$$

Similarly $\rho(x, X_t) < \epsilon$ for all x in X_α. Therefore $\rho(X_t, X_\alpha) \leq \epsilon$.

5. LOCALLY COMPACT SPACES

Many important metric spaces, such as the euclidean spaces $\mathbf{R}^n$, that are not compact have a property almost as good.

Definition 18 A nonvoid metric space X is *locally compact* if every bounded subset of X is contained in a compact subset. A function $f : X \to Y$ from X to a metric space Y is *continuous* if it is continuous (i.e., uniformly continuous) on every compact subset of X, or equivalently on every bounded subset of X. A continuous function $g\colon X \to \mathbf{R}$ such that the set

$$X_\alpha \equiv \{x \in X \colon g(x) \leq \alpha\}$$

is bounded for all α in $\mathbf{R}$ is called a *compactifier* for X.

If x_0 is any point in the locally compact space X, the function $x \to \rho(x,x_0)$ is a compactifier.

The term "compactifier" comes from the fact that if g is any compactifier, then X_α is either void or compact for all except countably many α. This follows from Theorem 8.

For maps $f\colon X \to Y$ between locally compact spaces the notion of metric equivalence is too restrictive. More useful is the concept of a *homeomorphism*.

Definition 19   A continuous map f from a locally compact space X to a locally compact space Y, with an inverse $g\colon Y \to X$ which is also continuous, is called a *homeomorphism*. In case ρ and ρ' are two metrics on the same set X, such that the spaces (X,ρ) and (X,ρ') are both locally compact and such that the identity map $x \to x$ is a homeomorphism from (X,ρ) to (X,ρ'), the metrics ρ and ρ' are called *homeomorphic* metrics on X.

If $f\colon X \to Y$ is a continuous map, and if K is any bounded subset of X, then there exists a compact set V with $K \subset V$. Since f is continuous on V, $f(V)$ is totally bounded, and therefore bounded. Hence $f(K)$ is bounded. Thus f takes bounded sets into bounded sets. In particular the property of boundedness is intrinsic (preserved under homeomorphisms).

Certain spaces which are not necessarily locally compact become locally compact under a new metric. As an example, let $g\colon X \to \mathbf{R}$ be a function from a metric space X to $\mathbf{R}$ such that (i) the sets X_α are compact for certain arbitrarily large values of α, and (ii) g is uniformly continuous on each of the sets X_α. Then X need not be locally compact because not every bounded set need be a subset of some X_α, but X is locally compact under the new metric

$$(x,y) \to \rho(x,y) + |g(x) - g(y)|$$

and g is a compactifier. This construction can be used when X is the metric complement $-Z$ of a located subset Z of a locally compact space Y. In this case the function g can be taken to be $x \to h(x) + \rho(x,Z)^{-1}$, where h is a compactifier for Y.

We now show that every locally compact space can be obtained by this construction as the metric complement of a single point in a compact space.

Theorem 9 *Let (X,ρ) be a locally compact space. Then there exists a compact space (Y,d), called the* one-point compactification *of X, a point ∞ in Y, and an inclusion map $i\colon X \to Y$ such that $i(X)$ is the metric complement of the set $\{\infty\}$ in Y and such that the metric ρ_0 on X defined by*

$$\rho_0(x,y) \equiv d(i(x),i(y)) + |d(i(x),\infty)^{-1} - d(i(y),\infty)^{-1}|$$

is homeomorphic to ρ.

Proof Let $\{x_n\}$ be a countable dense subset of X. For each n in Z^+ define the bounded continuous function f_n on X by

$$f_n(x) \equiv \min\{1,\rho(x,x_n)\}$$

Let Z be the product space

$$Z \equiv \prod_{n=1}^{\infty} [0,1]$$

with product metric d defined in the usual way. Then X is a subset of Z, with the inclusion map i defined by $i(x) \equiv \{f_n(x)\}$. For each x and y in X,

$$
\begin{aligned}
(5.1) \quad d(i(x),i(y)) &= \sum_{n=1}^{\infty} 2^{-n}|f_n(x) - f_n(y)| \\
&= \sum_{n=1}^{\infty} 2^{-n}|\min\{1,\rho(x,x_n)\} - \min\{1,\rho(y,x_n)\}| \\
&\leq \sum_{n=1}^{\infty} 2^{-n}|\rho(x,x_n) - \rho(y,x_n)| \\
&\leq \sum_{n=1}^{\infty} 2^{-n}\rho(x,y) = \rho(x,y)
\end{aligned}
$$

In particular, i is uniformly continuous.

To show that $i(X)$ is totally bounded (in the metric d), for each N in Z^+ we define

$$
\begin{aligned}
X_N &\equiv \{x \in X : \rho(x,x_1) < \alpha_N\} \\
\text{and} \quad Y_N &\equiv \{x \in X : \rho(x,x_1) \geq N - 1\}
\end{aligned}
$$

where the real number α_N in $(N - 1, N)$ is chosen to make X_N compact. Clearly $X = X_N \cup Y_N$, and the diameter of $i(Y_N)$ approaches 0 as $N \to \infty$. Therefore $i(X)$ is the union of a totally bounded set $i(X_N)$ and a set $i(Y_N)$ of arbitrarily small diameter. It follows that the set $i(X)$ is totally bounded. The closure Y of $i(X) \cup \{\infty\}$ is therefore compact, where ∞ is the point in Z all of whose coordinates are 1. Since $\{x_n\}$ is dense in X, for each compact set $K \subset X$ we have

$$(5.2) \qquad d(i(K), \infty) > 0$$

It follows that the identity map $(K, \rho) \to (K, \rho_0)$ is uniformly continuous.

Consider conversely any subset A of X which is bounded in the metric ρ_0. Then $d(i(x), \infty)$ is bounded below by some $c > 0$, for all x in A. Therefore there exists n in Z^+ such that for each x in A at least one of the numbers $\rho(x, x_1), \ldots, \rho(x, x_n)$ is less than 1. Therefore A is bounded in the metric ρ. To show that the identity map $(A, \rho_0) \to (A, \rho)$ is uniformly continuous, consider any ϵ in $(0, \frac{1}{2})$. Choose the integer N so large that for each z in A there exists an integer $n \leq N$ with $\rho(z, x_n) \leq \epsilon$. Let x and y be any points in A with $\rho_0(x, y) \leq 2^{-N-1}$. Then $|f_n(x) - f_n(y)| \leq \frac{1}{2}$ whenever $n \leq N$. Take an integer $n \leq N$ such that $\rho(x, x_n) \leq \epsilon$. Since $\epsilon < \frac{1}{2}$, we have $f_n(x) = \rho(x, x_n)$. Therefore $f_n(y) \leq \frac{1}{2} + f_n(x) < 1$, and thus $f_n(y) = \rho(y, x_n)$ also. We compute

$$\begin{aligned}
\rho(x, y) &\leq \rho(x, x_n) + \rho(y, x_n) \\
&\leq 2\rho(x, x_n) + |\rho(y, x_n) - \rho(x, x_n)| \\
&\leq 2\epsilon + 2^N \rho_0(x, y)
\end{aligned}$$

It follows that $\rho(x, y) \leq 3\epsilon$ if $\rho_0(x, y) \leq \epsilon 2^{-N}$. Hence the identity map $(A, \rho_0) \to (A, \rho)$ is uniformly continuous.

Thus ρ and ρ_0 are homeomorphic metrics on X.

It remains to show that $i(X)$ is the metric complement $-\{\infty\}$ of $\{\infty\}$ in Y. It is enough to show that $-\{\infty\} \subset i(X)$, since $i(X) \subset -\{\infty\}$, by (5.2). Consider therefore y in $-\{\infty\}$, so that $d(y, \infty) > 0$. By the definition of Y there exists a sequence $\{x_n\}$ of points in X such that $\{i(x_n)\}$ converges to y. Therefore $\{x_n\}$ is a Cauchy sequence in the metric ρ_0. Since the metrics ρ and ρ_0 are equivalent on bounded sets, $\{x_n\}$ is Cauchy in the metric ρ. Let x_0 be the limit of $\{x_n\}$. Then $i(x_0) = \lim_{n \to \infty} i(x_n) = y$. Thus $y \in i(X)$, as was to be proved.

Next we generalize to locally compact spaces some results previously obtained for compact spaces.

Proposition 12 *The product X of finitely many locally compact spaces $X_1, \ldots, X_n$ is locally compact.*

Proof Let A be a bounded subset of X. By Definition 3 of the product metric,

$$A \subset A_1 \times \cdots \times A_n$$

where A_i is some bounded subset of X_i $(1 \leq i \leq n)$. Let B_i be any compact subset of X_i with $A_i \subset B_i$. Then $B \equiv B_1 \times \cdots \times B_n$ is a compact subset of X, and $A \subset B$. Therefore X is locally compact.

Proposition 13 *A locally compact subset Y of a metric space X is closed and located, and a closed located subset Y of a locally compact space X is locally compact.*

Proof Assume first that Y is a locally compact subspace of the metric space X. Let y_0 be any point in the closure of Y. Choose $c > 0$ so that

$$Y_c \equiv \{y \in Y : \rho(y,y_0) \leq c\}$$

is compact. Then y_0 is in the closure of Y_c, and Y_c is closed by Proposition 9. Therefore $y_0 \in Y_c \subset Y$. It follows that Y is closed.

To see that Y is located, let x be any point in X. Choose y_0 and c as above so that, in addition, $c > 2\rho(x,y_0)$. Then if y is any point of Y with $\rho(x,y) \leq \rho(x,y_0)$, we have

$$\rho(y,y_0) \leq \rho(x,y) + \rho(x,y_0) \leq 2\rho(x,y_0) < c$$

and thus $y \in Y_c$. Therefore $\rho(x,Y)$ exists and equals $\rho(x,Y_c)$. Thus Y is located.

Assume next that Y is a closed located subset of a locally compact space X. Let y_0 be any point of Y. For each α in $\mathbf{R}$ write

$$X_\alpha = \{x \in X : \rho(x,y_0) \leq \alpha\}$$

Consider any $c > 0$ for which X_{4c} is compact. Choose r in $(c,2c)$ so that the set

$$V \equiv \{x \in X : r \leq \rho(x,y_0) \leq 4c\}$$

is compact. Then the set

$$U \equiv V \cup (Y \cap X_{4c})$$

is located in X_{4c}, since if $x \in X_{4c}$ and $\rho(x,y_0) < 2c$, then $\rho(x, Y \cap X_{4c})$ exists and equals $\rho(x,Y)$, and if $x \in X_{4c}$ and $\rho(x,y_0) > r$, then $\rho(x,V) = 0$. The closure $\bar{U}$ of U is therefore a compact subset of X_{4c}, by Proposition 7, and thus there exists α in (c,r) such that $\bar{U} \cap X_\alpha$ is compact. But

$$\bar{U} \cap X_\alpha = \{y \in Y : \rho(y,y_0) \leq \alpha\}$$

Since c can be arbitrarily large, Y is locally compact.

The problem of extending a continuous function to a larger space arises often and in many forms. The following solution to a special case of this problem is the famous *Tietze extension theorem*, or rather as much of it as is constructively valid.

Theorem 10 *Let Y be a locally compact subset of a metric space X, and $I \subset \mathbf{R}$ a compact proper interval. Let*

$$f \colon Y \to I$$

be continuous. Then there exists a function $h \colon X \to I$ which is uniformly continuous on bounded subsets of X such that $h(y) = f(y)$ for all y in Y.

Proof There is no loss of generality in assuming $I = [-1,1]$. By adjoining two additional points to X, we may assume also that there exist points x^+ and x^- in Y with $f(x^+) = 1$ and $f(x^-) = -1$. (This will serve to make certain sets which will be defined below nonvoid.) Let g be a compactifier for Y, with $g(x^+) = g(x^-) = 0$, so that there exists a sequence $\{\alpha_n\}$ of nonnegative constants with $\alpha_n \to \infty$ as $n \to \infty$ such that the set $\{y \in Y : g(y) \leq \alpha_n\}$ is compact for each n in Z^+. By Theorem 8 there exists an a in $(\frac{1}{4},\frac{1}{3})$ such that for each n in Z^+ the sets

$$A_n \equiv \{y \in Y : f(y) \leq -a,\, g(y) \leq \alpha_n\}$$
and
$$B_n \equiv \{y \in Y : f(y) \geq a,\, g(y) \leq \alpha_n\}$$

are compact. Therefore the sets

$$A \equiv \{y \in Y : f(y) \leq -a\}$$
and
$$B \equiv \{y \in Y : f(y) \geq a\}$$

are locally compact. We define the function f_1 on X by

$$f_1(x) \equiv (\rho(x,A) + \rho(x,B))^{-1}(a\rho(x,A) - a\rho(x,B))$$

for x in X. Since $(\rho(x,A) + \rho(x,B))^{-1}$ is bounded away from 0 on bounded subsets of X, we see that f_1 is uniformly continuous on bounded subsets of X. Clearly $|f_1(x)| \leq a < \frac{1}{3}$ for all x in X. For all y in A we have

$$|f(y) - f_1(y)| = |f(y) + a| \leq \tfrac{3}{4}$$

The same inequality holds for all y in B. If $-\frac{1}{3} \leq f(y) \leq \frac{1}{3}$, then

$$|f(y) - f_1(y)| \leq |f(y)| + |f_1(y)| \leq \tfrac{2}{3} < \tfrac{3}{4}$$

Therefore $|f(y) - f_1(y)| \leq \frac{3}{4}$ for all y in Y.

Now apply the same construction to the function $f - f_1$ which was applied to f. The result is a function $f_2 \colon X \to \mathbf{R}$ which is uniformly continuous on bounded subsets of X such that

$$|f_2(x)| \leq \tfrac{1}{3} \cdot \tfrac{3}{4} \qquad (x \in X)$$

and
$$|f(y) - f_1(y) - f_2(y)| \leq (\tfrac{3}{4})^2 \qquad (y \in Y)$$

Continuing in this way, we construct a series $\sum\limits_{n=1}^{\infty} f_n$ of functions from X to $\mathbf{R}$, each of which is uniformly continuous on bounded subsets of X, which converges uniformly on X to a function f_0 such that $f_0(y) = f(y)$ for all y in Y. By Proposition 4, we see that f_0 is uniformly continuous on bounded subsets of X. The function

$$h \equiv \min\{1, \max\{-1, f_0\}\}$$

therefore has the desired properties.

PROBLEMS

1. Prove Proposition 4.

2. A neighborhood space X satisfies the *second axiom of countability* if there exists a sequence $\{U_n\}$ of neighborhoods such that for each x in X and each neighborhood U with $x \in U$ there exists n in Z^+ with $x \in U_n \subset U$. Show that a metric space X is separable if and only if the associated neighborhood space satisfies the second axiom of countability.

3. Show that a located subset of a separable metric space is separable.

4. Show that a dense subset of a separable metric space is separable.

5. Let $\{F_n\}$ be a sequence of located subsets of a complete metric space X. Show that $G \equiv \bigcap_{n=1}^{\infty} - F_n$ can be metrized to be a complete metric space.

6. Call a sequence $\{x_n\}$ in a metric space X *fickle* if for each $\epsilon > 0$ there exists N in Z^+ such that whenever $n_1 < n_2 < \cdots < n_N$ are positive integers at least one of the numbers $\rho(x_{n_i}, x_{n_{i+1}})$ $(1 \leq i \leq N - 1)$ is less than ϵ. Define two fickle sequences $\{x_n\}$ and $\{y_n\}$ to be *equal* if $\{x_1, y_1, x_2, y_2, \ldots\}$ is a fickle sequence. Show that this is an equivalence relation. Show that a fickle sequence is equal to every subsequence.

7. For each metric space X let the *fickle completion* X_f of X consist of all fickle sequences with equality defined as above. Define operations of addition and multiplication on the fickle completion $\mathbf{R}_f$ of $\mathbf{R}$ so that the natural map $i \colon \mathbf{R} \to \mathbf{R}_f$ is a homomorphism. Define an order $x \leq y$ on $\mathbf{R}_f$. (Problem 7 hints at an attempt to enlarge the real number system so that every bounded monotone sequence converges.)

8. For each metric space X define a *fickle metric* $\rho_f \colon X_f \times X_f \to \mathbf{R}_f^{0+}$ on X_f.

9. Prove the converse of Ascoli's theorem.

10. Let S consist of all differentiable functions on $[0,1]$ with $\|f\| \leq c_0$, $\|f'\| \leq c_1$, where c_0 and c_1 are positive constants. Show that S is a totally bounded subset of $C([0,1])$.

11. Let S consist of all N times differentiable functions on $[0,1]$ with $\|f^{(n)}\| \leq c_n$ $(0 \leq n \leq N)$, where $c_0, \ldots, c_N$ are positive constants. Show that S is a totally bounded subset of $C([0,1])$.

12. Show that the set $\mathfrak{F}$ of all compact subsets of a compact space X is compact.

13. Show that if Y is a complete located subset of a metric space X and x a point of X with $x \neq y$ for all y in Y, then the distance from x to Y is positive.

14. Show that there exists a metric d on the set M of irrational real numbers such that a subset of M is locally compact with respect to d

if and only if it is locally compact with respect to the metric ρ on $\mathbf{R}$, so that d and ρ are homeomorphic metrics on such sets, and such that if $\{x_n\}$ is any sequence of irrational numbers converging in the metric ρ to a rational number, then there exists $\epsilon > 0$ such that for infinitely many n there exists $m > n$ with $d(x_m,x_n) \geq \epsilon$.

15. Let $f\colon [0,1] \to \mathbf{R}$ be continuous, with $f(0) = 0$ and $f(1) = 1$. Show that the set $\{x\colon f(x) = \alpha\}$ is compact for all except countably many α in $[0,1]$.

16. A compact space X is *locally connected* if for each $\epsilon > 0$ there exist finitely many compact sets, each connected in the sense of Prob. 12 of Chap. 3, of diameters at most ϵ whose union is X. If $f\colon X \to \mathbf{R}$ is continuous, show that $\{x\colon f(x) = \alpha\}$ is void or compact for all except countably many α in $\mathbf{R}$.

17. A *uniform space* is a set X and a set M of pseudometrics on X, such that $x_1 = x_2$ if and only if $\rho(x_1,x_2) = 0$ for all ρ in M. A function $f\colon X \to Y$ from a uniform space (X,M) to a uniform space (Y,N) is *uniformly continuous* if for each d in N and each $\epsilon > 0$ there exist $\rho_1, \ldots, \rho_n$ in M and $\delta > 0$ such that $d(f(x_1),f(x_2)) \leq \epsilon$ whenever $\rho_i(x_1,x_2) \leq \delta$ $(1 \leq i \leq n)$. In case f has an inverse which is also uniformly continuous, f is called a *metric equivalence*. Show that if M is countable, there exists a metric d on X such that the identity map $i\colon X \to X$ is a metric equivalence of (X,M) with $(X,\{d\})$.

18. Call a uniform space (X,M) *totally bounded* if X is totally bounded with respect to the pseudometric $\rho_1 + \cdots + \rho_n$, whenever $\rho_1, \ldots, \rho_n \in M$. Show that a uniformly continuous map takes totally bounded sets into totally bounded sets.

19. A *filter* $\mathfrak{F}$ in a set X is a collection of nonvoid subsets of X such that if $F_1 \in \mathfrak{F}$ and $F_2 \in \mathfrak{F}$, then $F_1 \cap F_2 \in \mathfrak{F}$, and if $F \subset G$ and $F \in \mathfrak{F}$, then $G \in \mathfrak{F}$. If (X,M) is a uniform space, a filter $\mathfrak{F}$ in X is a *Cauchy filter* if for each ρ in M and each $\epsilon > 0$ there exists F in $\mathfrak{F}$ with $\rho(x,y) < \epsilon$ whenever $x, y \in F$. A Cauchy filter $\mathfrak{F}$ is said to *converge* to a point x in X if for each ρ in M and each $\epsilon > 0$ there exists F in $\mathfrak{F}$ such that $\rho(x,y) < \epsilon$ for all y in F. If every Cauchy filter converges, X is said to be *complete*. Relate this definition to the definition of completeness for metric spaces.

20. Show that an arbitrary uniform space can be completed.

21. If $(X_\alpha,M_\alpha)_{\alpha \in A}$ is any family of uniform spaces, define the *product* (X,M) by taking $X \equiv \prod_{\alpha \in A} X_\alpha$ and letting M be the set of all pseudometrics

$$(\{x_\alpha\},\{y_\alpha\}) \to \rho(x_\beta,y_\beta) \qquad (\{x_\alpha\},\{y_\alpha\} \in X)$$

with $\beta \in A$ and $\rho \in M_\beta$. Show that (X,M) is complete if each (X_α,M_α) is complete, and totally bounded if each (X_α,M_α) is totally bounded.

NOTES

In classical analysis, two metrics are equivalent if they give rise to the same family of open sets. This notion is of no use constructively. The definition given in Sec. 1 is especially suited to compact spaces. Other notions are more suited to other spaces. (See, for example, Definition 19.)

The definition of a located set is due to Brouwer.

The definition of a nowhere-dense set can be extended to sets that are not necessarily located, as follows. A nowhere-dense set A in a metric space X is a subset of X such that there exists an open dense subset B of X with $x \neq y$ for all x in A and y in B. [In other words, (A,B) is a complemented set relative to the family of uniformly continuous real-valued functions on X.]

Definition 13 is due to Brouwer. The classical definition of a compact space as one for which every open cover has a finite subcover would not do. There would be no nontrivial example of a compact space!

In classical analysis, every nonvoid subset of a metric space is located. This is not true constructively. In fact, Proposition 10 would be false without the condition that X is located.

The extra hypothesis in Ascoli's theorem (Theorem 5), that for each $\epsilon > 0$ there exists an ϵ approximation of the type described, unfortunately makes it much harder to apply. For example, Theorem 6 would be trivial if this condition did not have to be checked.

The proof of Theorem 7 is another instance of minor complications caused by not being able to compare arbitrary real numbers. We are forbidden to assume that either $\rho(x_i,y) > r$ or $\rho(x_i,y) \leq r$.

Proposition 11 and Theorem 8 are partial substitutes for the classical result that a closed subset of a compact space is compact.

Definition 18 differs from the classical version, which states that a locally compact space is one in which every point has a compact neighborhood. The open interval $(0,1)$ is locally compact in this sense, but not constructively. Of course it can be given a new metric to make it locally compact constructively (as indicated in the discussion preceding Theorem 9).

The one-point compactification Y of a locally compact space X is much simpler to define classically than constructively. One reason is that constructively there is more to Y than the union of X and $\{\infty\}$.

Problems 6 to 8 indicate an attempt to beef up **R**. Such attempts have not led to significant results.

For reasons discussed in Appendix A, the concept of a uniform space (see Probs. 17 to 21) is not as useful as might be supposed.

Chapter **5**

COMPLEX ANALYSIS

The constructive development of elementary complex analysis, through Cauchy's integral formula, is a simple matter; the material seems to have a natural constructive cast. This is carried through in Secs. 1, 2, 3, and 4. Next it is shown that under certain conditions it is possible to find the zeros of an analytic function. Although results of this type entail much more careful estimates than their classical counterparts, nothing basically new is involved. The last section proves a constructive version of the Riemann mapping theorem, following the classical Koebe approach as presented by Ostrowski. Some care must be taken in finding the right definitions, since the Riemann mapping theorem is not constructively valid without additional restrictions on the domain.

112

The real numbers, for certain purposes, are too thin. Many beautiful phenomena become fully visible only when the complex numbers are brought to the fore.

1. THE COMPLEX PLANE

The complex numbers are obtained by adjoining a square root i of -1 to the real numbers.

Definition 1$\quad$A complex number is an expression of the form

$$z \equiv x + iy$$

where x and y are real numbers. Two complex numbers $z_1 \equiv x_1 + iy_1$ and $z_2 \equiv x_2 + iy_2$ are *equal* if and only if $x_1 = x_2$ and $y_1 = y_2$, and are *unequal*, $z_1 \neq z_2$, if and only if $x_1 \neq x_2$ or $y_1 \neq y_2$. The *sum* of z_1 and z_2 is

$$z_1 + z_2 \equiv (x_1 + x_2) + i(y_1 + y_2)$$

and the *product* is

$$z_1 z_2 \equiv (x_1 x_2 - y_1 y_2) + i(x_1 y_2 + x_2 y_1)$$

The *norm* or *absolute value* of the complex number $z \equiv x + iy$ is

$$|z| \equiv (x^2 + y^2)^{1/2}$$

Notice that the map $x + iy \to (x,y)$ is a bijection of the set $\mathbf{C}$ of complex numbers onto $\mathbf{R}^2$, and

$$|z_1 - z_2| = d((x_1,y_1), (x_2,y_2))$$

for all $z_1 \equiv x_1 + iy_1$ and $z_2 \equiv x_2 + iy_2$ in $\mathbf{C}$, where d is the metric (Chap. 4) on $\mathbf{R}^2$. The complex numbers are thus a locally compact metric space, with metric $(z_1,z_2) \to |z_1 - z_2|$, and the function $z \to |z|$ is a compactifier. The operations of addition and multiplication are continuous functions from $\mathbf{C} \times \mathbf{C}$ to $\mathbf{C}$ (that is, they are uniformly continuous on every bounded subset of $\mathbf{C} \times \mathbf{C}$). These operations obey the usual rules of arithmetic, which we do not list here. The additive identity is $0 + i0$, and the multiplicative identity is $1 + i0$. The map $x \to x + i0$ from $\mathbf{R}$ to $\mathbf{C}$ preserves sums, products, and distances. We therefore identify $\mathbf{R}$ with the subset $\{x + i0 : x \in \mathbf{R}\}$ of $\mathbf{C}$, by identifying each real number x with the complex number $x + i0$. The element $0 + i1$ of $\mathbf{C}$ will be written simply as i.

To each complex number $z \equiv x + iy$ we associate a complex number $z^* \equiv x - iy$ (called the *conjugate* of z). The map $z \to z^*$ preserves sums, products, and distances, as can be verified by computation. Also,

$$zz^* = x^2 + y^2 = |z|^2 \qquad (z \equiv x + iy \in \mathbf{C})$$

Consequently

$$|z_1 z_2|^2 = z_1 z_2 z_1^* z_2^* = |z_1|^2 |z_2|^2$$

for all z_1 and z_2, and thus

$$|z_1 z_2| = |z_1|\, |z_2|$$

The norm function also satisfies the triangle inequality

$$|z_1 + z_2| \le |z_1| + |z_2|$$

because of its relation to the metric.

Each complex number $z \neq 0$ has a multiplicative inverse $z^* |z|^{-2}$, since

$$z(z^* |z|^{-2}) = (zz^*)|z|^{-2} = |z|^2 |z|^{-2} = 1$$

2. DERIVATIVES

The notion of the derivative is analogous to the notion of the derivative of a function of a real variable.

Definition 2 Let f and g be continuous complex-valued functions on a compact set $K \subset \mathbf{C}$, and δ a function from $\mathbf{R}^+$ to $\mathbf{R}^+$ such that

$$(2.1) \qquad |f(w) - f(z) - g(z)(w - z)| \le \epsilon |w - z|$$

whenever $z,\ w \in K$, $\epsilon > 0$, and $|w - z| \le \delta(\epsilon)$. Then g is called a *derivative* of f on K, written $g = f'$, $g = Df$, or $g(z) = df(z)/dz$, and δ is called a *modulus of differentiability* for f on K. The function f is *differentiable* on K.

In order to discuss differentiation on an open set, a preliminary definition is needed.

Definition 3 A totally bounded set $K \subset \mathbf{C}$ is *well contained* in an open set $U \subset \mathbf{C}$, written

$$K \subset\subset U$$

if there exists $r > 0$ such that

$$K_r \equiv \{z: \rho(z,K) \leq r\} \subset U$$

Definition 4 A function $f: U \to \mathbf{C}$ defined on an open subset U of $\mathbf{C}$ is *continuous* if it is continuous on every compact set K which is well contained in U. If f and g are continuous on U and $f' = g$ on every compact set $K \subset\subset U$, then g is called the *derivative* of f on U, and f is said to be *differentiable* on U.

Many of the results obtained in Chap. 2 for differentiation in the real domain carry over to the complex numbers, for instance, the rules for differentiating sums, products, quotients, and polynomials. The chain rule,

$$(2.2) \qquad\qquad (g \circ f)' = f' \cdot (g' \circ f)$$

is valid whenever f is a differentiable function from a compact subset K of $\mathbf{C}$ into a compact subset L of $\mathbf{C}$ and g is a differentiable function on L. In case f is a differentiable function from an open subset U of $\mathbf{C}$ into an open subset V of $\mathbf{C}$, and g is a differentiable on V, equality (2.2) is valid under the extra hypothesis that $f(K) \subset\subset V$ for each compact $K \subset\subset U$. (This hypothesis is always satisfied: see Prob. 12.)

It is easily shown that if f and g are complex-valued functions on an open set $U \subset \mathbf{C}$ such that for each compact set $K \subset\subset U$ there exists a $\delta: \mathbf{R}^+ \to \mathbf{R}^+$ such that (2.1) holds, and such that either (a) f is continuous or (b) g is bounded, then f and g are continuous, and therefore $g = f'$.

There is another kind of derivative in the complex domain. Let f and g be continuous complex-valued functions on a compact set $K \subset \mathbf{C}$, and δ a function from $\mathbf{R}^+$ to $\mathbf{R}^+$ such that

$$|f(x_2 + iy) - f(x_1 + iy) - g(x_1 + iy)(x_2 - x_1)| \leq \epsilon|x_2 - x_1|$$

whenever $x_1 + iy \in K$, $x_2 + iy \in K$, $\epsilon > 0$, and $|x_2 - x_1| \leq \delta(\epsilon)$. Then g is called the *partial derivative of f with respect to x* on K (written $g = D_x f$, or $g = f_x$, or $g(z) = \partial f(z)/\partial x$), and δ is called a *modulus of x differentiability* for f on K. We say that $g = f_x$ on an open set U if

$g = f_x$ on every compact set $K \subset\subset U$. Partial derivatives with respect to y are defined similarly.

It is clear that a differentiable function f on a compact set K (respectively on an open set U) has partial derivatives f_x and f_y on K (respectively on U), and

$$f_x = f' \qquad f_y = if'$$

The following converse to this simple remark is a basic criterion for differentiability.

Theorem 1 *Let f be a continuous function on an open set U with partial derivatives f_x and f_y on U satisfying the* Cauchy-Riemann *equations*

$$f_y = if_x$$

Then f is differentiable on U, and $f' = f_x$.

Proof Let $K \subset\subset U$ be compact. Choose $r > 0$ so that K_r is compact and $K_r \subset\subset U$. Let $z_1 \equiv x_1 + iy_1$ and $z_2 \equiv x_2 + iy_2$ be any points of K, with $|z_2 - z_1| \leq r$. Then

$$
\begin{aligned}
f(z_2) &- f(z_1) - f_x(z_1)(z_2 - z_1) \\
&= f(x_2 + iy_2) - f(x_1 + iy_2) + f(x_1 + iy_2) - f(x_1 + iy_1) \\
&\qquad - \{f_x(z_1)(x_2 - x_1) + f_y(z_1)(y_2 - y_1)\} \\
&= f(x_2 + iy_2) - f(x_1 + iy_2) - f_x(x_1 + iy_2)(x_2 - x_1) \\
&\quad + f(x_1 + iy_2) - f(x_1 + iy_1) - f_y(x_1 + iy_1)(y_2 - y_1) \\
&\qquad + (x_2 - x_1)\{f_x(x_1 + iy_2) - f_x(x_1 + iy_1)\}
\end{aligned}
$$

If δ denotes the minimum of r, the modulus of continuity of f_x on K_r, and the moduli of x and y differentiability of f on K_r, it follows that

$$|f(z_2) - f(z_1) - f_x(z_1)(z_2 - z_1)| \leq 3\epsilon|z_2 - z_1|$$

whenever $z_1, z_2 \in K$ and $|z_2 - z_1| \leq \delta(\epsilon)$. Therefore f is differentiable on K and $f' = f_x$. It follows that f is differentiable on U, and $f' = f_x$.

A function $f\colon \mathbf{C} \to \mathbf{C}$ differentiable on $\mathbf{C}$ is called an *entire* function. The simplest examples are the polynomials $\sum_{k=0}^{n} a_k z^{n-k}$. Here are some other examples.

Definition 5 Define the function $\exp: \mathbf{C} \to \mathbf{C}$ by

$$\exp(z) \equiv \exp(x)(\cos y + i \sin y) \qquad (z \equiv x + iy \in \mathbf{C})$$

This function has the partial derivatives

$$\exp_x(z) = \exp(x)(\cos y + i \sin y)$$

and
$$\exp_y(z) = \exp(x)(-\sin y + i \cos y)$$

By Theorem 1 it is differentiable with derivative

$$\frac{d \exp(z)}{dz} = \exp_x(z) = \exp(z)$$

For all $z_1 \equiv x_1 + iy_1$ and $z_2 \equiv x_2 + iy_2$ in $\mathbf{C}$ we have

$$\begin{aligned}
\exp(z_1 + z_2) &= \exp(x_1 + x_2)(\cos(y_1 + y_2) + i \sin(y_1 + y_2)) \\
&= \exp(x_1) \exp(x_2)(\cos y_1 \cos y_2 - \sin y_1 \sin y_2 \\
&\qquad\qquad + i(\sin y_1 \cos y_2 + \cos y_1 \sin y_2)) \\
&= \exp(z_1) \exp(z_2)
\end{aligned}$$

The function exp thus satisfies the same functional equation that it satisfies in the real domain.

Another notation for $\exp(z)$ is e^z.

Other entire functions, with which we shall not be concerned, are the functions cos and sin given by

$$\cos z \equiv \frac{1}{2}(\exp(iz) + \exp(-iz))$$

and
$$\sin z \equiv \frac{1}{2i}(\exp(iz) - \exp(-iz))$$

3. INTEGRATION

We introduce the notion of a path, which is a domain along which integrations are performed.

Definition 6 A *path* γ is a continuous map from a proper compact interval $[a,b]$ into $\mathbf{C}$ which is *piecewise differentiable* in the sense that there exist real numbers $t_0 = a < t_1 < \cdots < t_n = b$ such that γ is differentiable on each of the intervals $[t_i, t_{i+1}]$ ($1 \le i \le n - 1$). These intervals are called the *intervals of differentiability* of γ. The interval

$[a,b]$ is called the *parameter interval* of γ, and $\gamma(a)$ and $\gamma(b)$ are the *left and right end points* of γ. In case these end points are equal the path γ is said to be *closed*. The *carrier* car γ of γ is the closure of the subset $\gamma([a,b])$ of **C**.

Since $[a,b]$ is totally bounded and γ is continuous, car γ is totally bounded. If Γ is a compact set with car $\gamma \subset \Gamma$ or an open set with car $\gamma \subset\subset \Gamma$, we say that γ *lies in* Γ or is a *path in* Γ.

Paths γ_1 and γ_2 are *equivalent* if, for the uniquely defined linear everywhere-increasing map λ of the parameter interval $[a_1,b_1]$ of γ_1 onto the parameter interval $[a_2,b_2]$ of γ_2, we have $\gamma_1 = \gamma_2 \circ \lambda$.

Path equivalence is an equivalence relation, and every path is equivalent to a path whose parameter interval can be chosen arbitrarily.

Definition 7   Let γ_1 and γ_2 be two paths such that the right end point of γ_1 equals the left end point of γ_2, with respective parameter intervals $[a_1,b_1]$ and $[a_2,b_2]$. Let λ be the increasing linear map of $[b_1,\ b_1 + b_2 - a_2]$ onto $[a_2,b_2]$. Then the unique path γ_3 with parameter interval $[a_1,\ b_1 + b_2 - a_2]$ such that $\gamma_3(t) = \gamma_1(t)$ for t in $[a_1,b_1]$ and $\gamma_3(t) = \gamma_2(\lambda(t))$ for t in $[b_1,\ b_1 + b_2 - a_2]$ is called the *sum* $\gamma_1 + \gamma_2$ of the paths γ_1 and γ_2.

It is clear that path addition is associative. Every path is a finite sum of differentiable paths.

The *negative* $-\gamma$ of a path γ is the path $\gamma \circ \lambda$, where λ is the linear map of the parameter interval of γ onto itself which is everywhere decreasing. The end points of $-\gamma$ are the end points of γ reversed.

Next we study a form, which is an object to be integrated along a path.

Definition 8   A *form* on a compact subset K (respectively, an open subset U) of **C** is an expression

$$\omega \equiv f\,dx + g\,dy$$

where f and g are continuous functions on K (respectively, on U). If $\gamma \equiv \alpha + i\beta$ (where α and β are real-valued functions) is a path, and if ω is a form on car γ, we define

$$\int_\gamma \omega \equiv \int_a^b \{f(\gamma(t))\alpha'(t) + g(\gamma(t))\beta'(t)\}\,dt$$

where the integral on the right is to be interpreted as a sum of integrals over the intervals of differentiability of γ.

If the sum $\gamma + \delta$ of the paths γ and δ is defined, and if ω is a form on car $(\gamma + \delta)$, then

$$\int_{\gamma+\delta} \omega = \int_{\gamma} \omega + \int_{\delta} \omega$$

If γ and δ are equivalent paths, then

$$\int_{\gamma} \omega = \int_{\delta} \omega$$

whenever the integrals are defined.

For any path γ and any form ω on car γ,

$$\int_{-\gamma} \omega = -\int_{\gamma} \omega$$

The magnitude of $\int_{\gamma} \omega$ can be estimated as follows. For each path $\gamma \equiv \alpha + i\beta$ the *length* $|\gamma|$ of γ is defined by

$$|\gamma| \equiv \int_{a}^{b} |\gamma'(t)| \, dt \equiv \int_{a}^{b} |\alpha'(t) + i\beta'(t)| \, dt$$

Clearly $|\gamma_1 + \gamma_2| = |\gamma_1| + |\gamma_2|$ whenever $\gamma_1 + \gamma_2$ exists. For a form $\omega \equiv f \, dx + g \, dy$ defined on a compact set K, the *norm* $\|\omega\|_K$ of ω on K is defined by

$$\|\omega\|_K \equiv \|(|f|^2 + |g|^2)^{1/2}\|_K$$

Cauchy's inequality gives the estimate

$$\left| \int_{\gamma} \omega \right| \leq \int_{a}^{b} \{f(\gamma(t))^2 + g(\gamma(t))^2\}^{1/2} (\alpha'(t)^2 + \beta'(t)^2)^{1/2} \, dt$$
$$\leq \|\omega\|_K |\gamma|$$

whenever γ lies in K.

If the function f has partial derivatives on an open set U, then df will denote the form $f_x \, dx + f_y \, dy$ on U. Thus dz is the form $dx + i \, dy$, and dz^* is the form $dx - i \, dy$.

If h is a continuous function and $\omega \equiv f \, dx + g \, dy$ is a form, both defined on the same set K, then $h\omega$ will denote the form $hf \, dx + hg \, dy$. Thus $h \, dz$ is the form $h \, dx + ih \, dy$. For each path γ in K we have the estimate

$$\left| \int_{\gamma} h \, dz \right| = \left| \int_{a}^{b} h(\gamma(t))(\alpha'(t) + i\beta'(t)) \, dt \right|$$
$$\leq \|h\|_K \int_{a}^{b} |\alpha'(t) + i\beta'(t)| \, dt = \|h\|_K |\gamma|$$

In case γ is a differentiable path and f is a differentiable function on car γ, the chain rule for differentiating composite functions gives

$$(3.1) \qquad \int_\gamma df = \int_\gamma f'(z)\, dz \equiv \int_a^b f'(\gamma(t))\gamma'(t)\, dt$$
$$= \int_a^b \frac{d}{dt} f(\gamma(t))\, dt = f(\gamma(b)) - f(\gamma(a))$$

This formula can be generalized to an arbitrary path γ, by breaking the parameter interval $[a, b]$ down into the intervals of differentiability of γ.

In case γ is a closed path, the right side of (3.1), and therefore the left side, vanishes.

A path $\gamma \equiv \alpha + i\beta$ is called *linear* if both α and β are linear functions on the parameter interval. There is a unique linear path $\gamma \equiv \mathrm{lin}\,(z_1, z_2)$ with parameter interval $[0,1]$ and end points z_1 and z_2, given by

$$\gamma(t) \equiv z_1 + t(z_2 - z_1)$$

We write

$$\int_{z_1}^{z_2} \omega \equiv \int_\gamma \omega$$

for the integral of a form ω along this path. The *polygonal path* $\gamma \equiv \mathrm{poly}\,(z_1, \ldots, z_n)$ whose vertices are the complex numbers $z_1, \ldots, z_n$ is defined by

$$\mathrm{poly}\,(z_1, \ldots, z_n) \equiv \mathrm{lin}\,(z_1, z_2) + \cdots + \mathrm{lin}\,(z_{n-1}, z_n)$$

and $\mathrm{lin}\,(z_i, z_{i+1})$ $(1 \le i \le n - 1)$ are called the *linear pieces* of γ.

If $z_1, \ldots, z_k$ are any complex numbers the closed convex set span $(z_1, \ldots, z_n)$ spanned by $z_1, \ldots, z_n$ is the closure of the set

$$\{\alpha_1 z_1 + \cdots + \alpha_n z_n : \alpha_j \ge 0,\ \alpha_1 + \cdots + \alpha_n = 1\}$$

It is easily seen to be compact. If $\gamma \equiv \mathrm{poly}\,(z_1, \ldots, z_n)$ is a polygonal path, we write span $\gamma \equiv$ span $(z_1, \ldots, z_n)$.

Definition 9   A continuous function f on an open set U is *analytic* on U if

$$(3.2) \qquad\qquad \int_\gamma f(z)\, dz = 0$$

whenever $\gamma \equiv \mathrm{poly}\ (z_1, z_2, z_3, z_1)$ is a triangular path in U whose span is well contained in U.

Our goal is to show that f is analytic if and only if it is differentiable, and to extend (3.2) to a large class of closed paths γ in U.

Proposition 1 *A differentiable function f on an open set U is analytic on U.*

Proof Let $\gamma_1 \equiv \mathrm{poly}\ (z_1, z_2, z_3, z_1)$ be a triangular path whose span is well contained in U. Let ϵ be any positive constant. Now,

$$\int_{\gamma_1} f(z)\ dz = \int_{\gamma_{11}} f(z)\ dz + \int_{\gamma_{12}} f(z)\ dz + \int_{\gamma_{13}} f(z)\ dz + \int_{\gamma_{14}} f(z)\ dz$$

where
$$\gamma_{11} \equiv \mathrm{poly}\ (z_1, \tfrac{1}{2}(z_1 + z_2), \tfrac{1}{2}(z_1 + z_3), z_1)$$
$$\gamma_{12} \equiv \mathrm{poly}\ (z_2, \tfrac{1}{2}(z_2 + z_3), \tfrac{1}{2}(z_1 + z_2), z_2)$$
$$\gamma_{13} \equiv \mathrm{poly}\ (z_3, \tfrac{1}{2}(z_1 + z_3), \tfrac{1}{2}(z_2 + z_3), z_3)$$
and
$$\gamma_{14} \equiv \mathrm{poly}\ (\tfrac{1}{2}(z_1 + z_2), \tfrac{1}{2}(z_2 + z_3), \tfrac{1}{2}(z_1 + z_3), \tfrac{1}{2}(z_1 + z_2))$$

It follows that

$$\left| \int_{\gamma_{1i}} f(z)\ dz \right| > \tfrac{1}{4} \left| \int_{\gamma_1} f(z)\ dz \right| - \tfrac{1}{4}\epsilon$$

for at least one value of i, or

$$\left| \int_{\gamma_1} f(z)\ dz \right| < 4 \left| \int_{\gamma_2} f(z)\ dz \right| + \epsilon$$

where $\gamma_2 \equiv \gamma_{1i}$. The length of γ_2 is $|\gamma_2| \equiv \tfrac{1}{2}|\gamma_1|$, and span $\gamma_2 \subset$ span γ_1.

Continuing in this way, we construct by induction a sequence $\{\gamma_n\}$ of triangular paths with

$$(3.3) \qquad\qquad \text{span } \gamma_{n+1} \subset \text{span } \gamma_n$$

$$(3.4) \qquad\qquad |\gamma_{n+1}| = \tfrac{1}{2}|\gamma_n| = 2^{-n}|\gamma_1|$$

and

$$(3.5) \qquad\qquad \left| \int_{\gamma_1} f(z)\ dz \right| < 4^{n-1} \left| \int_{\gamma_n} f(z)\ dz \right| + \epsilon$$

for all n in Z^+. By (3.3) and (3.4), there is a unique point z_0 common to the sets span γ_n.

Let δ be the modulus of differentiability of f on span γ_1. Choose the positive integer n so large that $|z - z_0| \leq \delta(\epsilon)$ whenever $z \in K \equiv$ span γ_n.

Then
$$|f(z) - f(z_0) - f'(z_0)(z - z_0)| \leq \epsilon|z - z_0|$$

for all z in K. The integral

$$\int_{\gamma_n} \{f(z_0) + f'(z_0)(z - z_0)\}\, dz$$

vanishes by (3.1) because the integrand $f(z_0) + f'(z_0)(z - z_0)$ is the derivative of the function $f(z_0)z + \frac{1}{2}f'(z_0)(z - z_0)^2$ of z. We have

$$\left| \int_{\gamma_1} f(z)\, dz \right| < 4^{n-1} \left| \int_{\gamma_n} f(z)\, dz \right| + \epsilon$$
$$= 4^{n-1} \left| \int_{\gamma_n} \{f(z) - f(z_0) - f'(z_0)(z - z_0)\}\, dz \right| + \epsilon$$
$$\leq 4^{n-1}|\gamma_n|\epsilon|z - z_0|_K + \epsilon \leq 4^{n-1}|\gamma_n|^2\epsilon + \epsilon$$
$$= |\gamma_1|^2\epsilon + \epsilon$$

Since ϵ is an arbitrary positive number it follows that $\int_{\gamma_1} f(z)\, dz = 0$. Therefore f is analytic in U.

Lemma 1 *Let U be an open convex subset of $\mathbf{C}$, z_0 a point of U, and f an analytic function on U. Then the function g on U defined by*

$$(3.6) \qquad\qquad g(z) \equiv \int_{z_0}^{z} f(z)\, dz$$

is differentiable, and $g' = f$.

Proof Since U is convex, span $(z_0, z_1, z_2) \subset\subset U$ whenever z_0, z_1, $z_2 \in U$. By Proposition 1,

$$\int_{z_0}^{z_1} f(z)\, dz + \int_{z_1}^{z_2} f(z)\, dz + \int_{z_2}^{z_0} f(z)\, dz = 0$$

or

$$(3.7) \qquad\qquad g(z_2) - g(z_1) = \int_{z_1}^{z_2} f(z)\, dz$$

Lemma 1 now follows from the following result.

Lemma 2 *Let f and g be continuous functions on an open set U such that (3.7) holds whenever z_1, $z_2 \in U$ and span $(z_1,z_2) \subset\subset U$. Then g is differentiable on U and $g' = f$.*

Proof Let $K \subset\subset U$ be compact. Choose $r > 0$ with $K_r \subset\subset U$. Let δ be a modulus of continuity for f on K_r. Then if $\epsilon > 0$, z_1, $z_2 \in K$, and $|z_2 - z_1| \leq \min \{r, \delta(\epsilon)\}$, we have

$$
\begin{aligned}
|g(z_2) - g(z_1) - f(z_1)(z_2 - z_1)| \\
= \left| \int_{z_1}^{z_2} (f(z) - f(z_1))\, dz \right| \\
\leq \sup \{|f(z) - f(z_1)| : z \in K_r, |z - z_1| \leq \delta(\epsilon)\}|z_2 - z_1| \\
\leq \epsilon|z_2 - z_1|
\end{aligned}
$$

Therefore $g' = f$ on K. Since K is an arbitrary compact set with $K \subset\subset U$, $g' = f$ on U.

Lemma 3 *If f is an analytic function on an open convex set U, then (3.2) holds for every closed path γ in U.*

Proof Let z_0 be any point of U. Define the function g by (3.6). By Lemma 1, we have $g' = f$ on U. Therefore (3.2) holds, by (3.1).

Lemma 4 *Let K be compact, U open, and $\epsilon > 0$ such that $K_\epsilon \subset\subset U$. Let γ_0 and γ_1 be closed paths in K with common parameter interval $[a,b]$, such that*
$$
|\gamma_0(t) - \gamma_1(t)| \leq \tfrac{1}{2}\epsilon \qquad (t \subset [a,b])
$$
Then

$$
\text{(3.8)} \qquad \int_{\gamma_0} f(z)\, dz = \int_{\gamma_1} f(z)\, dz
$$

for all analytic functions f on U.

Proof Choose points

$$
t_0 = a < t_1 < \cdots < t_n = b
$$

in $[a,b]$ such that

$$
|\gamma_k(t_i) - \gamma_k(t)| \leq \tfrac{1}{4}\epsilon \qquad (k = 0 \text{ or } 1, \ t_i \leq t \leq t_{i+1}, \ 0 \leq i \leq n - 1)
$$

Let $\gamma_k{}^i$ ($k = 0$ or 1, $0 \leq i \leq n - 1$) be the restriction of γ_k to the interval $[t_i, t_{i+1}]$. Write $\sigma^i \equiv \text{lin} \{\gamma_0(t_i), \gamma_1(t_i)\}$ $(0 \leq i \leq n)$. Then $\sigma^i + \gamma_1^i - \sigma^{i+1} - \gamma_0^i$ $(1 \leq i \leq n-1)$ is a closed path δ^i whose carrier lies in the closed sphere of radius $\frac{1}{2}\epsilon + \frac{1}{4}\epsilon = \frac{3}{4}\epsilon$ about the point $\gamma_0(t_i)$. Since the open sphere of radius ϵ about $\gamma_0(t_i)$ is convex and lies in U, by Lemma 3

$$(3.9) \qquad\qquad \int_{\delta^i} f(z)\, dz = 0$$

Summation of (3.9) for $0 \leq i \leq n-1$ gives

$$
\begin{aligned}
0 &= \sum_{i=0}^{n-1} \left\{ \int_{\sigma^i} f(z)\, dz + \int_{\gamma_1^i} f(z)\, dz - \int_{\sigma^{i+1}} f(z)\, dz - \int_{\gamma_0^i} f(z)\, dz \right\} \\
&= \int_{\gamma_1} f(z)\, dz - \int_{\gamma_0} f(z)\, dz + \int_{\sigma^0} f(z)\, dz - \int_{\sigma^n} f(z)\, dz \\
&= \int_{\gamma_1} f(z)\, dz - \int_{\gamma_0} f(z)\, dz
\end{aligned}
$$

(since the paths σ^0 and σ^n are the same), as was to be proved.

Definition 10   Closed paths γ_0 and γ_1 in a compact set K are *homotopic in K* if there exists a continuous function $\sigma \colon [0,1] \times [0,1] \to K$, such that for each t in $[0,1]$ the function $\sigma_t \colon [0,1] \to \mathbf{C}$ defined by $\sigma_t(x) \equiv \sigma(t,x)$ is a closed path, with σ_0 equivalent to γ_0 and σ_1 to γ_1. The function σ is called a *homotopy* of γ_0 and γ_1. In case U is open and $K \subset\subset U$, γ_0 and γ_1 are said to be *homotopic in U*. In case γ_1 is a constant path, γ_0 is said to be *null-homotopic*.

Our next result is the famous integral theorem of Cauchy.

Theorem 2   *Let the closed paths γ_0 and γ_1 in the open set U be homotopic in U. Then (3.8) is valid for all analytic functions f on U.*

Proof   Let $K \subset\subset U$ and let σ be a homotopy of γ_0 and γ_1 in K. Choose $\delta > 0$ such that $K_\delta \subset\subset U$. Choose points

$$t_0 = 0 < t_1 < \cdots < t_n = 1$$

so that $|\sigma(t_i,x) - \sigma(t_{i+1},x)| < \frac{1}{2}\delta$ $(0 \leq i \leq n-1, x \in [0,1])$. By Lemma 4,

$$\int_{\sigma_{t_i}} f(z)\, dz = \int_{\sigma_{t_{i+1}}} f(z)\, dz \qquad (0 \leq i \leq n-1)$$

Therefore (3.8) is valid, and our theorem is proved.

As a corollary, (3.2) holds whenever f is analytic in an open set U and the closed path γ in U is null-homotopic in U.

Definition 11   An open set $U \subset \mathbf{C}$ is *connected* if any two points of U can be joined by a path in U. If in addition every closed path in U is null-homotopic, then U is called *simply connected*.

For simply connected open sets Theorem 2 takes the following form.

Corollary 1   *If γ is any closed path in a simply connected open set U, and f is any function analytic in U, then (3.2) holds.*

Corollary 1 is the form in which Cauchy's integral theorem is usually stated. Another version is the following.

Corollary 2   *If γ_0 and γ_1 are paths in a simply connected open set U with common right and left end points, then (3.8) is valid for all f analytic in U.*

Proof   By Corollary 1, equality (3.2) is valid for the path $\gamma \equiv \gamma_0 - \gamma_1$. Therefore (3.8) is valid.

Theorem 3   *If f is analytic in the simply connected open set U, there exists a differentiable function g in U with $g' = f$, and g is unique to within an additive constant.*

Proof   Let z_0 be any point of U. For each z in U let γ_z be a path from z_0 to z. Define the function g on U by

$$g(z) \equiv \int_{\gamma_z} f(z)\,dz$$

By Corollary 2 to Theorem 2, any other choice of γ_z would give the same result. Also,

$$g(z_2) - g(z_1) = \int_{\gamma} f(z)\,dz$$

where γ is any path from z_1 to z_2. In particular,

$$g(z_2) - g(z_1) = \int_{z_1}^{z_2} f(z)\,dz$$

whenever span $(z_1, z_2) \subset\subset U$. From Lemma 2 it follows that g is analytic on U and that $g' = f$.

Let h be any function on U with $h' = f$, and consider any points z_1, z_2 in U and a path γ from z_1 to z_2. Then

$$g(z_2) - g(z_1) = \int_\gamma g'(z)\, dz = \int_\gamma f(z)\, dz = \int_\gamma h'(z)\, dz = h(z_2) - h(z_1)$$

by (3.1). Therefore $g - h$ is constant on U.

4. THE WINDING NUMBER

Further developments depend on the concept of how many times a closed path winds around a point.

Proposition 2   *Let γ be a path in the metric complement $\mathbf{C} - \{z_0\}$ of a point z_0. Then*

$$(4.1) \qquad \exp\left[\int_\gamma (z - z_0)^{-1}\, dz\right] = (z_2 - z_0)(z_1 - z_0)^{-1}$$

where z_1 and z_2 are the left and right end points of γ.

Proof   Since every path is a finite sum of differentiable paths, it is enough to consider the case in which γ is differentiable. For each u in the parameter interval $[a,b]$ of γ we write

$$\lambda(u) \equiv \int_a^u (\gamma(t) - z_0)^{-1}\gamma'(t)\, dt$$

so that $\lambda : [a,b] \to \mathbf{C}$ is continuous, $\lambda(b) = \displaystyle\int_\gamma (z - z_0)^{-1}\, dz$, and $\lambda(a) = 0$.

Also, $\lambda'(u) = (\gamma(u) - z_0)^{-1}\gamma'(u)$

Therefore

$$\frac{d}{du}\{(\gamma(u) - z_0)\exp[-\lambda(u)]\}$$
$$= (\gamma(u) - z_0)(-\lambda'(u))\exp[-\lambda(u)] + \gamma'(u)\exp[-\lambda(u)] = 0$$

Therefore the function $u \to (\gamma(u) - z_0)\exp[-\lambda(u)]$ is constant on $[a,b]$, and so equal to its value at a, which is $\gamma(a) - z_0$. In particular,

$$(\gamma(b) - z_0)\exp[-\lambda(b)] = \gamma(a) - z_0$$

which is equivalent to (4.1).

A complex number w is a *logarithm* of a complex number z if $\exp(w) = z$. Every complex number $z \neq 0$ has at least one logarithm w, and the numbers

$$(4.2) \qquad w + 2\pi i n \qquad (n \in Z)$$

are the totality of the logarithms of z. The fact that z has at least one logarithm follows from Proposition 2. Since

$$\exp(2\pi i) = \cos 2\pi + i \sin 2\pi = 1$$

the numbers (4.2) are all logarithms of z. If, conversely, ζ is any logarithm of z, then $\exp(\zeta - w) = 1$, and thus if $\zeta - w \equiv u + iv$ then $e^u(\cos v + i \sin v) = 1$; hence $u = 0$, $v = 2\pi n$ for some n in Z, and consequently ζ has the form (4.2).

If U is a simply connected open subset of $\mathbf{C} - \{0\}$, z_0 is any point of U, and a is any logarithm of z_0, by Theorem 3 there exists a unique differentiable function g in U with $g'(z) = z^{-1}$ and $g(z_0) = a$. Then $g(z) - a = \int_\gamma z^{-1}\, dz$, where γ is any path in U from z_0 to z. By Proposition 2,

$$\exp(g(z)) = z \qquad (z \in U)$$

The function g is called a *branch* of the logarithmic function in U.

Proposition 2 has the following corollary.

Proposition 3 *If γ is a closed path in $\mathbf{C} - \{z_0\}$, the quantity*

$$0(\gamma, z_0) \equiv (2\pi i)^{-1} \int_\gamma (z - z_0)^{-1}\, dz$$

is an integer, called the winding number *of γ with respect to z_0.*

Proof By Proposition 2,

$$\exp(2\pi i 0(\gamma, z_0)) = 1$$

Therefore $0(\gamma, z_0)$ is an integer.

Since $0(\gamma, z_0)$ depends continuously on z_0, we have $0(\gamma, z_1) = 0(\gamma, z_2)$ whenever z_1 and z_2 can be joined by a path whose carrier is well contained in the metric complement of the carrier of γ.

An important example is the path γ defined by

$$\gamma(t) \equiv z_0 + r \exp(int) \qquad (0 \le t \le 2\pi)$$

where $r > 0$. (In case $n = 1$, γ is called the *circle of radius r* about the point z_0.) The winding number of γ is

$$0(\gamma, z_0) \equiv (2\pi i)^{-1} \int_0^{2\pi} r^{-1} \exp(-int) inr \exp(int) \, dt = n$$

This accords with the intuitive meaning of winding. By the above remark

$$0(\gamma, z) = 0(\gamma, z_0) = n$$

whenever $|z - z_0| < r$.

As another example, let z_0 be any point in $\mathbf{C}$ and x_1, x_2 elements of $\mathbf{R}^+$. Let γ_1 be any path from $z_0 + x_1$ to $z_0 - x_2$ which lies in the simply connected open set

$$U_1 \equiv \mathbf{C} - \{z_0 + iy : y \le 0\}$$

Let γ_2 be any path from $z_0 - x_2$ to $z_0 + x_1$ which lies in

$$U_2 \equiv \mathbf{C} - \{z_0 + iy : y \ge 0\}$$

Let α be the circular path of radius $r = \frac{1}{2}(x_2 + x_1)$ with center $z_0 + \frac{1}{2}(x_1 - x_2)$. Then α passes through the points $z_0 + x_1$ and $z_0 - x_2$. The path α is the sum of the paths α_1 and α_2, where α_1 is a semicircular path from $z_0 + x_1$ to $z_0 - x_2$ lying in U_1 and α_2 is a semicircular path from $z_0 - x_2$ to $z_0 + x_1$ lying in U_2. By Corollary 2 of Theorem 2,

$$\begin{aligned}
0(\gamma_1 + \gamma_2, z_0) &= (2\pi i)^{-1} \left\{ \int_{\gamma_1} (z - z_0)^{-1} \, dz + \int_{\gamma_2} (z - z_0)^{-1} \, dz \right\} \\
&= (2\pi i)^{-1} \left\{ \int_{\alpha_1} (z - z_0)^{-1} \, dz + \int_{\alpha_2} (z - z_0)^{-1} \, dz \right\} \\
&= 0(\alpha, z_0) = 1
\end{aligned}$$

In particular, $0(\gamma, z_0) = 1$ whenever γ is triangular or rectangular path surrounding z_0.

The following result, *Cauchy's integral formula*, uses the winding number to establish a representation of a differentiable function by means of an integral.

Theorem 4 *Let f be a differentiable function on an open set U, let z_0 be a point of U, and let γ be a closed path in $U - \{z_0\}$ which is null-*

homotopic in U. Then

$$0(\gamma,z_0)f(z_0) \ = \ (2\pi i)^{-1} \int_\gamma f(z)(z - z_0)^{-1}\, dz$$

Proof The function $z \to (f(z) - f(z_0))(z - z_0)^{-1}$ extends to an analytic function h on U, as we shall see in the next lemma. Since $\int_\gamma h(z)\, dz = 0$,

$$(2\pi i)^{-1} \int_\gamma f(z)(z - z_0)^{-1}\, dz \ = \ (2\pi i)^{-1} \int_\gamma f(z_0)(z - z_0)^{-1}\, dz$$
$$= \ 0(\gamma,z_0)f(z_0)$$

Lemma 5 *Let f be a differentiable function on an open set U, and let z_0 be a point of U. Then there exists an analytic function h on U such that*

$$h(z) \ = \ (f(z) - f(z_0))(z - z_0)^{-1} \qquad (z \in U - \{z_0\})$$

Proof Let $K \subset\subset U$ be a compact set such that z_0 lies in the closure of $K - \{z_0\}$. Let ω be a modulus of continuity for f' and δ a modulus of differentiability for f on K. Whenever $\epsilon > 0$, $z_1, z_2 \in K - \{z_0\}$, and $|z_2 - z_1| \leq \min\{\delta(\epsilon),\omega(\epsilon)\}$, we have

$$|h(z_2) - h(z_1)| \leq |h(z_2) - f'(z_2)| + |h(z_1) - f'(z_1)| + |f'(z_2) - f'(z_1)|$$
$$\leq \epsilon + \epsilon + \epsilon = 3\epsilon$$

It follows that h is uniformly continuous on $K - \{z_0\}$. By Lemma 1 of Chap. 4, we see that h has a continuous extension to K. Since this is true for all K, we see that h has a continuous extension to U.

To show that h is analytic on U, consider points z_1, z_2, z_3 in U with $K \equiv \mathrm{span}\,(z_1,z_2,z_3) \subset\subset U$. Choose $r > 0$ such that $K_r \subset\subset U$. Either $\rho(z_0,K) > r/2$ or $\rho(z_0,K) < r$. In the first case $K \subset\subset U - \{z_0\}$. Since h is differentiable on $U - \{z_0\}$, it follows from Proposition 1 that

$$(4.3) \qquad\qquad \int_\gamma h(z)\, dz \ = \ 0$$

where $\gamma \equiv \mathrm{poly}\,(z_1,z_2,z_3,z_1)$. Assume therefore that $\rho(z_0,K) < r$. Then span $(z_0,z_1,z_2,z_3) \subset K_r \subset\subset U$. Then

$$\int_\gamma h(z)\, dz \ = \ \int_{\gamma_1} h(z)\, dz + \int_{\gamma_2} h(z)\, dz + \int_{\gamma_3} h(z)\, dz$$

where $\gamma_1 \equiv \mathrm{poly}\,(z_1,z_2,z_0,z_1)$, $\gamma_2 \equiv \mathrm{poly}\,(z_2,z_3,z_0,z_2)$, and $\gamma_3 \equiv \mathrm{poly}$

(z_3,z_1,z_0,z_3). Thus we have reduced the proof of (4.3) to the case in which z_0 is one of the vertices of γ, say $z_0 = z_1$. In this case there exist points w_1, w_2, and w_3 such that $|w_1 - z_1|$, $|w_2 - z_2|$, and $|w_3 - z_3|$ are arbitrarily small, and $\rho(z_0, \text{span }(w_1,w_2,w_3)) > 0$. Then

$$\int_\delta h(z) \, dz = 0$$

where $\delta \equiv \text{poly }(w_1,w_2,w_3,w_1)$. Since this integral is arbitrarily near to $\int_\gamma h(z) \, dz$, it follows that the latter integral vanishes. Therefore h is analytic on U.

Cauchy's integral formula is often used in conjunction with the following proposition.

Proposition 4 *Let h be a continuous function on the carrier of the path γ, and let n be a positive integer. Then the function f defined by*

$$f(z) \equiv (2\pi i)^{-1} \int_\gamma h(\zeta)(\zeta - z)^{-n} \, d\zeta$$

is differentiable on the metric complement $-$ car γ of car γ, and $f' = g$, where

$$g(z) \equiv (2\pi i)^{-1} n \int_\gamma h(\zeta)(\zeta - z)^{-n-1} \, d\zeta$$

Proof Let K be any compact set with $K \subset\subset -$ car γ. Write

$$C \equiv \sup \left\{ |h(\zeta)| \sum_{k=1}^n |\zeta - z|^{-k} : \zeta \in \text{car } \gamma, \, z \in K \right\}$$

Let ω be a common modulus of continuity of the continuous functions

$$(\zeta,z) \to (\zeta - z)^{-k} \qquad (1 \le k \le n)$$

on the product space car $\gamma \times K$. Consider $\epsilon > 0$ and z_1, z_2 in K with $|z_1 - z_2| \le \omega(\epsilon)$. Then

$$\begin{aligned}
&|f(z_2) - f(z_1) - g(z_1)(z_2 - z_1)| \\
&\quad = (2\pi)^{-1} \left| \int_\gamma h(\zeta) \{ (\zeta - z_2)^{-n} - (\zeta - z_1)^{-n} \right. \\
&\qquad\qquad\qquad\qquad \left. - n(\zeta - z_1)^{-n-1}(z_2 - z_1) \} \, d\zeta \right|
\end{aligned}$$

$$= (2\pi)^{-1}|z_2 - z_1| \left| \int_\gamma h(\zeta) \left\{ \sum_{k=1}^n (\zeta - z_1)^{-k}(\zeta - z_2)^{-n-1+k} \right. \right.$$
$$\left. \left. - n(\zeta - z_1)^{-n-1} \right\} d\zeta \right|$$

$$= (2\pi)^{-1}|z_2 - z_1| \left| \int_\gamma h(\zeta) \sum_{k=1}^n (\zeta - z_1)^{-k}\{(\zeta - z_2)^{-n-1+k} \right.$$
$$\left. - (\zeta - z_1)^{-n-1+k}\} d\zeta \right|$$

$$\leq (2\pi)^{-1}|z_2 - z_1|\,|\gamma|C\epsilon$$

It follows that $f' = g$ on K, and therefore on $-\mathrm{car}\ \gamma$.

Lemma 6 *If a function f is continuous on an open set U and differentiable on every closed sphere $Sc(z,r) \subset\subset U$, then f is differentiable on U.*

Proof For each open sphere S whose closure is well contained in U let g_S be the derivative of f on S. Then $g_{S_1}(z) = g_{S_2}(z)$ whenever $z \in S_1 \cap S_2$. Thus there exists a unique function $g: U \to \mathbf{C}$ such that $g(z) = g_S(z)$ whenever z is a point of an open sphere S whose closure is well contained in U. Let K be any compact set well contained in U. Choose $r > 0$ with $K_r \subset\subset U$, and let $w_1, \ldots, w_n$ be an $r/2$ approximation to K. Let δ be a common modulus of differentiability of f on the closed spheres $Sc(w_i,r)$ $(1 \leq i \leq n)$, with $\delta \leq r/2$. Consider points z_1 and z_2 in K and $\epsilon > 0$ with $|z_2 - z_1| \leq \delta(\epsilon)$. Then $|z_1 - w_i| < r/2$ for some i, and thus $|z_2 - w_i| < r/2 + \delta(\epsilon) \leq r$. Thus z_1 and z_2 belong to $S(w_i,r)$, and

$$|f(z_2) - f(z_1) - g(z_1)(z_2 - z_1)| \leq \epsilon|z_2 - z_1|$$

Therefore $g = f'$ on K, and hence on U.

Theorem 5 *Let f be differentiable on an open set U. Then f has derivatives of all orders in U, and for any path γ in U which is null-homotopic in U, any z in $U - \mathrm{car}\ \gamma$, and any $n \geq 0$,*

$$(4.4) \qquad 0(\gamma,z)f^{(n)}(z) = (2\pi i)^{-1}n! \int_\gamma f(\zeta)(\zeta - z)^{-n-1}\, d\zeta$$

Proof To show that f has derivatives of all orders, it is enough to show that f' is differentiable, since by the same token f'' and so on will be differentiable. Consider a closed sphere $Sc(z_0,r) \subset\subset U$. Choose $t > r$ with $Sc(z_0,t) \subset\subset U$. Let γ be the circular path of radius t about

z_0. By Theorem 4,

$$f(z) = (2\pi i)^{-1} \int_\gamma f(\zeta)(\zeta - z_0)^{-1}\, d\zeta$$

whenever $|z - z_0| < t$. By Proposition 4, we see that f' is differentiable on $Sc(z_0, r)$. By Lemma 6, we see that f' is differentiable on U.

The formula (4.4) now follows from Theorem 4 and Proposition 4.

Corollary 1 *A function f analytic in an open set U is differentiable in U.*

Proof When U is simply connected, by Theorem 3 there exists a differentiable function g in U with $g' = f$, and thus f is differentiable in U by Theorem 5. The general case follows from this special case by means of Lemma 6.

Corollary 2 *Let $\{f_n\}$ be a sequence of differentiable functions on an open set U, converging uniformly on every compact set $K \subset\subset U$ to a function f on U. Then f is differentiable on U.*

Proof If γ is any triangular path with span $\gamma \subset\subset U$, then

$$\int_\gamma f(z)\, dz = \lim_{n \to \infty} \int_\gamma f_n(z)\, dz = 0$$

Therefore f is analytic, and consequently differentiable, on U.

Corollary 3 *If f is differentiable on $Sc(z_0, r_0)$, then*

$$|f^{(n)}(z_0)| \leq n!\, r_0^{-n} \sup\,\{|f(z)| : |z - z_0| = r_0\} \qquad (n \geq 0)$$

Proof For $0 < r < r_0$ let γ be the circular path about z_0 of radius r.

Then
$$\begin{aligned}
|f^{(n)}(z_0)| &= (2\pi)^{-1} n! \left| \int_\gamma f(\zeta)(\zeta - z_0)^{-n-1}\, d\zeta \right| \\
&\leq (2\pi)^{-1} n! |\gamma| r^{-n-1} \sup\,\{|f(z)| : |z - z_0| = r\} \\
&= n!\, r^{-n} \sup\,\{|f(z)| : |z - z_0| = r\}
\end{aligned}$$

Letting $r \to r_0$ gives the result.

Corollary 4 *If f is differentiable on an open set U, and if z_0 is an interior point of U, then the function $z \to (f(z) - f(z_0))(z - z_0)^{-1}$ on $U - \{z_0\}$ extends to a differentiable function on U.*

Proof   The result follows from Lemma 5 and Corollary 1 above.

Power series are uniquely suited to the representation of differentiable functions in the complex domain. The complex variable version of Taylor's theorem reads as follows.

Theorem 6   *Let f be differentiable in an open sphere $S \equiv S(z_0,r)$. Then the series*

$$(4.5) \qquad \sum_{n=0}^{\infty} a_n (z - z_0)^n$$

with coefficients a_n given by

$$(4.6) \qquad a_n \equiv (n!)^{-1} f^{(n)}(z_0)$$

converges uniformly to f on every compact set $K \subset\subset S$. Conversely, if the series (4.5) converges uniformly to f on every compact set $K \subset\subset S$, then a_n is given by (4.6).

Proof   Consider a compact set $K \subset\subset S$. Let $T \equiv Sc(z_0,r_0)$ be a closed sphere, with $0 < r_0 < r$ and $K \subset T$. Write $t = \frac{1}{2}(r + r_0)$. Let γ be the circular path of radius t about z_0. Then for all z in T,

$$
\begin{aligned}
f(z) &= (2\pi i)^{-1} \int_{\gamma} f(\zeta)(\zeta - z)^{-1}\, d\zeta \\
&= (2\pi i)^{-1} \int_{\gamma} f(\zeta)(\zeta - z_0)^{-1}\{1 - (z - z_0)(\zeta - z_0)^{-1}\}^{-1}\, d\zeta \\
&= (2\pi i)^{-1} \int_{\gamma} f(\zeta) \left\{ \sum_{n=0}^{\infty} (\zeta - z_0)^{-n-1}(z - z_0)^n \right\} d\zeta \\
&= (2\pi i)^{-1} \sum_{n=0}^{\infty} \left\{ \int_{\gamma} f(\zeta)(\zeta - z_0)^{-n-1}\, d\zeta \right\} (z - z_0)^n
\end{aligned}
$$

By Theorem 5 this means that (4.5), with a_n given by (4.6), converges to f uniformly on T, and therefore on K.

If, conversely, (4.5) converges uniformly to f on T, we get

$$
\begin{aligned}
(n!)^{-1} f^{(n)}(z_0) &= (2\pi i)^{-1} \int_{\gamma} f(\zeta)(\zeta - z_0)^{-n-1}\, d\zeta \\
&= \sum_{k=0}^{\infty} a_k (2\pi i)^{-1} \int_{\gamma} (\zeta - z_0)^{k}(\zeta - z_0)^{-n-1}\, d\zeta = a_n
\end{aligned}
$$

and thus (4.6) is valid.

Proposition 10 of Chap. 2 is valid in the complex domain. Hence a power series $\sum\limits_{n=0}^{\infty} a_n(z - z_0)^n$ converges absolutely and uniformly on every compact set $K \subset\subset S(z_0,r_0)$ if there exists N in Z^+ such that $|a_{n+1}| \leq r_0^{-1}|a_n|$ whenever $n \geq N$.

5. ESTIMATES OF SIZE, AND THE LOCATION OF ZEROS

It is a famous principle of classical analysis, called the *maximum principle*, that an analytic function attains its maximum on the boundary. Here is the constructive version of this result.

Definition 12 If K and B are compact sets, with $B \subset K$, such that

$$Sc(z,\rho(z,B)) \subset K \qquad (z \in K)$$

then B is called a *border* for K.

Proposition 5 *If B is a border for K, then*

$$\|f\|_K = \|f\|_B$$

for all differentiable functions f on K.

Proof Let ϵ be any positive number. Choose $\delta > 0$ so that $K \cap B_\delta$ is compact and $|f(z)| \leq \|f\|_B + \epsilon$ whenever $z \in K \cap B_\delta$. Consider z_0 in K, with $\rho(z_0,B) > \delta/2$. Let γ be the circle of radius r about z_0, where r is chosen so that $\rho(z_0,B) - \delta/2 < r < \rho(z_0,B)$. Then γ is null-homotopic in $K - B$, and there exists some point z_1 in γ with $\rho(z_1,B) < \delta/2$. Let the path γ_1 be the arc of the circle γ extending in length an amount $\delta/2$ on either side of z_1. Then $|\gamma_1| = \delta$, and γ_1 lies in B_δ, and thus $|f(z)| \leq \|f\|_B + \epsilon$ for all z in car γ_1. Let γ_2 be the complementary arc, so that $\gamma = \gamma_1 + \gamma_2$. Then

$$
\begin{aligned}
|f(z_0)| &= (2\pi)^{-1} \left| \int_\gamma f(z)(z - z_0)^{-1}\,dz \right| \\
&\leq (2\pi)^{-1} \left| \int_{\gamma_1} f(z)(z - z_0)^{-1}\,dz \right| + (2\pi)^{-1} \left| \int_{\gamma_2} f(z)(z - z_0)^{-1}\,dz \right| \\
&\leq (2\pi)^{-1}|\gamma_1|(\|f\|_B + \epsilon)r^{-1} + (2\pi)^{-1}|\gamma_2|\,\|f\|_K r^{-1} \\
&= (2\pi r)^{-1}\{\delta(\|f\|_B + \epsilon) + (2\pi r - \delta)\|f\|_K\}
\end{aligned}
$$

This inequality also holds for z_0 in $K \cap B_\delta$, since the right side is greater than or equal to $\|f\|_{B_\delta}$. It therefore holds for all z_0 in K. Hence

$$\|f\|_K \leq (2\pi r)^{-1}\{\delta(\|f\|_B + \epsilon) + (2\pi r - \delta)\|f\|_K\}$$

or

$$\|f\|_K \leq \|f\|_B + \epsilon$$

Since this is true for all $\epsilon > 0$, the proposition is proved.

For each compact set K and each continuous function f on K write

$$\lambda(f,K) \equiv \inf\ \{|f(z)| : z \in K\}$$

We have the following corollary to Proposition 5.

Corollary *Let K be a compact set, and let B be a border for K. Let f be a differentiable function on K with $\lambda(f,B) > 0$. Then either $\lambda(f,K) = \lambda(f,B)$ or $\lambda(f,K) = 0$.*

Proof Either $\lambda(f,K) > 0$ or $\lambda(f,K) < \lambda(f,B)$. In the first case the function f^{-1} is differentiable on K, and

$$\lambda(f,K)^{-1} = \|f^{-1}\|_K = \|f^{-1}\|_B = \lambda(f,B)^{-1}$$

and thus $\lambda(f,K) = \lambda(f,B)$. In the second case, for each n in Z^+ either $\lambda(f,K) < n^{-1}$ or $\lambda(f,K) > 0$; and $\lambda(f,K) > 0$ is ruled out because $\lambda(f,K) = \lambda(f,B)$ whenever $\lambda(f,K) > 0$. Therefore $\lambda(f,K) < n^{-1}$ for all n in Z^+, hence $\lambda(f,K) = 0$.

Lemma 7 *Let h be a differentiable function on a closed sphere $S \equiv Sc(w,r)$. Let δ be a constant, with $0 \leq \delta < r$. Let $z_1, \ldots, z_n$ be points of S with $|z_k - w| \leq \delta\ (1 \leq i \leq n)$. Write*

$$f(z) \equiv h(z)(z - z_1) \cdots (z - z_n) \qquad (z \in S)$$

Then

$$|f(z)| \leq (|z - w| + \delta)^n(r - \delta)^{-n}\|f\|_S$$

Proof If $|z - w| = r$, then $|z - z_k| \geq r - \delta\ (1 \leq k \leq n)$. Thus $\|h\|_S = \|h\|_\Gamma \leq (r - \delta)^{-n}\|f\|_\Gamma = (r - \delta)^{-n}\|f\|_S$, where Γ is the bounding circle $\{z : |z - w| = r\}$ of S. Then

$$|f(z)| \leq |z - z_1| \cdots |z - z_n|\ \|h\|_S \leq (|z - w| + \delta)^n(r - \delta)^{-n}\|f\|_S$$

for all z in S.

The case $n = 1$ and $\delta = 0$ of Lemma 7 is known as *Schwarz's lemma*.

Lemma 8 *Let r, δ, d, and α be positive constants, with $r > 3\delta$, and n a positive integer. Let f be differentiable on a closed sphere $S \equiv Sc(w,r)$. Let $z_1, \ldots, z_n$ be distinct points of S with*

$$|f(z_k)| \leq \alpha \quad |z_k - w| \leq \delta \quad \text{and} \quad |z_j - z_k| \geq d \quad (1 \leq j < k \leq n)$$

Write $T \equiv Sc(w, r - 3\delta)$. Then

$$\|f\|_T \leq \|f\|_S (r-2\delta)^n (r - \delta)^{-n} + 2\alpha n (r + \delta)^{n-1} d^{-n+1}$$

Proof Consider a closed sphere S, a function f, and points $z_1, \ldots, z_n$ as described. Let g be the polynomial of degree $n - 1$ with the same values as f at the points $z_1, \ldots, z_n$; that is,

$$g(z) \equiv \sum_{k=1}^{n} f(z_k) \prod_{k} (z_k - z_j)^{-1} (z - z_j)$$

where $\prod_{k}$ means the product taken from $j = 1$ to $j = n$ with the factor corresponding to $j = k$ left out. Then

$$(5.1) \qquad \|g\|_S \leq \alpha n (r + \delta)^{n-1} d^{-n+1}$$

since $|z - z_j| \leq |z - w| + |z_j - w| \leq r + \delta$ for z in S and $1 \leq j \leq n$. The function $f - g$ vanishes at $z_1, \ldots, z_n$. By Corollary 4 of Theorem 5,

$$f(z) - g(z) \equiv (z - z_1) \cdots (z - z_n) h(z) \qquad (z \in S)$$

for some differentiable function h on S. It follows from Lemma 7 that

$$(5.2) \qquad \|f - g\|_T \leq (r - 2\delta)^n (r - \delta)^{-n} \|f - g\|_S$$

By (5.1) and (5.2),

$$(5.3) \quad \begin{aligned} \|f\|_T &\leq \|f - g\|_T + \|g\|_T \leq (r - 2\delta)^n (r - \delta)^{-n} \|f - g\|_S + \|g\|_S \\ &\leq (r - 2\delta)^n (r - \delta)^{-n} (\|f\|_S + \|g\|_S) + \|g\|_S \\ &\leq \|f\|_S (r - 2\delta)^n (r - \delta)^{-n} + 2\alpha n (r + \delta)^{n-1} d^{-n+1} \end{aligned}$$

Our next lemma could be proved in a more general framework (this is left to the reader).

Lemma 9   *Let K be a compact subset of $\mathbf{C}$ such that K_δ is compact, for a given $\delta > 0$. If $\rho(z,K) > \delta$ then*

$$(5.4) \qquad\qquad \rho(z,K_\delta) \leq \rho(z,K) - \delta$$

Proof   For each w in K we have $\rho(z, Sc(w,\delta)) = |z - w| - \delta$. Since $Sc(w,\delta) \subset K_\delta$, it follows that $\rho(z,K_\delta) \leq |z - w| - \delta$. Taking the infimum over all w in K gives (5.4).

Lemma 10   *Let $K \subset \mathbf{C}$ be compact, and let B be a border for K. Let f be a differentiable function on K with $\lambda(f,B) > \lambda(f,K) = 0$. Then for each $t > 0$ there exists a closed sphere $S \equiv Sc(w,t_0) \subset K$ of radius $t_0 \leq t$ such that $\lambda(f,\Gamma) > \lambda(f,S) = 0$, where Γ is the bounding circle*

$$\Gamma \equiv \{z \in \mathbf{C}: |z - w| = t_0\}$$

of S.

Proof   Let r be the diameter of K. Choose positive constants ϵ and δ, with $\delta < \min\{t,\tfrac{1}{3}r\}$, such that

$$(5.5) \qquad\qquad |f(z)| > \epsilon \qquad (z \in K \cap B_{4\delta})$$

Choose the positive integer n so large that

$$\|f\|_K (r - 2\delta)^n (r - \delta)^{-n} \leq \frac{\epsilon}{2}$$

Write $d \equiv n^{-1}\delta$, and choose $\alpha > 0$ so small that

$$2\alpha n (r + \delta)^{n-1} d^{-n+1} \leq \frac{\epsilon}{2}$$

Choose w in K with $|f(w)| < \min\{\alpha,\epsilon\}$. Since $|f(w)| < \epsilon$, the inequality $\rho(w,B) < 4\delta$ would contradict (5.5). Hence $\rho(w,B) \geq 4\delta$. Write

$$\Gamma_k \equiv \{z \in \mathbf{C}: |z - w| = kd\} \qquad (1 \leq k \leq n)$$

Since $\rho(w,B) \geq 4\delta > kd$, we have $\Gamma_k \subset K$. Either

$$\lambda(f,\Gamma_k) > |f(w)|$$

for at least one value of k, $1 \leq k \leq n$, or else

$$(5.6) \qquad\qquad \lambda(f,\Gamma_k) < \alpha \qquad (1 \leq k \leq n)$$

In the former case we define $t_0 \equiv kd$, $\Gamma \equiv \Gamma_k$, and $S \equiv Sc(w,kd)$. Then

$$\lambda(f,\Gamma) > |f(w)| \geq \lambda(f,S)$$

By the Corollary to Proposition 5, we have $\lambda(f,S) = 0$. Thus the requirements are all satisfied.

It remains to consider case (5.6). Then there exist points z_k in Γ_k $(1 \leq k \leq n)$ with $|f(z_k)| < \alpha$. If $0 \leq k < j \leq n$, then

$$|z_k - z_j| \geq |z_j - w| - |z_k - w| = (j - k)d \geq d$$

Also, $|z_k - w| = kd \leq \delta$.

Write $r_0 \equiv \rho(w,B)$. Since $Sc(w,r_0) \subset K$, it follows from Lemma 8 that $|f(z)| \leq \epsilon$ whenever $|z - w| \leq r_0 - 3\delta$. By Lemma 9, $\rho(w,B_{4\delta}) \leq r_0 - 4\delta$. Hence there exists z in $B_{4\delta}$ with $|z - w| \leq r_0 - 3\delta$. Thus, on the one hand, $|f(z)| \leq \epsilon$, and on the other, $|f(z)| > \epsilon$ [because of (5.5)]. The possibility (5.6) is thus ruled out, and the lemma holds in all cases.

Lemma 11   *Under the hypotheses of Lemma 10 there exists z in K with $f(z) = 0$.*

Proof   Using Lemma 10, construct by induction a sequence $K \supset S_1 \supset S_2 \supset \cdots$ of closed spheres with $\lambda(f,\Gamma_k) > \lambda(f,S_k) = 0$, where Γ_k is the boundary of S_k, so that the radius of S_k is at most k^{-1}. These spheres have a common point z with $f(z) = 0$.

Theorem 7   *Let K be a compact set, and let B be a border for K. Let f be a differentiable function on K with $\lambda(f,B) > \lambda(f,K) = 0$. Then there exists a finite sequence $z_1, \ldots, z_n$ of points in K, a differentiable function g on K, and a constant $c > 0$ such that*

$$f(z) = (z - z_1) \cdots (z - z_n)g(z)$$

and $|g(z)| \geq c > 0$ for all z in K.

Proof   Choose $\delta > 0$ and $M > 0$ so that $|f(z)| > M$ for all z in $B_{3\delta}$. Let r be the diameter of K. Choose m in Z^+ so that

$$(r - \tfrac{3}{2}\delta)^m (r - \delta)^{-m}\|f\|_K < M$$

Since K is totally bounded, there exists N in Z^+ such that for any family of N points in K some subfamily of m points have mutual distances less than δ. By Lemma 11 and Corollary 4 of Theorem 5, there exists z_1 in K and a differentiable function g_1 on K with

$$f(z) = (z - z_1)g_1(z) \qquad (z \in K)$$

Either $\lambda(g_1,K) = 0$ or $\lambda(g_1,K) = \lambda(g_1,B) \neq 0$. In the latter case we take $g \equiv g_1$ and are finished. Otherwise there exists z_2 in K and a differentiable function g_2 on K with

$$f(z) = (z - z_1)(z - z_2)g_2(z) \qquad (z \in K)$$

Repeat this process inductively to obtain at the nth stage points $z_1, \ldots, z_n$ in K and a differentiable function g_n on K with

$$f(z) = (z - z_1) \cdots (z - z_n)g_n(z) \qquad (z \in K)$$

unless the process stops at some previous stage. If the process stops at the nth stage because $\lambda(g_n,K) > 0$, we are through. Otherwise we continue.

Either the process proceeds for N stages or it terminates at some stage $n < N$. Consider the former case. By the choice of N, there exist m of the points $z_1, \ldots, z_N$ whose mutual distances are less than δ. Call these points $w_1, \ldots, w_m$. Then

$$(5.7) \qquad f(z) = (z - w_1) \cdots (z - w_m)h(z) \qquad (z \in K)$$

for some differentiable function h on K. Since $f(w_1) = 0$ and $|f(z)| > M$ for all z in $B_{3\delta}$, we have

$$r_0 \equiv \rho(w_1,B) > 3\delta$$

Since B is a border for K,

$$(5.8) \qquad\qquad\qquad Sc(w_1,r_0) \subset K$$

By Lemma 9, there exists z in $B_{3\delta}$ with

$$(5.9) \qquad\qquad\qquad |z - w_1| < r_0 - \tfrac{5}{2}\delta$$

Since $|w_i - w_1| \leq \delta \ (1 \leq i \leq m)$, we get, using (5.7), (5.8), (5.9), and Lemma 7,

$$\begin{aligned}
|f(z)| &\leq (|z - w_1| + \delta)^m (r_0 - \delta)^{-m}\|f\|_K \\
&\leq (r_0 - \tfrac{3}{2}\delta)^m (r_0 - \delta)^{-m}\|f\|_K \\
&\leq (r - \tfrac{3}{2}\delta)^m (r - \delta)^{-m}\|f\|_K < M
\end{aligned}$$

This contradicts the fact that $z \in B_{3\delta}$, and rules out the possibility that the process continues for N stages. Therefore at some stage $n < N$ the process stops, and the requirements of the theorem are satisfied by the function $g \equiv g_n$ and the associated points $z_1, \ldots, z_n$.

Definition 13 A polynomial $z \to a_0 z^n + \cdots + a_n$ has *degree at least* k if $a_{n-j} \neq 0$ for some $j \geq k$.

Theorem 8 *If the polynomial* $p(z) = a_0 z^n + \cdots + a_n$ *has degree at least* k, *there exist complex numbers* $z_1, \ldots, z_k$ *and a polynomial* q *with*

$$p(z) \equiv (z - z_1) \cdots (z - z_k) q(z) \qquad (z \in \mathbf{C})$$

Proof Choose $j \geq k$ with $a_{n-j} \neq 0$. Choose $r > 0$ with

$$|a_{n-j}| r^j > \sum_{m=0}^{j-1} |a_{n-m}| r^m + |p(0)|$$

Then either $a_{n-m} \neq 0$ for some $m > j$ or else

$$(5.10) \qquad |a_{n-j}| r^j > \sum_{m=0}^{j-1} |a_{n-m}| r^m + \sum_{m=j+1}^{n} |a_{n-m}| r^m + |p(0)|$$

In the latter case

$$\inf \{ |p(z)| : |z| = r \} > |p(0)|$$

and there exists a complex number z_1 with $p(z_1) = 0$ by Lemma 11. In the former case we replace j with m and perform the same construction. Eventually this leads to a value of j for which (5.10) is true, and therefore to a complex number z_1 with $p(z_1) = 0$.

Using the process of polynomial division, we find a polynomial $q_1(z) \equiv b_0 z^{n-1} + \cdots + b_{n-1}$ and a constant c in $\mathbf{C}$ with

$$p(z) = (z - z_1) q_1(z) + c \qquad (z \in \mathbf{C})$$

Since $p(z_1) = 0$, we have $c = 0$. Now $|a_0|, \ldots, |a_{n-k}|$ will be arbitrarily small if $|b_0|, \ldots, |b_{n-k}|$ are all arbitrarily small. Therefore $b_{n-1-j} \neq 0$ for some $j \geq k - 1$. Thus q_1 has degree at least $k - 1$. Replacing p with q_1 and repeating the construction, we find

$$p(z) = (z - z_1)(z - z_2) q_2(z)$$

where q_2 has degree at least $k - 2$. By induction we finally construct
the desired numbers $z_1, \ldots, z_k$ and the desired polynomial $q \equiv q_k$.

6. THE RIEMANN MAPPING THEOREM

In this section we prove the beautiful theorem of Riemann that under
very general conditions any two simply connected open sets in $\mathbf{C}$ have
the same analytic structure. The first step is to derive some integral
formulas which are important in their own right.

Proposition 6 *Let the function f be differentiable in $Sc(z_0,R)$. Then for
each $z = z_0 + re^{i\theta}$ ($\theta \in \mathbf{R}$, $0 \leq r < R$) we have*

$$(6.1) \quad f(z) = (2\pi)^{-1} \int_0^{2\pi} f(z_0 + Re^{i\varphi}) \frac{(R^2 - r^2)\, d\varphi}{R^2 - 2rR \cos(\theta - \varphi) + r^2}$$

Proof It is enough to consider the case $z_0 = 0$ and $R = 1$. Let γ be
the circle of radius 1 about the point 0. By Cauchy's integral formula,

$$(6.2) \qquad f(z) = (2\pi i)^{-1} \int_\gamma f(\zeta)(\zeta - z)^{-1}\, d\zeta$$
$$= (2\pi)^{-1} \int_\gamma f(\zeta)\zeta(\zeta - z)^{-1}(i\zeta)^{-1}\, d\zeta$$

and by Cauchy's integral theorem,

$$(6.3) \qquad 0 = (2\pi i)^{-1} \int_\gamma f(\zeta)z^*(1 - z^*\zeta)^{-1}\, d\zeta$$
$$= (2\pi)^{-1} \int_\gamma f(\zeta)z^*(\zeta^* - z^*)^{-1}(i\zeta)^{-1}\, d\zeta$$

Addition of (6.2) and (6.3) gives

$$f(z) = (2\pi)^{-1} \int_\gamma f(\zeta) \left\{ \frac{(1 - zz^*)}{1 - z\zeta^* - z^*\zeta + zz^*} \right\} (i\zeta)^{-1}\, d\zeta$$

Writing $\zeta = e^{i\varphi}$ and $z = re^{i\theta}$, we get

$$f(z) = (2\pi)^{-1} \int_0^{2\pi} f(e^{i\varphi}) \left\{ \frac{(1 - r^2)\, d\varphi}{1 - 2r \cos(\theta - \varphi) + r^2} \right\}$$

which is just (6.1) for the case $z_0 = 0$ and $R = 1$.

A consequence of Proposition 6 is a bound for an analytic function
inside a circle.

Proposition 7 *If f is differentiable in $S \equiv Sc(z_0,R)$, and if $|z - z_0| \equiv r < R$, then*

$$(6.4) \qquad |f(z)| \leq |f(z_0)| \frac{R - r}{R + r} + \|f\|_S \frac{2r}{R + r}$$

Proof There is no loss of generality in taking $z_0 = 0$ and $R = 1$. Write z in the form $z = re^{i\theta}$, and abbreviate

$$A \equiv \frac{1 - r^2}{1 - 2r \cos(\theta - \varphi) + r^2} \qquad B \equiv \frac{1 - r}{1 + r}$$

Then

$$A \geq \frac{1 - r^2}{(1 + r)^2} = B$$

By Proposition 6, $(2\pi)^{-1}\int A \, d\varphi = 1$. Also

$$\begin{aligned}
2\pi f(z) &= \left| \int_0^{2\pi} f(e^{i\varphi})A \, d\varphi \right| \\
&\leq \left| \int_0^{2\pi} f(e^{i\varphi})B \, d\varphi \right| + \left| \int_0^{2\pi} f(e^{i\varphi})(A - B) \, d\varphi \right| \\
&\leq 2\pi |f(0)| B + 2\pi \|f\|_S (1 - B)
\end{aligned}$$

which is (6.4) for the case $z_0 = 0$ and $R = 1$.

A weak form of Proposition 7 carries over to arbitrary connected open sets.

Corollary *Let U be a connected open set, $K \subset\subset U$ a compact set, z_0 a point of K, and r a constant with $0 \leq r < 1$. Then there exists a constant c with $0 < c < 1$ such that, for each differentiable function f on U, with $|f(z_0)| \leq r$ and $|f(z)| \leq 1$ for all z in U, we have $\|f\|_K \leq c$.*

Proof Since U is connected, for each w in U there exist points $w_1 \equiv z_0, w_2, \ldots, w_n = w$ in U and closed spheres $Sc(w_k,r_k) \subset U$ $(1 \leq k \leq n)$ such that $|w_{k+1} - w_k| < r_k$ $(1 \leq k \leq n - 1)$. Using Proposition 7, we see by induction on k that there exist constants $c_1, \ldots, c_n$ in $(0,1)$ such that $|f(w_k)| \leq c_k$ for all f satisfying the hypothesis. In particular, $|f(w)| \leq c_n$.

Now choose $\epsilon > 0$ so that $K_{2\epsilon} \subset U$ and let $z_1, \ldots, z_n$ be an ϵ approximation to K. By the above, there exists c_0 in $(0,1)$ such that $|f(z_k)| \leq c_0$ $(1 \leq k \leq m)$ whenever f satisfies our conditions. Consider z in K and choose z_k with $|z - z_k| < \epsilon$. Since $Sc(z_k,2\epsilon) \subset U$, we see by

Proposition 7 that

$$|f(z)| \leq \frac{2\epsilon - \epsilon}{2\epsilon + \epsilon} |f(z_k)| + \frac{2\epsilon}{2\epsilon + \epsilon} < \frac{1}{3} c_0 + \frac{2}{3} \qquad (z \in K)$$

The following proposition and its corollary give estimates that will be needed later.

Proposition 8 *Let f be differentiable in the closed sphere $S \equiv Sc(0,R)$, with $f(0) > 0$ and $1 - \epsilon \leq |f(z)| \leq 1$ for all z in S. Then*

$$|f(z) - 1| \leq (1 - rR^{-1})^{-1}(2\epsilon)^{1/2}$$

whenever $r \equiv |z| < R$.

Proof It is enough to consider the case $R = 1$. Whenever $|w| \leq 1$, we have

$$|w - 1|^2 = 1 - w - w^* + ww^* \leq 2 - w - w^*$$

This gives

$$\int_0^{2\pi} |f(e^{i\theta}) - 1|^2 \, d\theta \leq \int_0^{2\pi} (2 - f(e^{i\theta}) - f(e^{i\theta})^*) \, d\theta$$
$$= 2\pi(2 - f(0) - f(0)^*) \leq 4\pi\epsilon$$

The Cauchy Schwarz inequality for integrals (page 79) implies that

$$\int_0^{2\pi} |f(e^{i\theta}) - 1| \, d\theta \leq \left\{ \int_0^{2\pi} |f(e^{i\theta}) - 1|^2 \, d\theta \int_0^{2\pi} d\theta \right\}^{1/2}$$
$$\leq 2\pi(2\epsilon)^{1/2}$$

Whenever $r \equiv |z| < 1$, it follows that

$$|f(z) - 1| = (2\pi)^{-1} \left| \int_0^{2\pi} (f(e^{i\theta}) - 1)(e^{i\theta} - z)^{-1} e^{i\theta} \, d\theta \right|$$
$$\leq (2\pi)^{-1}(1 - r)^{-1} \int_0^{2\pi} |f(e^{i\theta}) - 1| \, d\theta$$
$$\leq (1 - r)^{-1}(2\epsilon)^{1/2}$$

Corollary *Let U and V be open sets with $S(0,R) \subset U \cap V$ and $U \cup V \subset S(0,1)$. Let $f \colon U \to V$ and $g \colon V \to U$ be analytic functions, with $f(0) = g(0) = 0$ and $f'(0) > 0$, such that $g \circ f \colon U \to U$ and $f \circ g \colon$*

V → V are the identity maps. Then

$$(6.5) \qquad |f(z) - z| \leq 3(R^2 - r)^{-1}(1 - R)^{1/2}$$

whenever $r \equiv |z| < R^2$.

Proof By Lemma 7, we have $|f(w)| \leq R^{-1}|w|$ and $|g(w)| \leq R^{-1}|w|$ whenever $|w| \leq R$. Since $|z| < R^2$, it follows that $|f(z)| \leq R$, and thus

$$|z| = |g(f(z))| \leq R^{-1}|f(z)|$$

Thus
$$R^2 \leq |Rz^{-1}f(z)| \leq 1$$

for $0 < |z| < R^2$. Using Proposition 8, we get

$$\begin{aligned}
|f(z) - z| &\leq |f(z) - R^{-1}z| + |R^{-1}z - z| \\
&\leq |Rz^{-1}f(z) - 1| + 1 - R \\
&\leq (1 - rR^{-2})^{-1}(2(1 - R^2))^{1/2} + 1 - R \\
&\leq 2(1 - rR^{-2})^{-1}(1 - R)^{1/2} + 1 - R \\
&< 3(R^2 - r)^{-1}(1 - R)^{1/2}
\end{aligned}$$

We are interested in whether two open subsets of **C** have the same structure. Here is a formal definition of this concept.

Definition 14 Two open subsets U_1 and U_2 of **C** are *equivalent* if there are differentiable functions $f_1: U_1 \to U_2$ and $f_2: U_2 \to U_1$ such that $f_2 \circ f_1: U_1 \to U_1$ and $f_1 \circ f_2: U_2 \to U_2$ are the identity maps, and such that $f_1(K_1) \subset\subset U_2$ (respectively, $f_2(K_2) \subset\subset U_1$) for all compact $K_1 \subset\subset U_1$ (respectively, $K_2 \subset\subset U_2$). The map f_1 is an *equivalence* of U_1 and U_2, and f_2 is the inverse equivalence.

The self-equivalences of $S(0,1)$ with itself are easily found.

Proposition 9 *Let U be the open sphere $S(0,1)$. For each a in **C** with $|a| < 1$ define the function $h_a: U \to$ **C** by*

$$h_a(z) = (z - a)(1 - a^*z)^{-1}$$

Then h_a is an equivalence of U with itself, and every equivalence of U with itself has the form αh_a, with $|a| < 1$ and $|\alpha| = 1$.

Proof   Since $|a| < 1$, we see that h_a is differentiable in $Sc(0,1)$. A simple computation shows that h_{-a} is inverse to h_a on $Sc(0,1)$. Therefore h_a is an equivalence of U with itself.

Conversely, consider an arbitrary equivalence f of U with itself. Write $b \equiv f(0)$. Then $h \equiv h_b \circ f$ is an equivalence of U with itself, and $h(0) = 0$. Choose the complex number α so that $|\alpha| = 1$ and $\alpha^* h'(0) > 0$. Then $g \equiv \alpha^* h_b \circ f$ is an equivalence of U with itself, and $g'(0) > 0$. By the above corollary, g is the identity map. Therefore f is the inverse of $\alpha^* h_b$, so that

$$\begin{aligned}
f(z) &= (\alpha z + b)(1 + \alpha b^* z)^{-1} \\
&= \alpha(z + \alpha^* b)(1 + \alpha b^* z)^{-1} \\
&= \alpha h_a(z)
\end{aligned}$$

where $a \equiv -\alpha^* b$.

Definition 15   An open set U is *mappable* if it is bounded and simply connected and if there exists a compact set K, called the *border* of U, and a point z_0 in $\mathbf{C} - K$ such that U consists of all points in $\mathbf{C} - K$ which can be joined to z_0 by a path lying in $\mathbf{C} - K$. The mappable set U is *sequestered* if $U \subset S(0,1)$ and $K \subset\subset S(0,1)$.

Our goal is to prove the Riemann mapping theorem, that every mappable set is equivalent to $S(0,1)$. To this end, we need several lemmas. The first lemma needs no proof.

Lemma 12   *If U is a sequestered set, if $|a| < 1$, and if $|\alpha| = 1$, then the set $V \equiv \alpha h_a(U)$ is a sequestered set and αh_a is an equivalence of U with V.*

Lemma 13   *Let K be the border of the sequestered set U, and let 0 belong to K. Let s be a branch of the square root function on U:*

$$s(z) \equiv \exp\left(\tfrac{1}{2}\log z\right) \qquad (z \in U)$$

where $\log$ is a branch of the logarithm on U. Then $U_0 \equiv s(U)$ is a sequestered set with border $K_0 \equiv \{w: w^2 \in K\}$, and $s: U \to U_0$ is an equivalence.

Proof   Clearly the map s_0 defined by $s_0(z) \equiv z^2$ of U_0 into U is inverse to s. Consider any compact set $L \subset\subset U$. Then

$$\begin{aligned}
(6.6) \qquad \rho(s(L),K_0) &\geq \inf\{|z - w|: z^2 \in L,\, w^2 \in K\} \\
&\geq \inf\{\tfrac{1}{2}|z^2 - w^2|: z^2 \in L,\, w^2 \in K\} \\
&= \tfrac{1}{2}\inf\{|z - w|: z \in L,\, w \in K\} > 0
\end{aligned}$$

Similarly if the set $L_0 \subset\subset \mathbf{C} - K_0$ is compact, then

$$(6.7) \quad \begin{aligned}
\rho(s_0(L_0),K) &= \inf\{|z^2 - w| : z \in L_0, w \in K\} \\
&= \inf\{|z^2 - w^2| : z \in L_0, w \in K_0\} \\
&\geq \inf\{|z - w| : z \in L_0, w \in K_0\} \cdot \\
&\qquad\qquad\qquad \inf\{|z + w| : z \in L_0, w \in K_0\} \\
&= \inf\{|z - w| : z \in L_0, w \in K_0\}^2 > 0
\end{aligned}$$

To show that U_0 is mappable, fix a point z_0 in U. We must show that U_0 consists of all w in $\mathbf{C} - K_0$ which can be connected to $s(z_0)$ by a path in $\mathbf{C} - K_0$. To this end consider a point z in U. Then z can be connected to z_0 by a path γ in U. By (6.6), we see that $s \circ \gamma$ is a path in $\mathbf{C} - K_0$ connecting $s(z_0)$ to $s(z)$.

Conversely, consider a path γ_0 in $\mathbf{C} - K_0$ connecting $s(z_0)$ to a point w. Then, by (6.7), we see that $s_0 \circ \gamma_0$ is a path in $\mathbf{C} - K$ connecting z_0 to $s_0(w)$. It follows that $s_0(w) \in U$. Therefore $w = s(s_0(w)) \in U_0$.

Thus U_0 is mappable with border K_0. Since $U_0 \subset \{z : |z| < 1\}$ and $K_0 \subset\subset \{z : |z| < 1\}$, it follows that U_0 is sequestered.

By (6.6), if $L \subset\subset U$ is compact, then $s(L) \subset\subset U_0$. By (6.7), if $L_0 \subset\subset U_0$ is compact, then $s_0(L_0) \subset\subset U$. Thus $s : U \to U_0$ is an equivalence.

Definition 16   Let U be a sequestered set and K be the border of U. Assume that $0 \in U$ and let a be a point of K with $|a| \leq \frac{1}{2}(1 + d)$, where $d \equiv \rho(0,K)$. Choose α with $|\alpha| = 1$ so that $\alpha a < 0$. By Lemma 12, we see that αh_a is an analytic equivalence of U with a sequestered set U_0, that $0 = \alpha h_a(a)$ is in the border K_0 of U_0, and that $|a| = \alpha h_a(0) \in U_0$. Let s be that branch of the square root function on U_0 for which $s(|a|) > 0$. Then the function

$$\varphi_U \equiv \alpha^* h_b \circ s \circ (\alpha h_a)$$

on U, where $b \equiv |a|^{1/2}$, is called the *canonical map* for the sequestered set U.

Lemma 14   *Let U be a sequestered set with border K, such that $0 \in U$. Write $d \equiv \rho(0,K)$. Then φ_U is an equivalence of U with a sequestered set U^* with border K^*, such that $\varphi_U(0) = 0$,*

$$\varphi_U'(0) > 1 + \tfrac{1}{32}(1 - d)^2$$

and $d \leq d^* \equiv \rho(0,K^*)$.

Proof By Lemma 13, we see that s is an equivalence of U_0 with $s(U_0)$. By Lemma 11, we see that $\alpha^* h_b$ is an equivalence of $s(U_0)$ with some sequestered set U^*. Therefore φ_U is an equivalence of U with U^*. Now,

$$\varphi_U(0) = \alpha^* h_b(s(|a|))$$
$$= \alpha^* h_b(b) = 0$$

Also,   (i) $\alpha h_a'(0) = \alpha(1 - (-a)(-a^*)) = \alpha(1 - |a|^2)$

       (ii) $s'(|a|) = \frac{1}{2}|a|^{-1/2}$

and   (iii) $\alpha^* h_b'(|a|^{1/2}) = \alpha^* \left(\dfrac{1 - |a|}{(1 - |a|)^2} \right) = \dfrac{\alpha^*}{1 - |a|}$

Therefore $\varphi_U'(0) = \frac{1}{2}|a|^{-1/2}(1 + |a|)$

$$= 1 + \tfrac{1}{2}|a|^{-1/2}(1 - |a|^{1/2})^2 > 1 + \tfrac{1}{2}(1 - |a|^{1/2})^2$$
$$\geq 1 + \tfrac{1}{2}(1 - (\tfrac{1}{2}(1 + d))^{1/2})^2$$
$$\geq 1 + \tfrac{1}{2}(1 - (\tfrac{3}{4} + \tfrac{1}{4}d))^2 = 1 + \tfrac{1}{32}(1 - d)^2$$

Now, the inverse $\psi \colon U^* \to U$ of φ_U is the composition of the inverse of $\alpha^* h_b$, the map $z \to z^2$, and the inverse of αh_a. Each of these maps is defined on $S(0,1)$ and takes $S(0,1)$ into itself. Therefore $|\psi(z)| \leq |z|$ whenever $|z| < 1$. Since $\psi(K^*) = K$, it follows that $d \leq d^*$.

Lemma 15   _Let U_0 be a sequestered set, $0 \in U_0$, with border K_0. Let φ_0 be the canonical map of U_0 into $U_1 \equiv U_0^*$. Let φ_1 be the canonical map of U_1 into $U_2 \equiv U_1^*$. Continuing in this way, we define a sequence $\{U_n\}_{n=0}^{\infty}$ of sequestered sets, with $0 \in U_n$ for all n, and for each n a canonical map $\varphi_n \colon U_n \to U_{n+1}$. Let d_n be the distance of the border K_n of U_n to 0. Then $d_0 \leq d_1 \leq d_2 \leq \cdots$, $\lim\limits_{n \to \infty} d_n = 1$, and_

$$(6.8) \qquad\qquad 1 + \tfrac{1}{32}(1 - d_n)^2 < d_0^{-1/(n+1)}$$

Proof   We have $d_0 \leq d_1 \leq \cdots$ by Lemma 14. For each n in Z^+ write

$$\varphi_0{}^{n+1} \equiv \varphi_n \circ \cdots \circ \varphi_0 \colon U_0 \to U_{n+1}$$

By Corollary 3 of Theorem 5,

$$|(D\varphi_0{}^{n+1})(0)| \leq d_0^{-1}$$

On the other hand,

$$|(D\varphi_0{}^{n+1})(0)| = |\varphi_n'(0) \cdots \varphi_0'(0)|$$
$$> (1 + \tfrac{1}{32}(1 - d_n)^2) \cdots (1 + \tfrac{1}{32}(1 - d_0)^2)$$
$$\geq (1 + \tfrac{1}{32}(1 - d_n)^2)^{n+1}$$

This gives (6.8), which in turn implies $d_n \to 1$ as $n \to \infty$.

We now prove the Riemann mapping theorem.

Theorem 9   *For each sequestered set U_0 with $0 \in U_0$ the maps*

$$\varphi_0{}^n \equiv \varphi_{n-1} \circ \cdots \circ \varphi_0 \colon U_0 \to U_n$$

converge uniformly on compact subsets of U_0 to an equivalence φ of U_0 with $\{z \colon |z| < 1\}$.

Proof   For each $m \le n$ write

$$\varphi_m{}^n \equiv \varphi_{n-1} \circ \cdots \circ \varphi_m \colon U_m \to U_n$$

Let $L \subset\subset U_0$ be a compact set. By the Corollary to Proposition 7, there exists $c < 1$ such that $|\varphi_0{}^n(z)| \le c$ for all n in Z^+ and all z in L. Using the Corollary to Proposition 8, from the fact that $d_n \to 1$ as $n \to \infty$ we see that, for $m \le n$,

$$|\varphi_0{}^n(z) - \varphi_0{}^m(z)| = |\varphi_m{}^n(\varphi_0{}^m(z)) - \varphi_0{}^m(z)|$$

will be arbitrarily small for all z in L if m and n are sufficiently large. It follows that $\{\varphi_0{}^n\}$ converges uniformly on L, to a continuous function from L to $Sc(0,c)$. Since L is an arbitrary compact set with $L \subset\subset U_0$, the sequence $\{\varphi_0{}^n\}$ converges on U_0 to a differentiable function $\varphi \colon U_0 \to S(0,1)$, such that $\varphi(L) \subset\subset S(0,1)$ for each compact set $L \subset\subset U_0$.

To construct the inverse $\varphi_\infty \colon S(0,1) \to U_0$ of φ, consider compact sets K and L with $K \subset\subset \text{int } L$ and $L \subset\subset S(0,1)$. For each $m \le n$, let $\varphi_n{}^m \colon U_n \to U_m$ be the inverse to $\varphi_m{}^n$. Since $d_n \to 1$ as $n \to \infty$, the functions $\varphi_n{}^0$ are defined on L for all n sufficiently large. By the Corollary to Proposition 8, $\varphi_m{}^n$ converges uniformly on K as $m \to \infty$ and $n \to \infty$ to the identity function $z \to z$. Therefore we may choose the positive integer N so large that if $m, n \ge N$ then $\varphi_m{}^0$ and $\varphi_n{}^0$ are defined on L and $\varphi_m{}^n(K) \subset L$. For all z in K and all $m, n \ge N$ we have

$$|\varphi_m{}^0(z) - \varphi_n{}^0(z)| = |\varphi_n{}^0(\varphi_m{}^n(z)) - \varphi_n{}^0(z)|$$

Hence the sequence $\{\varphi_n{}^0\}$ converges uniformly on K. Since $\varphi_N{}^0$ is an equivalence, $\varphi_N{}^0(L) \subset\subset U_0$. Therefore

$$\varphi_n{}^0(K) = \varphi_N{}^0(\varphi_n{}^N(K)) \subset \varphi_N{}^0(L) \subset\subset U_0 \qquad (n \ge N)$$

Thus there exists a differentiable map $\varphi_\infty\colon S(0,1) \to U_0$ such that for each compact set $K \subset\subset S(0,1)$ the sequence $\{\varphi_n{}^0\}$ is defined on K whenever n is sufficiently large and converges uniformly on K to φ_∞. Moreover $\varphi_\infty(K) \subset\subset U_0$.

To show that $\varphi\colon U_0 \to S(0,1)$ is an analytic equivalence, it remains to show that $\varphi_\infty \circ \varphi\colon U_0 \to U_0$ and $\varphi \circ \varphi_\infty\colon S(0,1) \to S(0,1)$ are the identity maps. For each z in U_0 the points $\varphi_0{}^n(z)$ lie inside some compact set $M \subset\subset S(0,1)$ [since the sequence $\{\varphi_0{}^n(z)\}$ converges to a point in $S(0,1)$], and thus for each $\epsilon > 0$ we can choose N in Z^+ with

$$|\varphi_k{}^0(\varphi_0{}^n(z)) - \varphi_\infty(\varphi_0{}^n(z))| \leq \epsilon \qquad (k, n \in Z^+, k \geq N)$$

Taking $n = k$ gives

$$|z - \varphi_\infty(\varphi_0{}^k(z))| \leq \epsilon \qquad (k \geq N)$$

Letting $k \to \infty$ gives

$$|z - \varphi_\infty(\varphi(z))| \leq \epsilon$$

Since ϵ is arbitrary, $\varphi_\infty \circ \varphi$ is the identity map. Similarly, $\varphi \circ \varphi_\infty$ is the identity map.

Corollary *Every mappable open set U is equivalent to $S(0,1)$.*

Proof The corollary is immediate because every mappable set is equivalent to a sequestered set containing the point 0.

PROBLEMS

1. Is homotopy an equivalence relation on the set of all closed paths lying in a given compact set K?

2. Show that a closed path in an open set U is homotopic to a closed differentiable path in U.

3. Show that the convex set spanned by finitely many points z_1, $\ldots$, z_n need *not* be closed.

4. Prove that homotopic closed paths in a compact set K have the same winding number with respect to each point of $-K$.

5. Let f and g be complex-valued functions on an open set $U \subset \mathbf{C}$, such that for every compact set $K \subset\subset U$ there exists $\delta\colon \mathbf{R}^+ \to \mathbf{R}^+$ such that (2.1) holds. Show that f and g are continuous (and that therefore $f' = g$).

6.   Let G be the set of all analytic functions in $U \equiv \{z: |z| < 1\}$ which are bounded by 1, and let $z_1, \ldots, z_n$ be points in U. Show that the set $A \equiv \{(f(z_1), \ldots, f(z_n)): f \in G\} \subset \mathbf{C}^n$ is totally bounded.

7.   Let $0 < r_1 < r_2 < \cdots$ be an increasing sequence of positive real numbers converging to 1. Let $\mathfrak{A}$ be the set of all analytic functions on $U \equiv \{z: |z| < 1\}$, with metric ρ defined by

$$\rho(f,g) \equiv \sum_{n=1}^{\infty} 2^{-n} \min \{1, \sup \{|f(z) - g(z)|: |z| \leq r_n\}\}$$

For each sequence $0 < c_1 < c_2 < \cdots$ show that the set

$$\{f \in \mathfrak{A}: |f(z)| \leq c_n \text{ whenever } |z| \leq r_n (n = 1, 2, 3, \ldots)\}$$

is compact.

8.   Show that the set of differentiable functions on $D \equiv \{z: |z| \leq 1\}$ is located in $C(D,\mathbf{C})$.

9.   Define the *Riemann sphere* S to be the one point compactification of $\mathbf{C}$. Show that the map $z \to z^{-1}$ from $\mathbf{C} - \{0\}$ onto itself extends to a metric equivalence of S with itself.

10.   Call open sets U_1 and U_2 of $\mathbf{C}$ *friendly* if for each compact set $K \subset\subset U_1 \cup U_2$ there exist compact sets $K_1 \subset\subset U_1$ and $K_2 \subset\subset U_2$ with $K \subset K_1 \cup K_2$. If U_1 and U_2 are friendly, and $f: U_1 \cup U_2 \to \mathbf{C}$ is differentiable on each of the sets U_1 and U_2, show that f is differentiable on $U_1 \cup U_2$.

11.   If $U \subset \mathbf{C}$ is open, if $z_0 \in U$, and if $f: U - \{z_0\} \to \mathbf{C}$ is differentiable and bounded, show that f extends to a differentiable function on U.

12.   Let f be a differentiable function on an open set U, and $K \subset\subset U$ a compact set, such that for each z in U there is a path γ in U from z to a point z' for which $f(z) \neq f(z')$. Show that $f(U)$ is open and $f(K) \subset\subset f(U)$. Conclude that if $h: U_1 \to U_2$ is differentiable, where U_1 and U_2 are open, then $h(K_1) \subset\subset U_2$ for each compact $K_1 \subset\subset U_1$.

13.   A sequence $\{y_n\}$ of complex numbers in an open set U is *thin* in U if for each compact set $K \subset\subset U$ all except finitely many of the numbers $\rho(y_n, K)$ are positive. Let the set $U - \{y_n\}$ consist of all z in U with $z \neq y_n$ for all n. A continuous function $f: U - \{y_n\} \to \mathbf{C}$ is called *meromorphic* in U if for each compact $K \subset\subset U$ there exists a differentiable function h on K and a polynomial p with at least one nonzero coefficient such that $fp = h$ on $K \cap (U - \{y_n\})$. Show that such a function f is differentiable on $U - \{y_n\}$.

14. Show that sums and products of meromorphic functions in an open set U are meromorphic in U. Give conditions for the reciprocal $1/f$ of a meromorphic function f to be meromorphic.

15. Let $f: U - \{y_n\} \to \mathbf{C}$ be meromorphic on U. Call a finite set $A \subset Z^+$ *detached* if $\rho(y_n, y_m) > 0$ whenever $n \in A$ and $m \in Z^+ - A$. Let γ be any closed path in U such that $\rho(y_n, \mathrm{car}\ \gamma) > 0$ for all n, which is null-homotopic in U. Show that the set

$$A(m,\gamma) \equiv \{n \in Z^+ : 0(\gamma, y_n) = m\}$$

is detached, for each m in Z^+, and that $\overset{\infty}{\underset{m=1}{\cup}} A(m,\gamma)$ is finite.

16. For each detached set A define the *residue* $r(f,A)$ of f on A in such a way that

$$\int_\gamma f(z)\ dz = 2\pi i \sum m\, r(f, A(m,\gamma))$$

for each closed path γ, with car $\gamma \subset\subset U - \{y_n\}$, which is null-homotopic in U.

17. Let a and b be real numbers. Show that

$$(z - a)(z - b) = (z - \min\ \{a,b\})(z - \max\ \{a,b\})$$

for all z. Conclude that the factorization of a polynomial is *not* unique.

18. Show that a bounded open set which is analytically equivalent to $\{z : |z| < 1\}$ is mappable (this is an unsolved problem).

NOTES

It is not hard to show that the following statements are equivalent: (i) $K \subset\subset U$ for every compact set $K \subset U \equiv \{z : |z| < 1\}$; (ii) inf $\{f(x) : 0 \leq x \leq 1\} > 0$ for every continuous real-valued function f on $[0,1]$ such that $f(x) > 0$ for all x in $[0,1]$. No proof or counterexample is known for either statement.

Proposition 1 proves more than is actually stated. The proof uses only the fact that f is pointwise differentiable on U. From Corollary 1 to Theorem 5, it follows that a pointwise differentiable function on an open set U is differentiable, in striking contrast to the real case.

The notion of border (Definition 12) substitutes for the classical notion of boundary. Certain compact sets have a border but not a boundary.

Classically, the proof of Lemma 11 is quite simple: if there were no such z, then f^{-1} would be differentiable on K, contradicting the assumption $\lambda(f,K) = 0$.

The part of Definition 14 about compact sets $K_1 \subset\subset U_1$ and $K_2 \subset\subset U_2$ would be trivially satisfied classically.

Not every simply connected bounded open set U is analytically equivalent to $\{z:|z| < 1\}$. *Example:* $U \equiv U_1 \cup U_2 \cup \cdots$, with $U_k \equiv \{z:|z| < 1\}$ whenever $n_k \neq 0$, and $U_k \equiv \{z:|z - 1| < 1\}$ whenever $n_k = 0$, where $\{n_k\}$ is any sequence of integers such that we do not know whether there is a k with $n_k = 0$.

The reader should give a routine for computing the mapping function φ of a given mappable set U at a given point z_0 to within a given accuracy ϵ. He should also give a routine for computing a modulus of continuity for φ on a given compact set $K \subset\subset U$.

The mapping function φ is classically obtained as an analytic function from U into $\{z:|z| < 1\}$ which vanishes at 0 and has the maximum possible derivative $\varphi'(0)$ at 0. Constructively this approach does not work, because it is not a priori evident that such a function exists. Once the mapping function is known to exist, it can easily be shown to be such a function.

CHAPTER 6

MEASURE

A measure is defined to be a continuous linear functional $f \to \int f \, d\mu$ on the space $C(X)$ of all continuous real-valued functions f with compact support on a given locally compact space X. The goal is to define the measure $\mu(A)$ of certain sets A (called integrable sets) and investigate the properties of the resulting set function μ. The constructive treatment of this problem is simplified by taking A to be a complemented set (as defined in Chap. 3), rather than a set. In addition, the use of complemented sets leads to more concrete results. A complete theory is developed only in case the measure is positive, although certain results are obtained in the general case. Assume therefore that the measure is positive. To define the measure of a complemented set, we approximate it (i.e., its characteristic function) by elements of $C(X)$. When the correct constructive version of this approximation process has been given, the theory develops along classical lines. However, certain theorems are significant modifications of their classical prototypes. For instance, the theorem that every compact set is integrable fails, but Theorem 4 says that nevertheless there are sufficiently many integrable compact sets. Using this result, we show that (i) every measure on $\mathbf{R}$ is induced by a monotone function, and (ii) an arbitrary integrable set can be approximated from within by compact integrable sets.

153

Any constructive approach to mathematics will find a crucial test in its ability to assimilate the intricate body of mathematical thought called measure theory. This subject was born in 1904 with a book by Lebesgue and now pervades abstract analysis in a way so essential that there is little point in trying to give meaning to this branch of mathematics without laying a constructive foundation for measure theory.

It was recognized by Lebesgue, Borel, and other pioneers in abstract function theory that the mathematics they were creating relied, in a way almost unique at that time, on set-theoretic methods, leading to results whose constructive content was problematical. Today it is true more than ever that much of abstract analysis has no ready constructive interpretation. It is to be expected that much abstract analysis will be seen in a proper constructive light only as the result of an investigation which remakes analysis from the very beginning in accord with constructive principles. We are at the stage in such an investigation in which we can profitably take up the study of measure, basing our work on the theory of complemented sets from Chap. 3 and the metric space theory of Chap. 4.

In two respects this study is crucial. First, as already indicated, measure theory and its techniques underlie many parts of modern analysis. Second, measure theory provides the proper framework for a discussion of discontinuous functions, which in most cases seem to be realized best in a measure-theoretic framework, as functions defined almost everywhere with respect to an appropriate measure.

1. TEST FUNCTIONS, MEASURES, EXAMPLES

The original approach was to let a measure be a function defined for sets. More recently it has been popular to let it be a function defined for functions. This approach fits our philosophy, that functions (at least continuous ones) should be preferred to sets as the primary objects of investigation whenever there is a choice.

Definition 1 A continuous real-valued function f on a locally compact space X is *supported* by a compact set $K \subset X$ if $f(x) = 0$ for all x in the metric complement $-K$ of K. The set K is a *support* for f, and f has *compact support*. A function with compact support is a *test function*, and the set of all test functions is written $C(X)$. The *norm* $\|f\|$ of a test function f is defined to be

$$\|f\| \equiv \|f\|_K \equiv \sup \{|f(x)| : x \in K\}$$

where K is any support for f.

Definition 2   A *measure* μ on a locally compact space X is a linear function from $C(X)$ to **R**, whose value at an element f of $C(X)$ is written $\int f\,d\mu$, such that for each f in $\mathbf{C}(X)$ there exists $c_f \geq 0$ with

$$|\textstyle\int h\,d\mu| \leq c_f$$

whenever $h \in C(X)$ and $|h| \leq |f|$. If $\int f\,d\mu \geq 0$ whenever $f \geq 0$, the measure μ is called *positive*.

The notation $\int f(x)\,d\mu(x)$ is sometimes used in place of $\int f\,d\mu$.

The set of all measures on X will be written $M(X)$, and the set of all positive measures $M^+(X)$.

Test functions and measures have some useful properties relative to certain maps. A continuous map $\lambda\colon X \to Y$ from one locally compact space into another is *proper* if $\lambda^{-1}(B)$ is a bounded subset of X for each bounded subset B of Y. The proper map λ induces a map $\lambda^*\colon C(Y) \to C(X)$ defined by

$$\lambda^*(f) \equiv f \circ \lambda \qquad (f \in C(Y))$$

If B is a support for f, then any compact set in X containing $\lambda^{-1}(B)$ is a support for $\lambda^*(f)$. The map λ^* in turn induces a map $\lambda_*\colon M(X) \to M(Y)$ defined by

$$\int g\,d\lambda_*(\mu) \equiv \int \lambda^*(g)\,d\mu \qquad (\mu \in M(X),\, g \in C(Y))$$

If $\mu \in M^+(X)$, then $\lambda_*(\mu) \in M^+(Y)$.

Measures can be multiplied by continuous functions: if $h\colon X \to \mathbf{R}$ is continuous, and if $\mu \in M(X)$, the measure $h\mu$ on X is defined by

$$\int f\,d(h\mu) \equiv \int hf\,d\mu \qquad (f \in C(X))$$

A measure μ on X will be said to be *supported* by a locally compact set $Y \subset X$ if $\int f\,d\mu = 0$ whenever the function f in $C(X)$ vanishes at all points of Y. (As an example, the measure $h\mu$ is supported on Y if h vanishes on $-Y$). If this is the case, consider any function g in $C(Y)$. By the Tietze extension theorem, g has a continuous extension g_0 to X, which can be taken to have compact support. Since μ is supported by Y, the integral $\int g_0\,d\mu$ does not depend on the choice of g_0. Therefore the equation

$$\int g\,d\nu \equiv \int g_0\,d\mu$$

defines a measure ν on Y, called the *restriction* of μ to Y.

We conclude this section with some examples.

Example 1   The integers Z are locally compact, and $C(Z)$ consists of all functions $f: Z \to \mathbf{R}$ which vanish outside a finite subset of Z. Every function $\alpha: Z \to \mathbf{R}$ determines a measure μ_α on Z, defined by

$$\int f \, d\mu_\alpha \equiv \sum_{n=-\infty}^{\infty} \alpha(n) f(n) \qquad (f \in C(Z))$$

where, of course, the sum is actually finite. Conversely, if μ is a measure on Z, then $\mu = \mu_\alpha$, where the function α is defined by

$$\alpha(n) \equiv \int \sigma_n \, d\mu$$

where σ_n is that element of $C(Z)$ whose value at n is 1 and whose value elsewhere is 0. In case $\alpha(n) = 1$ for all n in Z, the measure μ_α is called *counting measure* on Z.

Example 2   Every point x in a locally compact space X gives rise to a measure δ_x in $M(X)$, called the *point mass* at x, defined by

$$\int f \, d\delta_x \equiv f(x)$$

Example 3   *Lebesgue measure* μ on $\mathbf{R}$ is defined by

$$\int f \, d\mu \equiv \int_a^b f(x) \, dx \qquad (f \in C(\mathbf{R}))$$

where $[a,b]$ is any compact interval supporting f.

Example 4   *Lebesgue measure* on $\mathbf{R}^n$ can be obtained from Lebesgue measure on $\mathbf{R}$ as a special case of a general construction for a measure on a product space:

Definition 3   Let $X_1, \ldots, X_n$ be locally compact spaces, and let μ_i be an element of $M(X_i)$ $(1 \leq i \leq n)$. Let f be a test function on the product space $X \equiv X_1 \times \cdots \times X_n$. Then

$$(x_2, \ldots, x_n) \to \int f(x_1, \ldots, x_n) \, d\mu_1(x_1)$$

is a test function f_1 on $X_2 \times \cdots \times X_n$. Similarly,

$$(x_3, \ldots, x_n) \to \int f_1(x_2, \ldots, x_n) \, d\mu_2(x_2)$$

is a test function f_2 on $X_3 \times \cdots \times X_n$. This process of repeated integrations leads after n steps to a real number,

$$\int f \, d\mu \equiv \int f_{n-1}(x_n) \, d\mu_n(x_n)$$

The measure $f \to \int f \, d\mu$ on X so defined is called the *product* $\mu \equiv \mu_1 \times \cdots \times \mu_n$ of $\mu_1, \ldots, \mu_n$.

In case $X_1 = \cdots = X_n = \mathbf{R}$ and each μ_i is Lebesgue measure on $\mathbf{R}$, then μ is called *Lebesgue measure* on $\mathbf{R}^n$.

Product measures are independent of the ordering of the factors of the product; in other words, the *iterated integration* defining the product measure can be done in any order.

Proposition 1 *Let $X_1, \ldots, X_n$ be locally compact spaces, and let μ_i be a measure on X_i ($1 \leq i \leq n$). Let σ be a permutation of the integers $1, \ldots, n$. For each test function f on the product $X \equiv X_1 \times \cdots \times X_n$ define the test function f_σ on $X_\sigma \equiv X_{\sigma(1)} \times \cdots \times X_{\sigma(n)}$ by $f_\sigma(x_{\sigma(1)}, \ldots, x_{\sigma(n)}) \equiv f(x_1, \ldots, x_n)$. Then*

$$(1.1) \qquad \int f \, d(\mu_1 \times \cdots \times \mu_n) = \int f_\sigma \, d(\mu_{\sigma(1)} \times \cdots \times \mu_{\sigma(n)})$$

Proof In case f is a product of the form

$$f(x_1, \ldots, x_n) \equiv h_1(x_1) \cdots h_n(x_n)$$

with $h_i \in C(X_i)$ ($1 \leq i \leq n$), then (1.1) is valid because both sides are equal to $\int h_1 \, d\mu_1 \cdots \int h_n \, d\mu_n$. Therefore (1.1) is valid whenever f is in the algebra $\mathfrak{A}$ of linear combinations of such products.

Consider now an arbitrary test function f on X. For $1 \leq i \leq n$ choose a test function h_i on X_i such that the element h of $\mathfrak{A}$ defined by

$$h(x_1, \ldots, x_n) \equiv h_1(x_1) \cdots h_n(x_n)$$

has the value 1 at all points of the support K of f. (For instance, let K_i be a compact subset of X_i ($1 \leq i \leq N$), with $K \subset K_1 \times \cdots \times K_n$, and write $h_i(x_i) \equiv \max \{0, 1 - \rho_i(x_i, K_i)\}$.) Then $fh = f$. Let L be a support of h. By the Stone-Weierstrass theorem, there exists g in $\mathfrak{A}$ such that $\|f - g\|_L$ is arbitrarily small. Therefore $\|f - gh\|$ is arbitrarily small. Since (1.1) is valid for all functions in $\mathfrak{A}$, in particular for the function gh, it is valid for f.

Example 5 Much as functions on Z give rise to measures on Z, certain functions on a compact interval I, called functions of *bounded variation*, give rise to measures on I.

Definition 4   A real-valued function α defined on a dense subset A of
an interval $I \equiv [a,b]$ is of *bounded variation* on I if there exists $c > 0$
such that

$$(1.2) \qquad \sum_{i=0}^{n-1} |\alpha(x_{i+1}) - \alpha(x_i)| < c$$

whenever $x_1 < x_2 < \cdots < x_n$ are points in A. The constant c is
called a *bound* for the variation of f on I.

Note that condition (1.2) is satisfied if α is monotone and bounded.

Let f be a continuous function on a proper compact interval $I \equiv [a,b]$,
and α a function of bounded variation on I, whose domain includes the
end points of I. The *integral* $\int_a^b f(x)\, d\alpha(x)$ is defined in analogy to the
Riemann integral of Chap. 2. Here is a sketch of how this is done;
details are left to the reader. Only partitions $P \equiv \{a_0 = a, a_1, \ldots, a_n = b\}$ formed from points of A are considered. To each such P we
associate sums of the form

$$S(f,P) \equiv \sum_{i=0}^{n-1} f(x_i)(\alpha(a_{i+1}) - \alpha(a_i))$$

where $a_i \leq x_i \leq a_{i+1}$. There exists a unique constant L, called the
integral of f over I with respect to α, and written

$$\int_a^b f(x)\, d\alpha(x) \qquad \text{or simply} \qquad \int f(x)\, d\alpha(x)$$

such that $|S(f,P) - L|$ is arbitrarily small when the mesh of P is
sufficiently small. The integral is not changed if A is replaced by a
subset which is dense in I and contains the end points of I. The integral
depends linearly on the test function f, and

$$\left| \int_a^b f(x)\, d\alpha(x) \right| \leq c\|f\|_I$$

where c is any bound for the variation of α on I. Therefore the equation

$$\int f\, d\mu_\alpha \equiv \int f(x)\, d\alpha(x) \qquad (f \in C(I))$$

defines a measure μ_α on I, which is positive whenever α is monotone-
nondecreasing.

A similar construction associates a measure μ on $\mathbf{R}$ to a function α,
defined on a dense subset of $\mathbf{R}$ and of bounded variation on every

proper compact interval, as follows. For each f in $C(\mathbf{R})$ choose a proper compact interval I supporting f whose end points are in the domain of α, and define $\int f\, d\mu_\alpha$ as above.

2. MEASURES OF SETS

By definition, a measure μ assigns a value $\int f\, d\mu$ to certain functions. We shall use a measure to assign a value to certain sets (certain complemented sets, to be exact).

The *complemented sets* in a locally compact space X are the complemented sets relative to the set F of all continuous real-valued functions on X, and the *Borel sets* are the Borel sets generated by F.

Definition 5 Let A be a complemented set in a locally compact space X, let μ be a positive measure on X, and let ϵ be a positive real number. A continuous function $f\colon X \to \mathbf{R}$, with $0 \leq f \leq 1$, is said to *approximate A to within ϵ* (relative to μ) if there exists a sequence $\{f_j\}$ of nonnegative test functions such that

$$(2.1) \quad (a) \quad \sum_{j=1}^{\infty} \int f_j\, d\mu \qquad \text{exists and is less than } \epsilon$$

$$(b) \quad x \in -A \qquad \text{whenever } f(x) + \sum_{j=1}^{N} f_j(x) \leq 1 - \delta$$
$$\text{for all } N \text{ in } Z^+ \text{ and some } \delta > 0$$

$$(c) \quad x \in A \qquad \text{whenever } 1 - f(x) + \sum_{j=1}^{N} f_j(x) \leq 1 - \delta$$
$$\text{for all } N \text{ in } Z^+ \text{ and some } \delta > 0$$

The interpretation of Definition 5 is that A is "nearly" the set where f is positive and $-A$ is "nearly" the set where f is less than 1. In other words, f is "nearly" the characteristic function of A. The sequence $\{f_j\}$, called the *error sequence*, represents the maximum extent to which f departs from the characteristic function of A. Condition $(2.1)(a)$ gives a numerical measure to this maximum extent and requires it to be less than ϵ.

As an example, let $X \equiv \mathbf{R}$ and let μ be Lebesgue measure. Write $A \equiv (Q,Q')$, where Q is the set of rational numbers and Q' consists of all x in $\mathbf{R}$ with $x \neq y$ for all y in Q. Order Q to get a sequence $\{q_n\}$. For each $\epsilon > 0$ and each n in Z^+ there exists $f_n{}^\epsilon$ in $C(\mathbf{R})$ with $0 \leq f_n{}^\epsilon \leq 1$, $f_n{}^\epsilon(q_n) = 1$, and $\int f_n{}^\epsilon\, d\mu < 2^{-n}\epsilon$. Thus for each $\epsilon > 0$ the function $f \equiv 0$ approximates A to within ϵ with error sequence $\{f_n{}^\epsilon\}$.

If f approximates A to within ϵ, clearly $1 - f$ approximates $-A$ to within ϵ, with the same error sequence.

For each $\epsilon > 0$ the function $f \equiv 0$ approximates the void set to within ϵ, with error sequence $\{f_k\}$ consisting of functions which are all zero.

A complemented set A which for each $\epsilon > 0$ has an approximation to within ϵ by a test function is called *integrable*.

The next lemma, which is basic, generalizes Cantor's result on the uncountability of $\mathbf{R}$.

Lemma 1 *Let μ be a positive measure on a locally compact space X, and g a nonnegative test function on X. Let $\{f_j\}$ be a sequence of nonnegative test functions such that $\displaystyle\sum_{j=1}^{\infty} \int f_j \, d\mu$ converges and is less than $\int g \, d\mu$. Then*

$$(2.2) \qquad \sum_{j=1}^{m} f_j(x) \leq g(x) - \epsilon \qquad (m \in Z^+)$$

for some x in X and some $\epsilon > 0$.

Proof We show, by induction on n, that there exists a sequence $\{\lambda_n\}$ of nonnegative test functions and a sequence $\{x_n\}$ of points of X such that for some $\epsilon > 0$ we have

$$(2.3) \qquad \rho(x_k,x_n) \leq k^{-2} + n^{-2}$$

$$(2.4) \qquad \lambda_n(y) = 0 \qquad (\rho(y,x_n) \geq n^{-2})$$

and

$$(2.5) \qquad \int (g - \epsilon)\lambda_n \, d\mu - \sum_{j=1}^{\infty} \int f_j\lambda_n \, d\mu > 0$$

for all positive integers n and k. Let λ_0 be any nonnegative test function, with $0 \leq \lambda_0 \leq 1$, such that $\lambda_0 g = g$. By hypothesis, there exists $\epsilon > 0$ such that (2.5) is valid for $n = 0$. Now, let n be any positive integer. Assume that for each positive integer $k < n$ the function λ_k and the point x_k have been constructed. Let K be any support for λ_{n-1}. Let $\{y_1, \ldots, y_m\}$ be a $\frac{1}{2}n^{-2}$ approximation to K. Choose a constant c with $\frac{1}{2}n^{-2} < c < n^{-2}$ so that the sets

$$K_i \equiv \{y \in K : \rho(y,y_i) \geq c\} \qquad (1 \leq i \leq m)$$

are compact. For each i $(1 \leq i \leq m)$ let $\lambda_n{}^i$ be the test function defined by

$$\lambda_n{}^i(y) \equiv \lambda_{n-1}(y)\rho(y,K_i) \Big[\sum_{j=1}^{m} \rho(y,K_j) \Big]^{-1}$$

Then

$$0 \leq \lambda_n{}^i \leq \lambda_{n-1} \qquad \sum_{i=1}^{m} \lambda_n{}^i = \lambda_{n-1} \qquad \text{and} \qquad \lambda_n{}^i(y) = 0$$

whenever $\rho(y,y_i) \geq c$. By (2.5), we have

$$\sum_{i=1}^{m} \left\{ \int (g - \epsilon)\lambda_n{}^i \, d\mu - \sum_{j=1}^{\infty} \int f_j \lambda_n{}^i \, d\mu \right\} > 0$$

Therefore

$$(2.6) \qquad \int (g - \epsilon)\lambda_n{}^i \, d\mu - \sum_{j=1}^{\infty} \int f_j \lambda_n{}^i \, d\mu > 0$$

for at least one value of i. Define $\lambda_n \equiv \lambda_n{}^i$ and $x_n \equiv y_i$. Condition (2.5) is satisfied because of (2.6). Condition (2.4) is satisfied because if $y \in K$ and $\rho(y,x_n) \equiv \rho(y,y_i) \geq n^{-2} > c$, then $y \in K_i$, and thus $\lambda_n(y) \equiv \lambda_n{}^i(y) = 0$. To check (2.3), consider $k < n$ and note that $\lambda_n(y) = 0$ if either $\rho(y,x_n) > n^{-2}$ or $\rho(y,x_k) > k^{-2}$. Now there exists y in X with $\lambda_n(y) = \lambda_n{}^i(y) > 0$, by (2.6). For such a point y both $\rho(y,x_n) \leq n^{-2}$ and $\rho(y,x_k) \leq k^{-2}$, and thus (2.3) is valid. This completes the construction of the sequences $\{\lambda_n\}$ and $\{x_n\}$.

By (2.3), we see that $\{x_n\}$ is a Cauchy sequence whose limit x satisfies the inequality $\rho(x,x_n) \leq n^{-2}$ for all n. Since

$$\rho(y,x_n) \geq \rho(y,x) - \rho(x,x_n) \geq \rho(y,x) - n^{-2}$$

for all y in X and all n, we see by (2.4) that $\lambda_n(y) = 0$ whenever $\rho(y,x) > 2n^{-2}$.

Assume that $\sum_{j=1}^{m} f_j(x) > g(x) - \epsilon$ for some positive integer m. Then $\sum_{j=1}^{m} f_j \lambda_n \geq (g - \epsilon)\lambda_n$ for all sufficiently large n. This gives

$$\int (g - \epsilon)\lambda_n \, d\mu - \sum_{j=1}^{\infty} \int f_j \lambda_n \, d\mu \leq 0$$

This contradiction shows that $\sum_{j=1}^{m} f_j(x) \leq g(x) - \epsilon$ for all m in Z^+, as was to be proved.

Our next theorem is the final step to the goal of assigning a measure to every integrable set.

Theorem 1 *Let the test functions f and g approximate the complemented set A to within ϵ_1 and ϵ_2, respectively. Then*

$$(2.7) \qquad \int |f - g|\, d\mu \leq \epsilon_1 + \epsilon_2$$

Proof Let $\{f_j\}$ and $\{g_j\}$ be error sequences for f and g, respectively. Assume that

$$\sum_{j=1}^{\infty} \left\{ \int f_j\, d\mu + \int g_j\, d\mu \right\} < \int |f - g|\, d\mu$$

By Lemma 1 there exists $\delta > 0$ and x in X such that

$$(2.8) \qquad \sum_{j=1}^{m} \{ f_j(x) + g_j(x) \} + 2\delta \leq |f(x) - g(x)|$$

for all m in Z^+. Now either $|f(x) - g(x)| < f(x) - g(x) + \delta$ or $|f(x) - g(x)| < g(x) - f(x) + \delta$. Without loss of generality, we may assume that the first alternative holds. Then (2.8) gives

$$(2.9) \quad 1 - f(x) + g(x) + \sum_{j=1}^{m} \{ f_j(x) + g_j(x) \} < 1 - \delta \qquad (m \in Z^+)$$

Therefore

$$(2.10) \qquad g(x) + \sum_{j=1}^{m} g_j(x) < 1 - \delta \qquad (m \in Z^+)$$

and

$$(2.11) \qquad 1 - f(x) + \sum_{j=1}^{m} f_j(x) < 1 - \delta \qquad (m \in Z^+)$$

By (2.10) and condition (*b*) of Definition 5, it follows that $x \in -A$. By (2.11) and condition (*c*) of Definition 5, it follows that $x \in A$. Since A

is a complemented set, it follows that $0 = 1$. By Lemma 5 of Chap. 2, it follows that

$$(2.12) \qquad \epsilon_1 + \epsilon_2 \geq \sum_{j=1}^{\infty} \left\{ \int f_j \, d\mu + \int g_j \, d\mu \right\} \geq \int |f - g| \, d\mu$$

as was to be proved.

Corollary *To each integrable set A corresponds a unique constant $\mu(A)$, called the measure of A, such that if f is any test function that approximates A to within ϵ then*

$$(2.13) \qquad \left| \int f \, d\mu - \mu(A) \right| \leq \epsilon$$

Proof For each positive integer k choose a test function f_k that approximates A to within k^{-1}. By Theorem 1, $\{ \int f_k \, d\mu \}$ is a Cauchy sequence, whose limit we call $\mu(A)$. By (2.7)

$$\left| \int f \, d\mu - \mu(A) \right| = \lim_{k \to \infty} \left| \int (f - f_k) \, d\mu \right| \leq \lim_{k \to \infty} (\epsilon + k^{-1}) = \epsilon$$

Examples of integrable sets are hard to find at this stage. The void set ϕ_0 is integrable, and its measure is 0. If X is compact, then X_0 is integrable, and its measure is $\int 1 \, d\mu$. More generally, if X is compact, then a complemented set A is integrable if and only if its complement is integrable, and $\mu(A) + \mu(-A) = \int 1 \, d\mu$, since a function f approximates A to within ϵ if and only if $1 - f$ approximates $-A$ to within ϵ.

We look at how the measure function behaves with respect to the standard set-theoretic operations.

Lemma 2 *Let f and g be approximations, to within $\frac{1}{4}\epsilon$, to the complemented sets A and B, respectively, with respective error sequences $\{f_j\}$ and $\{g_j\}$. Then $f + g - fg$, fg, and $f - fg$, respectively, are approximations to within ϵ to the complemented sets $A \cup B$, $A \cap B$, and $A - B$, with common error sequence $\{h_j\}$ defined by*

$$h_j \equiv 2 \max \{f_j, g_j\}$$

Proof Write $h \equiv f + g - fg$. Clearly $0 \leq h \leq 1$.

The series $\sum_{j=1}^{\infty} \int h_j \, d\mu$ converges, and its sum is less than ϵ. The sequence $\{h_j\}$ therefore satisfies condition (2.1)(a).

To check condition $(2.1)(b)$, consider a point x in X with

$$h(x) + \sum_{j=1}^{m} h_j(x) \le 1 - \delta \qquad \text{for all } m \text{ in } Z^+$$

Then

$$f(x) + \sum_{j=1}^{m} f_j(x) \le 1 - \delta$$

and

$$g(x) + \sum_{j=1}^{m} g_j(x) \le 1 - \delta \qquad \text{for all } m \text{ in } Z^+$$

It follows that $x \in -A \cap -B = -(A \cup B)$. Therefore $(2.1)(b)$ is satisfied.

To check $(2.1)(c)$, consider a point x in X with $1 - h(x) + \sum_{j=1}^{m} h_j(x) \le 1 - \delta$ for all m in Z^+. This implies that

$$(2.14) \qquad -f(x) - g(x) \le - \sum_{j=1}^{m} h_j(x) - \delta \qquad (m \in Z^+)$$

Now, either $f(x) < g(x) + \frac{1}{2}\delta$ or $g(x) < f(x) + \frac{1}{2}\delta$. Consider the first alternative. From (2.14) we obtain

$$\begin{aligned}
1 - g(x) &< 1 - \tfrac{1}{2}(f(x) + g(x)) + \tfrac{1}{4}\delta \\
&\le 1 - \tfrac{1}{2}\Big(\sum_{j=1}^{m} h_j(x) + \delta \Big) + \tfrac{1}{4}\delta \\
&\le 1 - \sum_{j=1}^{m} g_j(x) - \tfrac{1}{4}\delta
\end{aligned}$$

for all m in Z^+. Therefore $x \in B$. Similarly $x \in A$ in case $g(x) < f(x) + \frac{1}{2}\delta$. Thus $x \in A \cup B$ in either case, and thus condition (c) is satisfied.

It therefore follows that $f + g - fg$ approximates $A \cup B$ to within ϵ, with error sequence $\{h_j\}$. The same result applied to the sets $-A$ and $-B$ shows that $(1 - f) + (1 - g) - (1 - f)(1 - g) = 1 - fg$ approximates $-A \cup -B$ to within ϵ with error sequence $\{h_j\}$. Therefore $1 - (1 - fg) = fg$ approximates $-(-A \cup -B) = A \cap B$ to within ϵ with the same error sequence. This result applied to the sets A and $-B$ shows that the function $f - fg = f(1 - g)$ approximates $A \cap -B = A - B$ to within ϵ with error sequence $\{h_j\}$.

Lemma 3 *Let A and B be complemented sets. Let f and g be approximations, to within $\frac{1}{4}\epsilon$, to A and to $A \cap B$, respectively, having error sequences $\{f_j\}$ and $\{g_j\}$, respectively. Then $h \equiv f - fg$ is an approximation to within ϵ to $A - B$, with the same error sequence $\{h_j\}$ as in Lemma 2.*

Proof By Lemma 2, we see that h approximates $A - (A \cap B)$ to within ϵ, with error sequence $\{h_j\}$. If

$$x \in A - (A \cap B) = A \cap (-A \cup -B) = (A \cap -A) \cup (A \cap -B)$$

then $x \in (A \cap -B) = A - B$. If

$$x \in -(A - (A \cap B)) = (-A \cup A) \cap (-A \cup B)$$

then $x \in -A \cup B = -(A - B)$. Therefore h approximates $A - B$ to within ϵ, with error sequence $\{h_j\}$.

Theorem 2 *Let A and B be integrable subsets of a locally compact space X relative to a positive measure μ on X. Then $A \cup B$, $A \cap B$, and $A - B$ are integrable, and*

$$(2.15) \qquad \mu(A) + \mu(B) = \mu(A \cup B) + \mu(A \cap B)$$

If E and F are complemented sets such that E and $E \cap F$ are integrable, then $E - F$ is integrable and

$$(2.16) \qquad \mu(E) = \mu(E \cap F) + \mu(E - F)$$

If $\{A_n\}$ is a sequence of integrable sets such that

$$\alpha \equiv \lim_{n \to \infty} \mu \left(\bigcup_{i=1}^{n} A_i \right)$$

exists, then $\bigcup_{i=1}^{\infty} A_i$ is integrable with measure α, while if

$$\beta \equiv \lim_{n \to \infty} \mu \left(\bigcap_{i=1}^{n} A_i \right)$$

exists, then $\bigcap_{i=1}^{\infty} A_i$ is integrable with measure β.

Proof The integrability of $A \cup B$, $A \cap B$, and $A - B$ follows from Lemma 2.

To prove (2.15), for each $\epsilon > 0$ choose ϵ approximations f and g to A and B, respectively. By Lemma 2, we see that $f + g - fg$ and fg are approximations, to within 4ϵ, to $A \cup B$ and $A \cap B$, respectively. From Theorem 1 and its corollary we get

$$
\begin{aligned}
|\mu(A) + \mu(B) &- \mu(A \cup B) - \mu(A \cap B)| \\
&\leq |\mu(A) - \smallint f \, d\mu| + |\mu(B) - \smallint g \, d\mu| \\
&\quad + |\mu(A \cup B) - \smallint (f + g - fg) \, d\mu| + |\mu(A \cap B) - \smallint fg \, d\mu| \\
&\leq \epsilon + \epsilon + 4\epsilon + 4\epsilon = 10\epsilon
\end{aligned}
$$

Letting $\epsilon \to 0$ gives (2.15).

Assume next that E and F are complemented sets such that E and $E \cap F$ are integrable. For each $\epsilon > 0$ let f approximate E to within ϵ, and let g approximate $E \cap F$ to within ϵ. By Lemma 3, we see that $f - fg$ approximates $E - F$ to within 4ϵ. Therefore $E - F$ is integrable, and a proof similar to the one just given proves (2.16).

Consider next a sequence $\{A_n\}$ of integrable sets such that the limit α exists. Write

$$
B_n \equiv \bigcup_{i=1}^{n} A_i \qquad (n \in Z^+)
$$

There is no loss of generality in assuming that $\sum_{k=n}^{\infty} \mu(B_{k+1} - B_k) \leq 2^{-n}$ for all n. In particular, $\mu(B_{n+1} - B_n) \leq 2^{-n}$. Consider an arbitrary positive integer N. For each $k > N$ let f^k be an approximation to $B_k - B_{k-1}$ to within 2^{-k}, with error sequence $\{f_j^k\}_{j=1}^{\infty}$. By the above corollary,

$$
\smallint f^k \, d\mu \leq \mu(B_k - B_{k-1}) + 2^{-k} \leq 2^{-k+1} + 2^{-k} < 2^{-k+2}
$$

Let f approximate B_N to within $(2N)^{-1}$, with error sequence $\{f_j\}$. Let $\{f_j'\}$ be the totality of all functions f_j $(1 \leq j < \infty)$, f^k $(N < k < \infty)$, and f_j^k $(N < k < \infty, 1 \leq j < \infty)$, arranged into a sequence. We shall show that f is an approximation to $A \equiv \bigcup_{i=1}^{\infty} A_i$ to within N^{-1}, with error sequence $\{f_j'\}$, for all sufficiently large N.

To check condition (2.1)(a), we compute

$$
\sum_{j=1}^{\infty} \int f_j' \, d\mu \leq (2N)^{-1} + \sum_{k=N+1}^{\infty} 2^{-k+2} + \sum_{k=N+1}^{\infty} 2^{-k} < N^{-1}
$$

if N is sufficiently large.

To check $(2.1)(b)$, consider a point x in X such that

$$f(x) + \sum_{j=1}^{m} f_j'(x) \le 1 - \delta \qquad (m \in Z^+)$$

Then $x \in -B_N$, and $x \in -(B_k - B_{k-1}) = B_{k-1} \cup -B_k$ for $k > N$. This implies that $x \in -B_k$ for all $k \ge N$, by an inductive argument. Therefore

$$x \in \bigcap_{i=N}^{\infty} -B_i = -A$$

and thus $(2.1)(b)$ is satisfied.

To check $(2.1)(c)$, consider x in X such that

$$1 - f(x) + \sum_{j=1}^{m} f_j'(x) \le 1 - \delta \qquad (m \in Z^+)$$

Then
$$1 - f(x) + \sum_{j=1}^{m} f_j(x) \le 1 - \delta \qquad (m \in Z^+)$$

so that $x \in B_N \subset A$. Therefore $(2.1)(c)$ is satisfied.

This completes the proof that f is an approximation to A to within N^{-1} for all sufficiently large N. Therefore A is integrable, and by (2.13),

$$|\mu(A) - \alpha| \le |\mu(A) - \textstyle\int f \, d\mu| + |\textstyle\int f \, d\mu - \mu(B_N)| + |\alpha - \mu(B_N)|$$
$$\le N^{-1} + (2N)^{-1} + 2^{-N}$$

for all N. Therefore $\mu(A) = \alpha$.

Consider finally a sequence $\{A_n\}$ of integrable sets such that β exists. A proof similar to the one just given shows that $\bigcap_{i=1}^{\infty} A_i$ is integrable and has measure β.

One of the most important uses of measure theory is to prove existence theorems, by means of the following proposition.

Proposition 2 *An integrable set A with $\mu(A) \ne 0$ has at least one point.*

Proof Let f approximate A to within $\frac{1}{4}\mu(A)$. Then

$$\textstyle\int f \, d\mu \ge \mu(A) - \tfrac{1}{4}\mu(A) = \tfrac{3}{4}\mu(A) > \sum_{j=1}^{\infty} \int f_j \, d\mu$$

By Lemma 1 there exists $\epsilon > 0$ and x in X such that

$$\sum_{j=1}^{m} f_j(x) \leq f(x) - \epsilon$$

or equivalently,

$$1 - f(x) + \sum_{j=1}^{m} f_j(x) \leq 1 - \epsilon \qquad (m \in Z^+)$$

It follows that $x \in A$.

To make further progress we must prove the existence of nontrivial integrable sets. First we show how to realize a locally compact set as a Borel set.

We identify a locally compact subset K of a locally compact space X with the Borel set

$$(\{x: \rho(x,K) \leq 0\}, \{x: \rho(x,K) > 0\}) = (K, -K)$$

and the metric complement $-K$ of K with its complement

$$(\{x: \rho(x,K) > 0\}, \{x: \rho(x,K) \leq 0\}) = (-K, K)$$

It will be clear from the context whether K is being thought of as a compact set or as a Borel set.

We next show that for a positive measure on $\mathbf{R}$ there are many integrable compact sets.

3. MEASURES ON R

The structure of a positive measure μ on $\mathbf{R}$ is revealed by the following result.

Theorem 3 *Let μ be a positive measure on $\mathbf{R}$. Then all except countably many points x of $\mathbf{R}$ are* steady, *in the sense that for each $\epsilon > 0$ there exist f in $C(\mathbf{R})$ and $\delta > 0$ such that $0 \leq f \leq 1$, and $f(y) = 1$ whenever $|y - x| \leq \delta$, and $\int f \, d\mu \leq \epsilon$.*

Proof Clearly it is sufficient to show that all except countably many points x in a given open proper interval (a,b) are steady. To this end let f_0 be any test function, $0 \leq f_0 \leq 1$, with $f_0(u) = 1$ for all u in $[a,b]$, and let the interval $[a_0,b_0] \supset [a,b]$ be a support for f_0. Let S be the subset of $\mathbf{R}^2$ consisting of all (u,v) such that $a_0 \le u \le b_0$ and $v = \int f \, d\mu$, for some test function f with $0 \le f \le f_0$ such that $f(t) = 0$ whenever $t \ge u$. Then S is totally bounded. To see this, consider any $\epsilon > 0$. Let $u_0 = a_0 < u_1 < \cdots < u_n = b_0$ be points in $[a_0, b_0]$ such that $u_i - u_{i-1} < \epsilon$ $(1 \le i \le n)$. For $1 \le i \le n$ let f_i be a test function, $0 \le f_i \le f_0$, such that $f_i(x) = f_0(x)$ for $x \le u_{i-1}$ and $f_i(x) = 0$ for $x \ge u_i$. We may assume $f_n = f_0$. For $1 \le i \le n$ choose points

$$0 = v_i{}^0 < v_i{}^1 < \cdots < v_i{}^N = \int f_i \, d\mu \text{ (where } N \text{ depends on } i)$$

such that $v_i{}^j - v_i{}^{j-1} < \epsilon$ $(1 \le j \le N)$. The points $(u_i, v_i{}^j)$, with $1 \le i \le n$ and $1 \le j \le N$, lie in S. If (u,v) is any point in S, then $u_i \le u \le u_{i+2}$ for some value of i with $0 \le i \le n - 2$. We have $v = \int f \, d\mu$, where f satisfies the above conditions. These conditions imply that $f \le f_k$, where $k \equiv \min \{i + 3, n\}$. Thus $v = \int f \, d\mu \le \int f_k \, d\mu$. Hence $|v - v_k{}^j| < \epsilon$ for some value of j. Since $|u - u_k| \le 3\epsilon$, it follows that $\rho((u,v), (u_k, v_k{}^j)) < 4\epsilon$. Hence the points $\{(u_i, v_i{}^j) : 1 \le i \le n, 1 \le j \le N\}$ are a 4ϵ approximation to S. Thus S is totally bounded. The closure T of S is therefore compact.

By Theorem 8 of Chap. 4, for all except countably many x in (a,b) the set $T_x \equiv \{(u,v) \in T : u \le x\}$ is compact, and for each $\epsilon > 0$ and each admissible x there exists $\delta > 0$ such that $\rho(T_y, T_z) < \epsilon$ whenever y and z are admissible points of (a,b) with $|y - x| < \delta$ and $|z - x| < \delta$.

We shall complete the proof by showing that each admissible point x is steady. To this end consider a constant $\epsilon > 0$ and choose δ as above. Let y and z be admissible points of (a,b) with $x - \delta < y < x < z < x + \delta$. Choose the test function g_1 with $0 \le g_1 \le f_0$ such that $g_1(u) = f_0(u)$ for $u \le \frac{1}{2}(x + z)$ and $g_1(u) = 0$ for $u \ge z$. Then $(z, \int g_1 \, d\mu) \in S_z$. Thus there exists an element $(y, \int g_2 \, d\mu)$ in S_y whose distance to $(z, \int g_1 \, d\mu)$ is less than ϵ. The function $f \equiv g_2 - g_1$ satisfies the inequalities $0 \le f \le 1$, $f(u) = 1$ if $y < u < \frac{1}{2}(x + z)$, and $\int f \, d\mu = \int (g_2 - g_1) \, d\mu < \epsilon$. Therefore x is a steady point.

Theorem 3 permits us to prove the existence of many integrable compact sets relative to a positive measure on *any* locally compact space.

Theorem 4 *Let μ be a positive measure on the locally compact space X, and let $h : X \to \mathbf{R}$ be a proper map. Then there exists a set S consisting of all except countably many points in $\mathbf{R}$ such that for all u and v in S with $u < v$ the set*

$$X(u,v) \equiv \{x \in X : u \le h(x) \le v\}$$

is compact and integrable, and for each $\epsilon > 0$ there exists $\delta > 0$ such that any f in $C(X)$ with $0 \leq f \leq 1$, with $f(x) = 1$ for $u + \delta \leq h(x) \leq v - \delta$ and with $f(x) = 0$ for $h(x) \leq u - \delta$ or $h(x) \geq v + \delta$, approximates $X(u',v')$ to within ϵ for arbitrary admissible u' and v' with $u' < v'$, $|u - u'| < \delta$, and $|v - v'| < \delta$.

Proof   Define the measure ν on $\mathbf{R}$ by

$$\textstyle\int f \, d\nu = \int f \circ h \, d\mu$$

for all f in $C(\mathbf{R})$. By Theorem 3, all except countably many real numbers are steady for ν. If u and v are steady for ν, then for each $\epsilon > 0$ there exists $\delta > 0$ and f_1 and f_2 in $C(\mathbf{R})$ with $0 \leq f_1, f_2 \leq 1$ such that (a) $f_1(t) = 1$ whenever $|t - u| \leq \delta$ and $f_2(t) = 1$ whenever $|t - v| \leq \delta$, and (b) $\int f_1 \, d\nu < \epsilon/2$ and $\int f_2 \, d\nu < \epsilon/2$. Each function f satisfying the given conditions approximates $X(u',v')$ to within ϵ, with error sequence $\{f_1 \circ h, f_2 \circ h, 0, 0, \ldots\}$. In particular, it approximates $X(u,v)$ to within ϵ. Hence $X(u,v)$ is integrable.

Corollary   *Let μ be a positive measure on a locally compact space X, and let g be a test function with $0 \leq g \leq 1$. Then for all except countably many $a > 0$, the sets*

$$X^a \equiv \{x : g(x) \geq a\} \qquad and \qquad X_a \equiv \{x : g(x) \leq a\}$$

are locally compact and the sets X^a and $-X_a$ are integrable and have the same measure. Moreover,

$$(3.1) \qquad\qquad \textstyle\int g \, d\mu \geq a\mu(X^a)$$

and

$$(3.2) \qquad\qquad \textstyle\int g \, d\mu \leq \|g\|\mu(X^a) + \int g_0 \, d\mu$$

where g_0 is any nonnegative test function with $g(x) \leq g_0(x)$ for all x in $-X^a$.

Proof   Consider any $c > 0$. Let $h: X \to \mathbf{R}$ be a proper map, with $h(x) = g(x)$ whenever $g(x) \geq c$, and $h(x) \leq c$ whenever $g(x) \leq c$. Then for all $a > c$ we have

$$X^a = \{x : h(x) \geq a\}$$

It follows from Theorem 4 that for all except countably many $a > c$

(and hence for all except countably many $a > 0$) the set X^a is compact, the set X_a is locally compact, and the set X^a is integrable. Moreover, for each admissible a and for each $\epsilon > 0$ there exists $\delta > 0$ such that any f in $C(X)$ with $0 \leq f \leq 1$, $f(x) = 1$ for $g(x) \geq a + \delta$, and $f(x) = 0$ for $g(x) \leq a - \delta$ approximates X^b to within ϵ, for each admissible b with $|a - b| < \delta$. If $a < b < a + \delta$, we have $X^b \subset -X_a \subset X^a$. Any function f that satisfies the above conditions therefore approximates $-X_a$ to within ϵ. Hence $-X_a$ is integrable and has the same measure as X^a.

To prove (3.1), note that for each $\epsilon > 0$ there exists f as described, with $g \geq af$.

To prove (3.2), note that $g - g_0 \leq 0$ on $-X^a$ and $g - g_0 \leq \|g\|$ on X^a. For each $\epsilon > 0$ choose a function f as described, with $f(x) = 1$ for all x in X^a. Then $g - g_0 \leq \|g\|f$. This gives (3.2).

Another consequence of Theorem 4 is that every positive measure on **R** comes from a monotone function.

Theorem 5   *Let μ be a positive measure on* **R**. *Then there exists a monotone-nondecreasing function* $\alpha: S \rightarrow$ **R** *defined on the set S of steady points of μ such that $\mu = \mu_\alpha$.*

Proof   Let x_0 be any steady point of μ. Let x be any steady point with $x \neq x_0$. If $x > x_0$, write $\alpha(x) = \mu([x_0,x])$, and if $x < x_0$, write $\alpha(x) = -\mu([x,x_0])$. Then α is monotone-nondecreasing, and $\alpha(x)$ will be arbitrarily near to 0 if x is sufficiently near to x_0. Thus α can be uniquely extended to a monotone-nondecreasing function on S, with $\alpha(x_0) = 0$. For arbitrary points x and y in S with $x < y$ we have $\alpha(y) - \alpha(x) = \mu([x,y])$. We must show that $\mu = \mu_\alpha$, or that

$$\int f \, d\mu = \int f(x) \, d\alpha(x)$$

for all test functions f. Let $I \equiv [a,b]$ be a proper compact interval which supports f, such that $a \in S$, $b \in S$, and $f(x)$ vanishes for all x sufficiently near to a. Let ω be the modulus of continuity of f on I. Let ϵ be any positive constant.

Choose a partition $P \equiv \{a_0 = a, a_1, \ldots, a_n = b\}$ of I, composed of points of S, of mesh less than $\omega(\epsilon)$, such that $f(a_1) = 0$ and

$$(3.3) \qquad\qquad |\int f(x) \, d\alpha(x) - S(f,P)| \leq \epsilon$$

with
$$S(f,P) \equiv \sum_{i=1}^{n} f(a_i)(\alpha(a_i) - \alpha(a_{i-1}))$$

Choose test functions $f_1, \ldots, f_n$ such that (a) $0 \leq f_i \leq 1$, (b) $\sum_{i=1}^{n} f_i(x) = 1$ for all x in I, (c) $f_i(x) = 0$ if $|x - a_i| \geq \omega(\epsilon)$, or if $2 \leq i \leq n$ and $x < a$, or if $1 \leq i \leq n - 1$ and $x > b$, and (d) $|\int f_i \, d\mu - (\alpha(a_i) - \alpha(a_{i-1}))| \leq n^{-1}\epsilon$ $(1 \leq i \leq n)$. This can be done because the points a_i are steady. Define the test function g by

$$g \equiv \sum_{i=1}^{n} f(a_i) f_i$$

Then

$$(3.4) \quad \left| \int g \, d\mu - S(f,P) \right| = \left| \sum_{i=1}^{n} f(a_i) \left[\int f_i \, d\mu - (\alpha(a_i) - \alpha(a_{i-1})) \right] \right|$$
$$\leq n \cdot n^{-1}\epsilon \cdot \|f\| = \epsilon\|f\|$$

If $x < a$, then $f(a_1) = 0$ and $f_i(x) = 0$ for $2 \leq i \leq n$, and thus $g(x) = 0$. Similarly $g(x) = 0$ for $x > b$. Therefore to estimate $|f(x) - g(x)|$ it is enough to consider the case $x \in I$. Since $\sum_{i=1}^{n} f_i(x) = 1$, and $f_i(x) = 0$ whenever $|x - a_i| \geq \omega(\epsilon)$, and $|f(x) - f(a_i)| \leq \epsilon$ whenever $|x - a_i| \leq \omega(\epsilon)$, we have

$$|f(x) - g(x)| = \left| \sum_{i=1}^{n} f_i(x)(f(x) - f(a_i)) \right|$$
$$\leq \sum_{i=1}^{n} f_i(x)\epsilon = \epsilon$$

Thus $\|f - g\| \leq \epsilon$. Since both f and g are supported by $[a,b]$, it follows that

$$(3.5) \qquad |\int f \, d\mu - \int g \, d\mu| \leq c\epsilon$$

where c is a constant that does not depend on ϵ. From (3.3), (3.4), and (3.5) we get

$$|\int f \, d\mu - \int f(x) \, d\alpha(x)| \leq \epsilon(1 + \|f\| + c)$$

Since ϵ is arbitrary, $\int f \, d\mu = \int f(x) \, d\alpha(x)$, as was to be proved.

For a general (not necessarily positive) measure on **R**, Theorem 4 is not valid. There does exist a partial result.

Definition 6   A compact set K in a locally compact space X is said to be *weakly integrable* with respect to a measure μ on X, with $\mu(K) = c$, if there exists a sequence $\{f_n\}$ of test functions with $0 \leq f_n \leq 1$, $f_n(x) = 1$ for all x in K, $f_n(x) = 0$ if $\rho(x,K) \geq n^{-1}$, such that $\int f_n \, d\mu \to c$ as $n \to \infty$.

Lemma 4   *If the compact set $K \subset X$ is weakly integrable relative to the measure μ on X, then $\mu(K)$ is uniquely defined.*

Proof   Let the sequence $\{f_n\}$ and the constant c satisfy the conditions of Definition 6. Let the sequence $\{g_n\}$ and the constant c' also satisfy those conditions. We must show $c = c'$. To this end we choose inductively positive integers $1 = n_1 < n_2 < n_3 < \cdots$ such that $\|h_k\| < 3$ for all k in Z^+, where $h_k \equiv f_{n_1} - g_{n_1} + f_{n_2} - g_{n_2} + \cdots + f_{n_k} - g_{n_k}$. To see that this is possible, assume that $n_1, \ldots, n_{k-1}$ have been chosen. Then h_{k-1} vanishes on K. If n_k is sufficiently large, we shall therefore have $|h_{k-1}(x)| < 1$ whenever $\rho(x,K) \leq 2n_k^{-1}$. Then if x is any point in X, either $\rho(x,K) < 2n_k^{-1}$ and therefore $|h_k(x)| \leq |h_{k-1}(x)| + |f_{n_k}(x)| + |g_{n_k}(x)| < 3$, or $\rho(x,K) > n_k^{-1}$ and therefore $|h_k(x)| = |h_{k-1}(x)| < 3$. Since the functions h_k all have a common support L, the sequence $\{\int h_k \, d\mu\}$ is bounded. Therefore

$$c - c' = \lim_{k \to \infty} \int (f_{n_k} - g_{n_k}) \, d\mu = \lim_{k \to \infty} \int (h_k - h_{k-1}) \, d\mu = 0$$

Theorem 6   *For each measure μ on $\mathbf{R}$ there exists an integrable Borel set B relative to Lebesgue measure on $\mathbf{R}$, of Lebesgue measure 0, such that $[u,v]$ is weakly integrable whenever $u, v \in -B$ and $u < v$.*

Proof   It is enough to show that if I is any compact proper interval, there exists an integrable set $A \subset I$ relative to Lebesgue measure on I, of Lebesgue measure 0, such that $[u,v]$ is weakly integrable whenever $u, v \in I - A$ and $u < v$. We may take $I = [0,1]$. Also, there is no loss of generality in assuming that $|\int f \, d\mu| \leq \|f\|$ whenever f is supported by I.

Consider integers n and k with $0 \leq k \leq n^4 - 1$. Let S_n^k consist of all f in $C(\mathbf{R})$ such that $\|f\| \leq 1$, and $f(x) = 0$ whenever $x \leq kn^{-4}$ or $x \geq (k+1)n^{-4}$, and $|f(x) - f(y)| \leq (2n)^6|x - y|$ for all x and y. Each of the sets S_n^k is a compact subset of $C(I)$, by Theorem 6 of Chap. 4. Therefore the supremum

$$a_n^k \equiv \sup \{\textstyle\int f \, d\mu : f \in S_n^k\}$$

of the continuous function $f \to \int f \, d\mu$ on S_n^k exists. For each n partition the set $\{k : 0 \leq k \leq n^4 - 1\}$ into disjoint subsets U_n and V_n, such that $a_n{}^k > n^{-2}$ whenever $k \in U_n$ and $a_n{}^k < 2n^{-2}$ whenever $k \in V_n$. For each k in U_n choose $f_n{}^k$ in $S_n{}^k$ with

$$\int f_n{}^k \, d\mu > n^{-2}$$

Define
$$f_n \equiv \sum_{k \in U_n} f_n{}^k$$

Then $\|f_n\| \leq 1$, and

$$1 \geq \int f_n \, d\mu > \alpha_n n^{-2}$$

where α_n is the cardinality of U_n. Therefore $\alpha_n < n^2$. Let A_n be the union of the intervals $[kn^{-4}, (k + 1)n^{-4}]$ for k in U_n, and the intervals $[kn^{-4} - n^{-6}, kn^{-4} + n^{-6}] \cap I$ for $0 \leq k \leq n^4$. Then the Lebesgue measure of A_n is at most $n^{-2} + (n^4 + 1)2n^{-6} \leq 5n^{-2}$. It follows that

$$A \equiv \bigcap_{N=1}^{\infty} \bigcup_{n=N}^{\infty} A_n$$

has zero Lebesgue measure.

Consider u in $I - A$. Then $u \in \bigcap_{n=N}^{\infty} (I - A_n)$, for some $N = N(u)$ in Z^+, and thus $u \in I - A_n$ whenever $n \geq N$. For each $n \geq N$, we see that u therefore belongs to an interval

$$I_n \equiv [k(u,n)n^{-4} + n^{-6}, (k(u,n) + 1)n^{-4} - n^{-6}]$$

for which $a_n{}^{k(u,n)} < 2n^{-2}$

Consider points u and v in $I - A$, with $u < v$. For each $n \geq$ max $\{N(u), N(v)\}$ let g_n be the continuous function which has the value 1 at each point of $[u,v]$, which vanishes at each point of $(-\infty, u - n^{-6}] \cup [v + n^{-6}, \infty)$, and which is linear on the intervals $[u - n^{-6}, u]$ and $[v, v + n^{-6}]$. Clearly $|g_n(y) - g_n(z)| \leq n^6|y - z|$ for all y and z. Therefore $g_n - g_{n+1}$ is the sum of functions g' in $S_n{}^{k(u,n)}$ and g'' in $S_n{}^{k(v,n)}$. Therefore

$$\left| \int (g_n - g_{n+1}) \, d\mu \right| \leq a_n{}^{k(u,n)} + a_n{}^{k(v,n)} < 4n^{-2}$$

Thus $\{\int g_n \, d\mu\}$ is a Cauchy sequence, whose limit we call c. The sequence $\{g_n\}$ and the constant c satisfy the conditions of Definition 6, relative to the compact set $[u,v] \subset \mathbf{R}$ and the measure μ on $\mathbf{R}$. Therefore $[u,v]$ is weakly integrable.

An analog of Theorem 4 can now be proved: If μ is a measure on a locally compact space X, and if $h: X \to \mathbf{R}$ is proper, then $\{x: u \leq h(x) \leq v\}$ is compact and weakly integrable for all u and v belonging to the complement $-B$ of a Borel set B in $\mathbf{R}$ of Lebesgue measure 0. The proof is left to the reader.

Theorem 5 has an analog for an arbitrary measure μ on $\mathbf{R}$: There exists a function α of bounded variation defined on the complement of a Borel set of Lebesgue measure 0 such that $\mu = \mu_\alpha$.

It would be interesting to take Theorem 6 as a point of departure for a detailed study of a general (not necessarily positive) measure, if possible. Since this has not been done, further developments will be confined to positive measures.

4. APPROXIMATION BY COMPACT SETS

We shall show that any integrable set can be approximated from within by a compact integrable set. Call a complemented set $B \equiv (B^1, B^2)$ *closed* if B^1 is closed, and *bounded* if $B \subset L$ [that is, $B \subset (L, -L)$] for some compact set L.

Lemma 5 *Let $A \equiv (A^1, A^2)$ be an integrable set of positive measure relative to a positive measure μ on a locally compact space X. Then there exists a closed bounded integrable set $B \equiv (B^1, B^2)$ with $B^1 \subset A^1$ such that $\mu(A) - \mu(B)$ is smaller than any given positive constant.*

Proof Let $\epsilon < \frac{1}{2}$ be a positive constant. Let the test function f approximate A to within $\frac{1}{4}\epsilon^2$ with error sequence $\{f_j\}$. For each n in Z^+ choose $N(n)$ with

$$\sum_{j=N(n)}^{\infty} \int f_j \, d\mu < \epsilon^2 2^{-2n-2}$$

We may assume $N(1) < N(2) < \cdots$. Write $N(0) \equiv 1$. Then the sequence $\{g_j\}$ of test functions defined by

$$g_j \equiv \sum_{k=N(j-1)}^{N(j)-1} f_k \qquad (j \in Z^+)$$

is an error sequence, and $\int g_j \, d\mu < \epsilon^2 2^{-2j}$ for $j \geq 2$, and therefore for all j. By the Corollary to Theorem 4 there exists c in $(\frac{1}{2}\epsilon, \epsilon)$ such that each of the sets B_0 and $-B_j$ is integrable, where

$$B_0 \equiv \{x : f(x) \geq c\}$$

and $\qquad\qquad B_j \equiv \{x : g_j(x) \leq 2^{-j-2}c\} \qquad (j \in Z^+)$

Moreover,

$$\mu(-B_j) \leq c^{-1}2^{j+2}\textstyle\int g_j \, d\mu \leq \epsilon^{-1}2^{j+3}\textstyle\int g_j \, d\mu \leq \epsilon 2^{-j+3} \qquad (j \in Z^+)$$

Therefore the set $B \equiv B_0 - \bigcup_{j=1}^{\infty} - B_j = \bigcap_{j=0}^{\infty} B_j$ is integrable.

For each point x in $-B_0$ we have $f(x) \leq c < \epsilon < \frac{1}{2}$, and thus

$$f(x) < 2f(x)(1 - f(x))$$

Since f^2 approximates A to within ϵ^2, by Lemma 2, and f approximates A to within $\frac{1}{4}\epsilon^2$, we have $\int (f - f^2) \, d\mu \leq \frac{1}{4}\epsilon^2 + \epsilon^2 < \epsilon$, by Theorem 1. By the Corollary to Theorem 4 it follows that

$$\int f \, d\mu - 2\epsilon \leq \int f \, d\mu - 2 \int (f - f^2) \, d\mu \leq \mu(B_0)$$

Therefore $\quad \mu(B) \geq \int f \, d\mu - 2\epsilon - \sum_{j=1}^{\infty} \mu(-B_j) \geq \int f \, d\mu - 10\epsilon$

Since f approximates A to within $\frac{1}{4}\epsilon^2$, we see that $\mu(A) - \mu(B)$ can be made smaller than any given positive constant by taking ϵ sufficiently small. Note that the Borel set $L \equiv B_0 \supset B$ is compact, and B therefore is bounded, and that $B^1 \equiv \bigcap_{j=0}^{\infty} B_j$ is closed, and B therefore is closed.

It remains to show that $B^1 \subset A^1$, or that $x \in A^1$ whenever $x \in B^1$. Consider x in B^1. Then

$$1 - f(x) + \sum_{j=1}^{\infty} g_j(x) \leq 1 - c + \sum_{j=1}^{\infty} 2^{-j-2}c \leq 1 - c + \tfrac{1}{4}c = 1 - \tfrac{3}{4}c$$

Therefore $x \in A^1$.

Our next lemma is interesting in its own right.

Lemma 6   *Let K be a locally compact subset of a locally compact space X, and let $A \equiv (A^1, A^2)$ be a complemented subset of X such that $K \subset A^1$. Then $K \subset A$.*

Proof We must show that each x in $-A$ is in $-K$, or equivalently, that $\rho(x,K) > 0$. Since $x \in -A$, we have $\rho(x,y) > 0$ for all y in A, and therefore for all y in K. The following lemma therefore implies that $\rho(x,K) > 0$.

Lemma 7 *Let K be a complete located subset of a metric space X and x a point in X with $x \neq y$ for all y in K. Then $\rho(x,K) > 0$.*

Proof Define a function $\lambda \colon Z^+ \to \{0,1\}$ such that $\rho(x,K) < n^{-1}$ whenever $\lambda(n) = 1$ and $\rho(x,K) > (2n)^{-1}$ whenever $\lambda(n) = 0$, with $\lambda(1) \geq \lambda(2) \geq \cdots$. Define a sequence $\{y_n\}$ of points of K as follows. Let z be any point of K. In case $\lambda(1) = 0$, write $y_n \equiv z$ for all n. Consider next the case $\lambda(1) = 1$. For each $n > 0$ with $\lambda(n) = 1$, let y_n be any point of K with $\rho(x,y_n) < n^{-1}$. For each n with $\lambda(n) = 0$, let $y_n \equiv y_m$, where m is the greatest integer for which $\lambda(m) = 1$. Then $\{y_m\}$ is a Cauchy sequence. In fact $\rho(y_n,y_m) \leq n^{-1} + m^{-1}$ for all m and n. It therefore converges to a point y in K for which $\rho(y_n,y) \leq n^{-1}$ for all n. Since $x \neq y$, we have $\rho(x,y) > 2n^{-1}$ for some n. Therefore

$$\rho(x,y_n) \geq \rho(x,y) - \rho(y,y_n) > n^{-1}$$

Therefore $\lambda(n) = 0$, and so $\rho(x,K) > (2n)^{-1} > 0$, as was to be proved.

Theorem 7 *Let $A \equiv (A^1, A^2)$ be an integrable set of positive measure relative to a measure μ on a locally compact space X. Then there exists a compact integrable set $K \subset A$ such that $\mu(A) - \mu(K)$ is arbitrarily small.*

Proof Let ϵ be any positive constant. By Lemma 5, there exists an integrable set $B \equiv (B^1, B^2)$ and a compact set L with $B \subset L$ such that B^1 is closed, $B^1 \subset A^1$, $\mu(A) - \mu(B) < \epsilon$, and $\mu(B) > 0$. Pick any x_0 in B. Write $B_0 \equiv B$ and $F_0 \equiv \{x_0\}$. We construct a decreasing sequence $\{B_k \equiv (B_k{}^1, B_k{}^2)\}$ of integrable sets and an increasing sequence $\{F_k\}$ of subfinite sets such that for each k in Z^+ the set $B_k{}^1$ is closed,

$$\mu(B_{k-1}) - \mu(B_k) < 2^{-k}\epsilon$$

and F_k is a k^{-1} approximation to $B_k{}^1$. To this end assume that B_0, $\ldots$, B_{k-1} and $F_0, \ldots, F_{k-1}$ have been constructed. Let the finite subset F of L be a $(4k)^{-1}$ approximation to L. Take c in $((4k)^{-1}, (2k)^{-1})$ so that $Sc(y,c)$ is compact and integrable for all y in F. Then $B_{k-1} \subset L \subset \bigcup_{y \in F} Sc(y,c)$. Write F as a union of sets F^1 and F^2, each of which is either finite or void, with $\mu(B_{k-1} \cap Sc(y,c)) > 0$ for all y in F^1 and $\mu(B_{k-1} \cap Sc(y,c)) < M^{-1}\epsilon 2^{-k}$ for all y in F^2, where M is the number of elements in F. For each y in F^1 there exists $z(y)$ in $B_{k-1} \cap Sc(y,c)$, by Proposition 2. Let F_k be the subfinite set consisting of the points $z(y)$ $(y \in F^1)$ and the points of F_{k-1}. Choose α in $(2c, k^{-1})$ so that the set

$$S \equiv \{x \in L : \rho(x, F_k) \leq \alpha\}$$

is compact and integrable. Then $Sc(y,c) \subset Sc(z(y), 2c) \subset S$ for all y in F^1. The set

$$B_k \equiv B_{k-1} \cap S$$

is integrable, $B_k{}^1$ is a closed set, and F_k is a k^{-1} approximation to B_k. Since $B_{k-1} \subset \bigcup_{y \in F} Sc(y,c)$ and $B_{k-1} \cap \bigcup_{y \in F^1} Sc(y,c) \subset B_k$, we have

$$\mu(B_{k-1}) - \mu(B_k) \leq \mu(B_{k-1} \cap \{\bigcup_{y \in F} Sc(y,c) - \bigcup_{y \in F^1} Sc(y,c)\})$$
$$\leq \sum_{y \in F^2} \mu(B_{k-1} \cap Sc(y,c)) < M(M^{-1} \epsilon 2^{-k}) = \epsilon 2^{-k}$$

This completes the induction.

Write $$K_0 \equiv (K, K') \equiv \bigcap_{k=1}^{\infty} B_k$$

Then $F_k \subset K$ for all k, and F_k is a k^{-1} approximation to K because it is a k^{-1} approximation to $B_k{}^1$. Therefore K is totally bounded. It is closed because each $B_k{}^1$ is closed. It is therefore compact. If $x \in -K$, then $\rho(x, K) > k^{-1}$ for some k in Z^+. This gives $\rho(x, F_k) \geq \rho(x, K) > k^{-1} > \alpha$. Therefore $x \in -S \subset -B_k \subset -K_0$. Thus $-K \subset K'$. Hence $K_0 \subset (K, -K)$. By Lemma 6, we also have $(K, -K) \subset K_0$. Therefore K, as a complemented set, equals K_0.

If $x \in K$, then $x \in B$, and thus $x \in A$. By Lemma 6 again, $K \subset A$. Finally, we compute

$$\mu(K) \geq \mu(B_0) - \sum_{k=1}^{\infty} (\mu(B_{k-1}) - \mu(B_k)) \geq \mu(B) - \sum_{k=1}^{\infty} 2^{-k}\epsilon \geq \mu(A) - 2\epsilon$$

This completes the proof.

The following proposition gives a useful method for proving integrability.

Proposition 3 *A complemented set A is integrable if and only if $A \cap K$ is integrable for each compact integrable set K and*

$$c \equiv \text{l.u.b.} \ \{\mu(A \cap K) \colon K \ \text{is compact and integrable}\}$$

exists, in which case $\mu(A) = c$.

Proof Assume that A is integrable. By Theorem 2, we see that $A \cap K$ is integrable for each compact integrable set K. Clearly $\mu(A) \geq c$. On the other hand, $\mu(A) \leq c$ by Theorem 7.

Assume next that $A \cap K$ is integrable for each compact integrable K, and that c exists. Let x_0 be any point of X. By Theorem 4, we may choose the numbers $a_1 < a_2 < \cdots$, with $a_n \to \infty$ as $n \to \infty$, so that sets

$$X_n \equiv \{x \colon \rho(x,x_0) \leq a_n\} \qquad (n \in Z^+)$$

are compact and integrable. If K is any compact integrable set, we have $K \subset X_n$ for some n. Therefore $c = \lim\limits_{n \to \infty} \mu(A \cap X_n)$. From Theorem 2, it follows that A is integrable and

$$c = \mu(\bigcup_{n=1}^{\infty} (A \cap X_n)) = \mu(A)$$

as was to be proved.

PROBLEMS

1. Give an example of a function of bounded variation on $[0,1]$ that can *not* be written as the difference of two monotone-nondecreasing functions.

2. Show that the complemented sets (respectively, Borel sets) of a locally compact space X are the same as the complemented sets (respectively, Borel sets) relative to the family of all functions $x \to \rho(x,x_0)$ $(x_0 \in X)$.

3. Show that $\int (f - f^2) \, d\mu \leq \epsilon$ whenever the test function f approximates the complemented set A to within ϵ.

4. Let δ be a *fixed* positive constant, with $\delta < \frac{1}{2}$. Show that a complemented set A is integrable if and only if for each $\epsilon > 0$ there exist f and $\{f_j\}$ satisfying the conditions of Definition 5, relative to the given constant δ.

5.   Let A be an integrable set and $\phi: X \to \mathbf{R}$ be continuous and bounded by 1. Show that there exists a constant $\int_A \phi \, d\mu$ such that $\left| \int \phi f \, d\mu - \int_A \phi \, d\mu \right| \leq \epsilon$ whenever the test function f approximates A to within ϵ.

6.   Let μ be a general (not necessarily positive) measure on a locally compact space. For each test function f let $B(f)$ consist of all $c \geq 0$ such that $|\int h \, d\mu| \leq c$ whenever $|h| \leq |f|$. Let g and $f_1, f_2, \ldots$ be test functions, and for each j let c_j be an element of $B(f_j)$ such that $\sum_{j=1}^{\infty} c_j$ converges and is less than $\int g \, d\mu$. Show that there exists x in X and $\epsilon > 0$ such that $\sum_{j=1}^{m} |f_j(x)| \leq |g(x)| - \epsilon$ for all m in $\mathbf{Z}^+$.

7.   Using Prob. 6, extend Definition 5 to general measures, and prove an analog of Theorem 2.

8.   Give an example of a compact subset of $[0,1]$ that is *not* integrable with respect to Lebesgue measure.

9.   Define the *distance* between integrable sets A and B to be $\rho(A,B) \equiv \mu(A - B) + \mu(B - A)$. Call A and B *equal* if $\rho(A,B) = 0$. Show that the integrable sets are a complete metric space.

10.   Construct an integrable subset A of $[0,1]$ relative to Lebesgue measure so that $\mu(I \cap A) > 0$ and $\mu(I - A) > 0$ for every proper subinterval I of $[0,1]$.

11.   Prove that for each integrable set A there exist Borel sets B and C with $B \subset A \subset C$ and $\mu(B) = \mu(A) = \mu(C)$.

12.   Let μ be a positive measure on a compact space X. Let a be a positive constant and n a positive integer with $na > \mu(X)$. Then there exists a subfinite set $S \equiv \{x_1, \ldots, x_n\}$ in X such that for each x in $-S$ there exists $f \geq 0$ in $C(X)$ with $f(x) = 1$ and $\int f \, d\mu < a$. (*Hint:* First consider the case $X \equiv [0,1]$.)

13.   Let μ be a positive measure on X. Show $\mu(\{x\}) = 0$ for all except countably many x in X.

14.   Show that a monotone-nondecreasing function $f: S \to \mathbf{R}$ defined on a dense set $S \subset \mathbf{R}$ is continuous at all except countably many x in S, in the sense that for each $\epsilon > 0$ there exists $\delta > 0$ such that $|f(y) - f(x)| < \epsilon$ whenever $y \in S$ and $|y - x| < \delta$. What can be said when f is of bounded variation rather than monotone?

15.   If μ is a positive measure on X, show that every weakly integrable compact set is integrable.

16.   Prove the analogs of Theorems 4 and 5 for general measures (as stated in the text).

17.   Let A be an integrable subset of $\mathbf{R}$, and ϵ a positive constant. Show that there exists a finite union I of bounded intervals with $\mu(A - I) + \mu(I - A) < \epsilon$.

NOTES

A function of bounded variation is classically defined at every point of the interval I under consideration. Constructively this is usually not possible.

It might seem more natural to have Definition 5 say that f is near to 1 at most points of A and near to 0 at most points of $-A$, in some appropriate sense. Such a definition would not do; the complemented set (ϕ,ϕ) would be approximable by any test function to within any $\epsilon > 0$.

If one approaches the problem of getting a set function from a general (not necessarily positive) measure by the techniques used for a positive measure (as indicated in Probs. 6 and 7), there seems to be no way to construct integrable sets. Therefore the circle of ideas suggested by Theorem 6 seems a more promising approach.

Lemma 5 is, of course, classically equivalent to Theorem 7, almost trivially, but constructively the gap is hard to fill.

It is natural to conjecture that Lemma 7 can be strengthened, to the effect that $\rho(S,K) > 0$ whenever S is a compact subset of a metric space X, K is a complete located subset of X, and $x \neq y$ for all x in S and y in K. In this regard, see the first note to Chap. 5.

Certain parts of measure theory in $\mathbf{R}^2$ have been developed by Brouwer (see Ref. [11]), by a different approach from the one used here.

CHAPTER 7

INTEGRATION

Chapter 6 showed how to define the measure of certain (complemented) sets, starting with a continuous linear functional. In this chapter we reverse the process. First, we define the notion of a measure space, by abstracting the fundamental properties of the set function $A \to \mu(A)$ constructed in Chap. 6. The goal is then to define the class of integrable functions, to define the integral, and to study its properties. For the most part, the theory develops along classical lines. In Sec. 1 we define a measure space, in Sec. 2 we construct the integral, and in Sec. 3 we study its convergence properties. Section 4 concerns the L_p spaces. In Sec. 5 we construct product measures and prove Fubini's theorem.

A measure μ, defined to be a function $f \to \int f\, d\mu$ of a test function f, gives rise to a function $A \to \mu(A)$ of an integrable set A. The set function $A \to \mu(A)$, in turn, gives rise to an *integral*, or a function $f \to \int f\, d\mu$, defined for a class of functions f much wider than the initial class of test functions. Our task is to construct this integral and study its properties.

1. MEASURABLE FUNCTIONS

The essential properties of the measure function $A \to \mu(A)$, constructed in Chap. 6, can be abstracted. This leads to the concept of a measure space.

Definition 1 Let F be a nonvoid family of real-valued functions on a set X, such that $\epsilon - f \in F$ whenever $\epsilon > 0$ and $f \in F$. Let $\mathfrak{F}$ be any family of complemented subsets of X (relative to F), closed with respect to countable unions, countable intersections, and complementation. Let $\mathfrak{M}$ be a subfamily of $\mathfrak{F}$ closed under finite unions, intersections, and differences. Let the function $\mu: \mathfrak{M} \to \mathbf{R}^{0+}$ satisfy the following conditions:

(a) There exists a sequence $S_1 \subset S_2 \subset \cdots$ of elements of $\mathfrak{M}$ such that $\bigcup\limits_{n=1}^{\infty} S_n = X_0$ and $\lim\limits_{n \to \infty} \mu(A \cap S_n) = \mu(A)$ for all A in $\mathfrak{M}$.

(b) If $A \in \mathfrak{F}$, and if there exist B and N in $\mathfrak{M}$ such that (i) $\mu(N) = 0$, (ii) $x \in A$ whenever $x \in B - N$, and (iii) $x \in -A$ whenever $x \in -B - N$, then $A \in \mathfrak{M}$ and $\mu(A) = \mu(B)$.

(c) If $A \in \mathfrak{M}$, if $B \in \mathfrak{F}$, and if $A \cap B \in \mathfrak{M}$, then $A - B \in \mathfrak{M}$ and $\mu(A) = \mu(A - B) + \mu(A \cap B)$.

(d) We have $\mu(A) + \mu(B) = \mu(A \cup B) + \mu(A \cap B)$ for all A and B in $\mathfrak{M}$.

(e) For each sequence $\{A_n\}$ of sets in $\mathfrak{M}$ such that $c \equiv \lim\limits_{n \to \infty} \mu(\bigcup\limits_{k=1}^{n} A_k)$ [respectively, $c \equiv \lim\limits_{n \to \infty} \mu(\bigcap\limits_{k=1}^{n} A_k)$] exists, the set $A \equiv \bigcup\limits_{k=1}^{\infty} A_k$ (respectively, $A \equiv \bigcap\limits_{k=1}^{\infty} A_k$) is in $\mathfrak{M}$, and $\mu(A) = c$.

(f) Each A in $\mathfrak{M}$ with $\mu(A) > 0$ is nonvoid.

Then the quintuple $(X, F, \mathfrak{F}, \mathfrak{M}, \mu)$ is called a *measure space*, μ is the *measure*, $\mathfrak{F}$ is the class of *Borel sets*, and $\mathfrak{M}$ is the class of *integrable sets*.

Definition 1 is justified by the following result.

Proposition 1 *Let μ be a measure on a locally compact space X, let F be the set of all continuous functions on X, let $\mathfrak{F}$ be the complemented sets* *relative to F, and let $\mathfrak{M}$ be the integrable sets. Then $(X,F,\mathfrak{F},\mathfrak{M},\mu)$ is a measure space.*

Proof Conditions (a), (c), (d), (e), and (f) were verified in Chap. 6. To verify (b), consider sets A, N, and B as described.   Let f approximate B to within ϵ, with error sequence $\{f_j\}$. Let g approximate N to within ϵ, with error sequence $\{g_j\}$. Then $\int g\, d\mu \le \mu(N) + \epsilon = \epsilon$. Let $\{h_j\}$ be an ordering of the totality of the functions g, g_j $(1 \le j < \infty)$, and f_j $(1 \le j < \infty)$. To prove (b), it is enough to show that f approximates A to within 3ϵ, with error sequence $\{h_j\}$. Note first that condition (2.1) (a) of Definition 5 of Chap. 6 is satisfied. To check (2.1) (b), assume that

$$f(x) + \sum_{j=1}^{m} h_j(x) < 1 - \delta \text{ for all } m \text{ in } Z^+. \text{ Then } x \in -B \text{ and } x \in -N,$$

or $x \in -B - N$. Hence $x \in -A$. To check (2.1) (c), assume that

$$1 - f(x) + \sum_{j=1}^{m} h_j(x) < 1 - \delta \text{ for all } m \text{ in } Z^+. \text{ Then } x \in B \text{ and } x \in -N,$$

or $x \in B - N$. Hence $x \in A$.

This example admits a slight modification. Again we take μ to be a measure on a locally compact space X, and take F to be the continuous functions on X. We take $\mathfrak{F}$ to be the Borel sets generated by F, and take $\mathfrak{M}$ to be the integrable Borel sets. The same proof shows that $(X,F,\mathfrak{F},\mathfrak{M},\mu)$ is a measure space.

A set $A \in \mathfrak{M}$ for which $\mu(A) = 0$ is called a *null set*.

Here are some consequences of Definition 1.

Proposition 2 *If $A \in \mathfrak{F}$, and if there exists a null set B with $A \subset B$, then A is a null set. The void set ϕ_0 is a null set. If $A \in \mathfrak{F}$ and if there exists B in $\mathfrak{M}$ so that $A - B$ and $B - A$ are null sets, then $A \in \mathfrak{M}$ and $\mu(A) = \mu(B)$.*

Proof If $A \in \mathfrak{F}$ and there exists a null set B with $A \subset B$, apply (b) of Definition 1, with $N \equiv B$, to conclude that $A \in \mathfrak{M}$ and $\mu(A) = 0$. By (a), we have $\phi_0 = -X_0 \in \mathfrak{F}$. Also, $\phi_0 \subset S_1 - S_1$, and $\mu(S_1 - S_1) = 0$ by (c). Hence $\phi_0 \in \mathfrak{M}$ and $\mu(\phi_0) = 0$. If $A \in \mathfrak{F}$ and there exists B in $\mathfrak{M}$ such that $A - B$ and $B - A$ are null sets, apply (b) of Definition 1, with $N \equiv (A - B) \cup (B - A)$, to conclude that $A \in \mathfrak{M}$ and $\mu(A) = \mu(B)$.

Next we lay down terminology for some special kinds of sets.

Definition 2   A Borel set A is *measurable* if $A \cap B \in \mathfrak{M}$ for all B in $\mathfrak{M}$. A Borel set A is a *full set* if $-A$ is a null set. Measurable sets A and B are *disjoint* if $A \subset -B$ (equivalently, $B \subset -A$).

Here are some consequences of Definitions 1 and 2.

Proposition 3   *The complement of a measurable set is measurable. The union, intersection, and difference of two measurable sets are measurable sets. A countable union of null sets is a null set. A countable intersection of full sets is a full set. A Borel subset of a null set is a null set. For each Borel set A we have $A = \bigcup_{n=1}^{\infty} A \cap S_n$. A Borel set A is a null set if $A \cap S_n$ is a null set for each n. For each measurable set A the set $A - A \equiv A \cap -A$ is a null set and the set $A \cup -A$ is a full set. If A and B are disjoint integrable sets, then $\mu(A \cup B) = \mu(A) + \mu(B)$.*

Proof   Let B be any measurable set. Then for every integrable set A the set $A - B$ is integrable by (c) of Definition 1. Therefore $-B$ is measurable. It is obvious that the union and intersection of two measurable sets are measurable sets. Since the complement of a measurable set is measurable, the difference of two measurable sets is measurable.

A countable union of null sets is a null set by (d) and (e) of Definition 1. A countable intersection of full sets is therefore a full set.

A Borel subset of a null set is a null set by Proposition 2.

If A is any Borel set, since $X_0 = \bigcup_{n=1}^{\infty} S_n$ we have $x \in A$ if and only if $x \in \bigcup_{n=1}^{\infty} A \cap S_n$, and $x \in -A$ if and only if $x \in \bigcap_{n=1}^{\infty} (-A \cup -S_n)$. Therefore $A = \bigcup_{n=1}^{\infty} A \cap S_n$. In particular, A is a null set if $A \cap S_n$ is a null set for each n.

If B is any integrable set, then $B - B$ is integrable by (c), and $\mu(B - B) = 0$. If A is any measurable set, then

$$(A - A) \cap S_n = A \cap -A \cap S_n \subset (A \cap S_n) \cap -(A \cap S_n)$$

is therefore a null set for each n. Hence $A - A$ is a null set, and $A \cup -A$ is a full set.

If A and B are disjoint integrable sets, then $A \cap B \subset A \cap -A$ is a null set. Therefore

$$\mu(A \cup B) = \mu(A) + \mu(B) - \mu(A \cap B) = \mu(A) + \mu(B)$$

Definition 3   Let f be a function from a subset D_f of X, called the *domain* of f, to **R**. If there exists a full set $A \equiv (A^1, A^2)$ with $A^1 \subset D_f$, then f is said to be defined *almost everywhere* on X.

Since the domain of the functions $f + g$ and fg is the intersection $D_f \cap D_g$ of the domains of f and g, sums and products of functions defined almost everywhere are defined almost everywhere.

In case f and g are defined almost everywhere on X, we say that $f \leq g$ a.e. if there exists a full set $A \equiv (A^1, A^2)$ with $A^1 \subset D_f \cap D_g$ such that $f(x) \leq g(x)$ for all x in A. In a similar way we define $f = g$ a.e.

For each Borel set $A \equiv (A^1, A^2)$ the *characteristic function* χ_A of A is that function from $A^1 \cup A^2$ to **R** defined by $\chi_A(x) \equiv 1$ for $x \in A^1$ and $\chi_A(x) \equiv 0$ for $x \in A^2$. If A is measurable, then χ_A is defined almost everywhere because $A \cup -A$ is full. For measurable sets A and B we have

$$\chi_{A \cup B} = \chi_A + \chi_B - \chi_A \chi_B \text{ a.e.}$$
$$\chi_{A \cap B} = \chi_A \chi_B \text{ a.e.}$$

and
$$\chi_{A-B} = \chi_A - \chi_A \chi_B \text{ a.e.}$$

Functions f and g are *equal almost everywhere* on a measurable set A (or $f = g$ a.e. on A) if there exists a full set B such that f and g are defined and equal on $A \cap B$. In a similar way we define $f \leq g$ a.e. on A and $f < g$ a.e. on A.

Measurable sets A and B are *equal almost everywhere*, written $A = B$ a.e., if $\chi_A = \chi_B$ a.e. (equivalently, if there exists a full set K such that $x \in K \cap A$ if and only if $x \in K \cap B$, and $x \in K - A$ if and only if $x \in K - B$). This is an equivalence relation on the class of measurable sets. Moreover, integrable sets A and B that are equal almost everywhere have the same measure. To see this, choose the full set K as above. Write $N \equiv -K$. If $x \in B - N = K \cap B$, then $x \in A$, and if $x \in -B - N = K - B$, then $x \in -A$. By (b) of Definition 1, we see that $\mu(A) = \mu(B)$.

If A and B are measurable sets such that $\chi_A \leq \chi_B$ a.e., we say that $A \subset B$ a.e. It is easily seen that $A = B$ a.e. if and only if $A \subset B$ a.e. and $B \subset A$ a.e. A proof similar to the one just given shows that $\mu(A) \leq \mu(B)$ whenever A and B are integrable sets with $A \subset B$ a.e.

Measurable sets A and B are *disjoint almost everywhere* if $A \subset -B$ a.e. (or equivalently, $B \subset -A$ a.e.).

Definition 4   A simple function χ on X is a function which is defined almost everywhere on X and equal almost everywhere to a linear combination $c_1 \chi_1 + \cdots + c_n \chi_n$ of characteristic functions $\chi_1, \ldots, \chi_n$ of integrable sets $A_1, \ldots, A_n$. The expression $c_1\chi_1 + \cdots + c_n\chi_n$ is called a *representation* of the simple function χ.

Sums and products of simple functions are again simple functions, by the above.

Definition 5 A function f defined almost everywhere is *measurable* if for each $\epsilon > 0$ and each integrable set A there exists a simple function χ and an integrable set $B \subset A$ such that $\mu(A - B) \leq \epsilon$ and $|f(x) - \chi(x)| \leq \epsilon$ for all x in B. The function χ is called a *simple ϵ approximation* to f on B.

Since sums and products of simple functions are simple, sums and products of measurable functions are measurable.

The measurable functions turn out to constitute just the right family of functions for most measure-theoretic purposes. A preliminary result in this direction is the following.

Proposition 4 *A continuous function $f: X \to \mathbf{R}$ on a locally compact space X is measurable relative to an arbitrary positive measure μ on X.*

Proof It is sufficient to consider the case $f \geq 0$. Let A be an integrable set and ϵ a positive constant. Let K be a compact integrable set with $K \subset A$ and $\mu(A - K) \leq \epsilon$. Choose constants $c_1 < c_2 < \cdots < c_n$ with $c_1 < \epsilon$, $c_n > \|f\|_K - \epsilon$, and $c_{i+1} - c_i \leq \epsilon$ $(1 \leq i \leq n - 1)$ such that the sets

$$Y_i \equiv \{x \in K : f(x) \geq c_i\} \qquad (1 \leq i \leq n)$$

are integrable. For $1 \leq i \leq n$ let χ_i be the characteristic function of Y_i. Let B be the full set

$$B \equiv (Y_1 \cup -Y_1) \cap \cdots \cap (Y_n \cup -Y_n)$$

and write

$$\chi \equiv c_1\chi_1 + (c_2 - c_1)\chi_2 + \cdots + (c_n - c_{n-1})\chi_n$$

Then $\chi(x) \leq f(x) \leq \chi(x) + \epsilon$ (all x in $B \cap K$). Therefore χ is a simple ϵ approximation to f on $B \cap K$. It follows that f is measurable.

Further progress in the study of measurable functions depends on a simple but crucial combinatorial lemma.

Theorem 1 *Let $A(1), \ldots, A(n)$ be integrable sets and $a_1, \ldots, a_n$ be real numbers. Let P consist of all free subsets S of $\{1, \ldots, n\}$. For each S in P let $A(S)$ be the integrable set*

$$A(S) \equiv \bigcap_{i \in S} A(i) \cap \bigcap_{i \in -S} -A(i)$$

Then the sets $\{A(S) : S \in P\}$ are disjoint, and for each i the union of the sets $\{A(S) : i \in S\}$ equals $A(i)$ almost everywhere. Moreover,

$$(1.1) \qquad \mu(A(i)) = \sum_{i \in S} \mu(A(S)) \qquad (1 \leq i \leq n)$$

and

$$(1.2) \qquad \sum_{i=1}^{n} a_i \chi_{A(i)} = \sum_{S \in P} \Big(\sum_{i \in S} a_i \Big) \chi_{A(S)} \text{ a.e.}$$

Proof If S and T are distinct elements of P, then $A(S) \subset A(i)$ and $A(T) \subset -A(i)$ for some i, so $A(S)$ and $A(T)$ are disjoint. Let x be any point of the full set $B \equiv (A(1) \cup -A(1)) \cap \cdots \cap (A(n) \cup -A(n))$. Then $x \in A(i)$ if and only if $x \in \bigcup_{i \in S} A(S)$, and $x \in -A(i)$ if and only if $x \in \bigcap_{i \in S} -A(S)$. Therefore $A(i)$ equals the union $\bigcup_{i \in S} A(S)$ almost everywhere. Since the sets $A(S)$ are disjoint, this implies (1.1). For each x in B we have

$$\sum_{i=1}^{n} a_i \chi_{A(i)}(x) = \sum_{i=1}^{n} a_i \sum_{i \in S} \chi_{A(S)}(x)$$
$$= \sum_{S \in P} \Big(\sum_{i \in S} a_i \Big) \chi_{A(S)}(x)$$

which implies (1.2).

Corollary 1 *If f and g are simple functions, the functions $|f|$, $\max\{f,g\}$, and $\min\{f,g\}$ are simple functions.*

Proof By Theorem 1, there exist disjoint integrable sets $B_1, \ldots, B_m$ and constants $b_1, \ldots, b_m$ such that

$$f = \sum_{i=1}^{m} b_i \chi_{B_i} \text{ a.e.}$$

Then

$$|f| = \sum_{i=1}^{m} |b_i| \chi_{B_i} \text{ a.e.}$$

Therefore $|f|$ is a simple function. Similarly, $|f - g|$ is a simple function. Hence $(f - g)^+ = \frac{1}{2}(f - g + |f - g|)$ is a simple function. It follows that $\max\{f,g\} = g + (f - g)^+$ and $\min\{f,g\} = -\max\{-f,-g\}$ are simple functions.

Corollary 2 *Absolute values and maxima and minima of measurable functions are measurable functions.*

Proof The proof follows directly from the definition of a measurable function and Corollary 1.

2. THE INTEGRAL

We are now in a position to define the integral.

Theorem 2 *Let*

$$f \equiv \sum_{i=1}^{n} a_i \chi_{A(i)} \text{ a.e.} \qquad g \equiv \sum_{j=1}^{m} b_j \chi_{B(j)} \text{ a.e.}$$

be simple functions, with $f \leq g$ a.e. Then

$$(2.1) \qquad \int f \, d\mu \leq \int g \, d\mu$$

where the integra's $\int f \, d\mu$ and $\int g \, d\mu$ are defined by

$$\int f \, d\mu = \sum_{i=1}^{n} a_i \mu(A(i)) \qquad \int g \, d\mu - \sum_{j=1}^{m} b_j \mu(B(j))$$

Proof By introducing coefficients equal to 0 we reduce to the case $m = n$ and $B(i) = A(i)$ $(1 \leq i \leq n)$. Define the sets $A(S)$ as in Theorem 1. By Theorem 1,

$$(2.2) \qquad \int f \, d\mu = \sum_{i=1}^{n} a_i \left(\sum_{i \in S} \right) \mu(A(S))$$
$$= \sum_{S \in P} \left(\sum_{i \in S} a_i \right) \mu(A(S))$$

and similarly,

$$(2.3) \qquad \int g \, d\mu = \sum_{S \in P} \left(\sum_{i \in S} b_i \right) \mu(A(S))$$

Now, $\qquad \left(\sum_{i \in S} b_i - \sum_{i \in S} a_i \right) \mu(A(S)) \geq 0 \qquad (S \in P)$

because if $\mu(A(S)) > 0$, then by (f) by Definition 1 there exists x in $A(S)$ with

$$\sum_{i \in S} a_i = f(x) \leq g(x) = \sum_{i \in S} b_i$$

This, together with (2.2) and (2.3), implies (2.1).

Corollary *If f and g are simple functions with $f = g$ a.e., then $\int f \, d\mu = \int g \, d\mu$.*

In particular, the integral $\int f \, d\mu$ of a simple function f does not depend on which linear combination of characteristic functions is used to represent it.

Definition 6 A nonnegative measurable function f is *integrable* if

$$c \equiv \text{l.u.b. } \{ \textstyle\int \chi \, d\mu : \chi \text{ is simple and } 0 \leq \chi \leq f \text{ a.e.} \}$$

exists, in which case $c \equiv \int f \, d\mu$ is called the *integral* of f. A measurable function f is *integrable* if both $f^+ \equiv \max \{f,0\}$ and $f^- \equiv -\min \{f,0\}$ are integrable, in which case $\int f \, d\mu \equiv \int f^+ \, d\mu - \int f^- \, d\mu$ is the *integral* of f. If $f \geq 0$, it is convenient to use the notation $\int f \, d\mu = \infty$ to indicate that there are simple functions χ ($0 \leq \chi \leq f$ a.e.) such that $\int \chi \, d\mu$ is arbitrarily large. In the same vein we define $\int f \, d\mu = \infty$ (respectively, $\int f \, d\mu = -\infty$) for an arbitrary measurable function f to mean that f^- is integrable and $\int f^+ \, d\mu = \infty$ (respectively, f^+ is integrable and $\int f^- \, d\mu = \infty$).

In case f is a simple function, the integral of f, as a measurable function, exists, and equals the integral $\int f \, d\mu$ of f as a simple function, since $\int \chi \, d\mu \leq \int f \, d\mu$ whenever χ is a simple function with $\chi \leq f$ a.e.

Another possible conflict in notation arises when f is a test function on a locally compact space X and μ is a measure on X. The following proposition resolves this conflict.

Proposition 5 *Let μ be a measure on a locally compact space X, and f a test function. Then the integral of f as a measurable function, which we temporarily denote by $I(f)$, exists and equals the integral $\int f \, d\mu$ of f as a test function.*

Proof Since it is sufficient to consider f^+ and f^-, we may assume that $f \geq 0$. Let ϵ be any positive constant and f_0 any test function with $0 \leq f_0 \leq 1$ and $f_0 f = f$.

By the proof of Proposition 4, there exists a simple function χ of the form

$$\chi \equiv c_1\chi_1 + (c_2 - c_1)\chi_2 + \cdots + (c_n - c_{n-1})\chi_n$$

with $c_1 < c_2 < \cdots < c_n$, $0 < c_1 < \epsilon$, and $\|f\| - \epsilon < c_n < \|f\|$, where (for $1 \leq i \leq n$) χ_i is the characteristic function of the compact integrable set

$$Y_i = \{x : f(x) \geq c_i\}$$

such that

$$(2.4) \qquad\qquad \chi(x) \leq f(x) \leq \chi(x) + \epsilon f_0$$

for all x in

$$B \equiv (Y_1 \cup -Y_1) \cap \cdots \cap (Y_n \cup -Y_n)$$

By Theorem 4 of Chap. 6, we can even choose the constants c_i ($1 \leq i \leq n$) such that for each i there exist test functions g_i and h_i, with $0 \leq g_i \leq h_i \leq 1$, such that $g_i(x) = 0$ for all x in $-Y_i$, $h_i(x) = 1$ for all x in Y_i, $|\mu(Y_i) - \int g_i \, d\mu| \leq \epsilon$, and $|\mu(Y_i) - \int h_i \, d\mu| \leq \epsilon$. Write

$$g \equiv c_1 g_1 + (c_2 - c_1)g_2 + \cdots + (c_n - c_{n-1})g_n$$
$$h \equiv c_1 h_1 + (c_2 - c_1)h_2 + \cdots + (c_n - c_{n-1})h_n$$

Then $g \leq f \leq h + \epsilon f_0$. Also,

$$|\int \chi \, d\mu - \int g \, d\mu| \leq \epsilon(c_1 + c_2 - c_1 + \cdots + c_n - c_{n-1}) = \epsilon c_n \leq \epsilon\|f\|$$

and

$$|\int \chi \, d\mu - \int h \, d\mu| \leq \epsilon\|f\|$$

Hence

$$(2.5) \qquad\qquad |\int f \, d\mu - \int \chi \, d\mu| < \epsilon(\|f\| + \int f_0 \, d\mu)$$

To show that $\int f \, d\mu$ is an upper bound for the set

$$\Gamma \equiv \{\int \chi' \, d\mu : \chi' \text{ is simple}, 0 \leq \chi' \leq f \text{ a.e.}\}$$

consider any element $\int \chi' \, d\mu$ of Γ. Then $0 \leq \chi' \leq \chi + \epsilon\chi_K$ a.e., by (2.4), where K is any integrable support for f_0. Hence

$$\int \chi' \, d\mu \leq \int \chi \, d\mu + \epsilon\mu(K) \leq \int f \, d\mu + \epsilon(\|f\| + \int f_0 \, d\mu + \mu(K))$$

Since ϵ is arbitrary, it follows that $\int \chi' \, d\mu \leq \int f \, d\mu$. Thus $\int f \, d\mu$ is an upper bound for Γ.

By (2.4) and (2.5), it is in fact the least upper bound. Therefore f is integrable and $I(f) = \int f \, d\mu$.

For further study of the integral we need some lemmas about nonnegative integrable functions. For convenience we introduce the notation $\tilde{\mu}(A) \equiv \mu(A) + 1$.

Lemma 1 *If f and g are nonnegative measurable functions with $f \leq g$ a.e., and if g is integrable, then f is integrable.*

Proof We must prove that for each $\epsilon > 0$ there exists a simple function χ' $(0 \leq \chi' \leq f$ a.e.$)$ such that

$$(2.6) \qquad \int \chi \, d\mu \leq \int \chi' \, d\mu + \epsilon$$

whenever χ is a simple function with $0 \leq \chi \leq f$ a.e. To this end choose a simple function χ^0 $(0 \leq \chi^0 \leq g$ a.e.$)$ with

$$\int \chi^0 \, d\mu \geq \int g \, d\mu - \frac{\epsilon}{4}$$

Let A be an integrable set such that χ^0 vanishes a.e. on $X - A$. Let $c > 0$ be an upper bound for χ^0. Since f is measurable, there exists an integrable subset B of A with $\mu(A - B) \leq c^{-1}\epsilon/4$ and a simple function χ' $(0 \leq \chi' \leq f$ a.e.$)$ which approximates f to within $\tilde{\mu}(A)^{-1}\epsilon/2$ on B. To show that χ' has the desired property, consider any simple function χ with $0 \leq \chi \leq f$ a.e. Then $0 \leq \chi_{-B}\chi + \chi_B\chi^0 \leq g$ a.e. Therefore

$$(2.7) \qquad \int \chi_{-B}\chi \, d\mu \leq \int g \, d\mu - \int \chi_B\chi^0 \, d\mu$$

$$\leq \int \chi^0 \, d\mu + \frac{\epsilon}{4} - \int \chi_B\chi^0 \, d\mu$$

$$= \int \chi_{-B}\chi^0 \, d\mu + \frac{\epsilon}{4} \leq \int c\chi_{A-B} \, d\mu + \frac{\epsilon}{4}$$

$$\leq cc^{-1}\frac{\epsilon}{4} + \frac{\epsilon}{4} = \frac{\epsilon}{2}$$

Since $\chi_B\chi \leq \chi_B f \leq \chi_B(\chi' + \tilde{\mu}(A)^{-1}\epsilon/2)$ a.e., we also have

$$(2.8) \qquad \int \chi_B\chi \, d\mu \leq \int \chi' \, d\mu + \frac{\epsilon}{2}$$

The desired inequality (2.6) now follows from (2.7) and (2.8).

Lemma 2 *The sum $f + g$ of nonnegative integrable functions f and g is integrable, and*

$$\int (f + g)\, d\mu = \int f\, d\mu + \int g\, d\mu$$

Proof We must show that $\int f\, d\mu + \int g\, d\mu$ is the least upper bound of the set

$$\Gamma \equiv \{\textstyle\int \chi\, d\mu : \chi \text{ is a simple function, } 0 \leq \chi \leq f + g \text{ a.e.}\}$$

Consider any element $\int \chi\, d\mu$ of Γ and any $\epsilon > 0$. Let $c > 0$ be a bound for χ, and let A be an integrable set such that χ vanishes almost everywhere on $X - A$. Choose an integrable subset B of A and simple functions χ^1 and χ^2 with $\mu(A - B) \leq c^{-1}\epsilon,\ 0 \leq \chi^1 \leq f$ a.e., $0 \leq \chi^2 \leq g$ a.e., so that χ^1 and χ^2 approximate f and g, respectively, to within $\tilde{\mu}(A)^{-1}\epsilon$ on B. Then

$$\chi_B \chi \leq \chi_B(f + g) \leq \chi_B(\chi^1 + \chi^2 + 2\epsilon\tilde{\mu}(A)^{-1}) \text{ a.e.}$$

Therefore $\quad \int \chi_B \chi\, d\mu \leq \int \chi_B \chi^1\, d\mu + \int \chi_B \chi^2\, d\mu + \mu(B)2\epsilon\tilde{\mu}(A)^{-1}$
$$\leq \int f\, d\mu + \int g\, d\mu + 2\epsilon$$
and $\qquad \int \chi_{-B}\chi\, d\mu = \int \chi_{A-B}\chi\, d\mu \leq c\mu(A - B) \leq \epsilon$

It follows that

$$\int \chi\, d\mu = \int \chi_B \chi\, d\mu + \int \chi_{-B}\chi\, d\mu \leq \int f\, d\mu + \int g\, d\mu + 3\epsilon$$

Since ϵ is any positive constant, we get $\int \chi\, d\mu \leq \int f\, d\mu + \int g\, d\mu$. Thus $\int f\, d\mu + \int g\, d\mu$ is an upper bound for Γ. Now, for an arbitrary $\epsilon > 0$ there exist simple functions χ_f and χ_g with $0 \leq \chi_f \leq f$ a.e. and $0 \leq \chi_g \leq g$ a.e., such that $\int f\, d\mu \leq \int \chi_f\, d\mu + \epsilon$ and $\int g\, d\mu \leq \int \chi_g\, d\mu + \epsilon$, and thus $\int f\, d\mu + \int g\, d\mu \leq \int (\chi_f + \chi_g)\, d\mu + 2\epsilon$ and $0 \leq \chi_f + \chi_g \leq f + g$ a.e.; it is therefore the least upper bound.

Corollary *If $0 \leq h \leq f$ a.e., where f is integrable and h is measurable, then $\int (f - h)\, d\mu = \int f\, d\mu - \int h\, d\mu$.*

Proof By Lemma 1, we see that h and $f - h$ are integrable. By Lemma 2, $\int h\, d\mu + \int (f - h)\, d\mu = \int f\, d\mu$.

Lemma 3 *If f and g are nonnegative integrable functions, then $f - g$ is integrable, and $\int (f - g)\, d\mu = \int f\, d\mu - \int g\, d\mu$.*

Proof Write $h \equiv \min \{f,g\}$. Then

$$(f - g)^+ = f - h \qquad \text{and} \qquad (f - g)^- = g - h$$

By the above corollary, $f - g$ is integrable, and

$$\int (f - g) \, d\mu = \int f \, d\mu - \int h \, d\mu - \left(\int g \, d\mu - \int h \, d\mu\right) = \int f \, d\mu - \int g \, d\mu$$

Theorem 3 *A measurable function f is integrable if and only if $|f|$ is integrable. If f and g are measurable functions with $0 \leq |f| \leq g$, and if g is integrable, so is f. The sum $f + g$ of integrable functions f and g is integrable, and $\int (f + g) \, d\mu = \int f \, d\mu + \int g \, d\mu$. For each a in $\mathbf{R}$ and each integrable function f the function af is integrable, and $\int af \, d\mu = a \int f \, d\mu$.*

Proof Consider any integrable function f. Then f^+ and f^- are integrable. By Lemma 2, so is $|f| = f^+ + f^-$. Conversely, if $|f|$ is integrable, so are f^+ and f^-, by Lemma 1. Consequently so is f.

Consider measurable functions $0 \leq |f| \leq g$ a.e., with g integrable. By Lemma 1, we see that $|f|$ is integrable. Therefore f is integrable.

Consider integrable functions f and g. Then

$$f + g = f^+ + g^+ - (f^- + g^-)$$

is integrable by Lemmas 2 and 3, and

$$\begin{aligned}
\int (f + g) \, d\mu &= \int (f^+ + g^+) \, d\mu - \int (f^- + g^-) \, d\mu \\
&= \int f^+ \, d\mu + \int g^+ \, d\mu - \int f^- \, d\mu - \int g^- \, d\mu \\
&= \int f \, d\mu + \int g \, d\mu
\end{aligned}$$

Consider next an integrable function f and a real number a. If $a > 0$, then af is integrable and $\int af \, d\mu = a \int f \, d\mu$, by Definition 6. In the general case we write $a = b - c$, with $b \equiv 1 + |a| > 0$ and $c \equiv 1 + |a| - a > 0$, and get

$$\int af \, d\mu = \int bf \, d\mu - \int cf \, d\mu = b \int f \, d\mu - c \int f \, d\mu = a \int f \, d\mu$$

3. CONVERGENCE OF INTEGRALS

Proximity in the class of measurable functions is most conveniently described by a notion of convergence. Here are three.

Definition 7 Let $\{f_n\}$ be a sequence of measurable functions, and f a function which is defined almost everywhere. For each integrable set

K and each $\epsilon > 0$ let $\Gamma(K,\epsilon)$ consist of all integrable sets A with $A \subset K$ and $\mu(K - A) \leq \epsilon$ such that the functions f_n and the function f are defined at every point of A. The sequence $\{f_n\}$ converges *almost uniformly* to f if for each integrable set K and each $\epsilon > 0$ there exists A in $\Gamma(K,\epsilon)$ such that $\{f_n\}$ converges uniformly to f on A. The sequence $\{f_n\}$ converges to f *almost everywhere* if for each integrable set K and each $\epsilon > 0$ there exists N in Z^+ and A in $\Gamma(K,\epsilon)$ such that $|f(x) - f_n(x)| \leq \epsilon$ for all x in A and all $n \geq N$. The sequence $\{f_n\}$ converges to f *in measure* if for each integrable set K and each $\epsilon > 0$ there exists N in Z^+ such that for each $n \geq N$ there exists A in $\Gamma(K,\epsilon)$ with $|f(x) - f_n(x)| \leq \epsilon$ for all x in A.

Clearly, almost-uniform convergence implies convergence almost everywhere, and convergence almost everywhere implies convergence in measure.

We next prove measurability of the limit function f, for the case of convergence in measure (and therefore for almost uniform convergence and convergence almost everywhere).

Proposition 6 *The limit in measure f of a sequence $\{f_n\}$ of measurable functions is measurable.*

Proof Let K be any integrable set and ϵ any positive number. By Definition 7, there exists A in $\Gamma(K,\epsilon/2)$ and n in Z^+ such that $|f(x) - f_n(x)| \leq \epsilon/2$ for all x in A. Since A is integrable, there exists an integrable set $B \subset A$ with $\mu(A - B) \leq \epsilon/2$ and a simple function χ which approximates f_n on B to within $\epsilon/2$. Then $\mu(K - B) \leq \epsilon$, and χ approximates f on B to within ϵ. Therefore f is measurable.

If $\{f_n\}$ converges to f and $\{g_n\}$ to g in measure (respectively, almost uniformly or almost everywhere) then $\{af_n + bg_n\}$ converges to $af + bg$ in the same sense (for arbitrary a and b in **R**). This remark enables us to show that limits in measure, and therefore almost-uniform limits and limits almost everywhere, are essentially unique.

Proposition 7 *If $\{f_n\}$ converges in measure to both f and f', then $f = f'$ a.e.*

Proof The sequence $\{h_n\}$, with $h_n = f_n - f_n = 0$ a.e., converges in measure to $f - f'$. This implies that if A is any integrable set and ϵ any positive constant, there exists an integrable set $B \subset A$ with $\mu(A - B) \leq \epsilon$ and $|f - f'| \leq \epsilon$ a.e. on B.

Now choose the sequence $\{S_n\}$ of integrable sets as in (a) of Definition 1. For each $\epsilon > 0$ and each pair (m,n) of positive integers we can choose $B_n{}^m \subset S_n$ with $\mu(S_n - B_n{}^m) \leq m^{-1}$ and $|f - f'| \leq \epsilon$ a.e. on $B_n{}^m$. Thus $|f - f'| \leq \epsilon$ a.e. on $B \equiv \underset{m,n}{\cup} B_n{}^m$. Since $S_n - B$ is a null set for each n, we see that $-B$ is a null set. Thus B is a full set. Hence $|f - f'| \leq \epsilon$ a.e., for each $\epsilon > 0$. Therefore $f = f'$ a.e.

Corresponding to each of the above notions of convergence is a notion of Cauchyness.

Definition 8 Let $\{f_n\}$ be a sequence of measurable functions. Define the classes $\Gamma(K,\epsilon)$ as in Definition 7, but without reference to the function f. The sequence $\{f_n\}$ is *Cauchy almost uniformly* if to each integrable set K and each $\epsilon > 0$ there exists A in $\Gamma(K,\epsilon)$ such that for each $\delta > 0$ there exists N in Z^+ such that $|f_m(x) - f_n(x)| \leq \delta$ for all x in A and m, $n \geq N$. The sequence $\{f_n\}$ is *Cauchy almost everywhere* if to each integrable set K and each $\epsilon > 0$ there exists A in $\Gamma(K,\epsilon)$ and N in Z^+ such that $|f_m(x) - f_n(x)| \leq \epsilon$ for all x in A and all m, $n \geq N$. The sequence $\{f_n\}$ is *Cauchy in measure* if to each integrable set K and each $\epsilon > 0$ there exists N in Z^+ such that for all m, $n \geq N$ there exists A in $\Gamma(K,\epsilon)$ with $|f_m(x) - f_n(x)| \leq \epsilon$ for all x in A.

It is obvious that a sequence $\{f_n\}$ of measurable functions which converges in one of the three possible senses is Cauchy in the corresponding sense. It is also obvious that a sequence which is Cauchy almost uniformly is Cauchy almost everywhere, and that a sequence which is Cauchy almost everywhere is Cauchy in measure.

To see that a sequence which is Cauchy in any of the three senses converges in the corresponding sense, we need a lemma.

Lemma 4 *Let the sequence $\{A_n\}$ of measurable sets have the property that $\mu(K) = \lim_{n \to \infty} \mu(K \cap A_n)$ for every integrable set K. Let the sequence $\{f_n\}$ of measurable functions converge uniformly to a function f:* $\overset{\infty}{\underset{n=1}{\cup}} A_n \to \mathbf{R}$ *on each A_n. Then $\{f_n\}$ converges almost uniformly to f.*

Proof Consider any integrable set K and any $\epsilon > 0$. There exists a value of n for which $\mu(K - K \cap A_n) \leq \epsilon$, by hypothesis. Since $\{f_n\}$ converges uniformly to f on $K \cap A_n$, it follows that $\{f_n\}$ converges almost uniformly to f.

Theorem 4 *A sequence $\{f_n\}$ which is Cauchy almost everywhere converges almost uniformly to some measurable function f. In addition $\{f_n(x)\}$ converges to $f(x)$ at each point x of some full set B. A sequence $\{f_n\}$ which is Cauchy in measure converges in measure to some measurable function f, and some subsequence of $\{f_n\}$ converges almost uniformly to f.*

Proof Let the sequence $\{f_n\}$ of measurable functions be Cauchy almost everywhere. Let the sequence $\{S_n\}$ have the property (a) of Definition 1. For arbitrary positive integers k and n choose $B_n{}^k$ in $\Gamma(S_n, 2^{-k}n^{-1})$ and $N_n{}^k$ in Z^+ such that $|f_i(x) - f_j(x)| \leq 2^{-k}$ whenever $x \in B_n{}^k$ and $i, j \geq N_n{}^k$. Then $A_n \equiv \bigcap_{k=1}^{\infty} B_n{}^k$ is integrable, $\mu(S_n - A_n) \leq n^{-1}$, and $\{f_n\}$ converges uniformly on each of the sets A_n to a function $f: \bigcup_{n=1}^{\infty} A_n \to \mathbf{R}$. By Lemma 4, we see that $\{f_n\}$ converges almost uniformly to f. It also converges pointwise to f on the full set $B \equiv \bigcup_{n=1}^{\infty} A_n$.

Consider next a sequence $\{f_n\}$ of measurable functions which is Cauchy in measure. Let $\{S_n\}$ be the sequence of (a) of Definition 1. We define (by induction on m) for each m in Z^+ a subsequence $\{g_n{}^m\}_{n=1}^{\infty}$ of $\{f_n\}$ such that $\{g_n{}^{m+1}\}_{n=1}^{\infty}$ is a subsequence of $\{g_n{}^m\}$ and such that for each m there exists an integrable set $A_m \subset S_m$, with $\mu(S_m - A_m) \leq m^{-1}$, such that $\{g_n{}^m\}_{n=1}^{\infty}$ converges uniformly on A_m. We prepare the induction by taking $g_n{}^m \equiv f_n$ for $m = 0$. Assume then that $\{g_n{}^{m-1}\}_{n=1}^{\infty}$ has been defined for a given value of $m \geq 1$. For each k in Z^+ choose $N(k)$ in Z^+ so that for each $i, j \geq N(k)$ there exists $B_k(i,j)$ in $\Gamma(S_m, 2^{-k}m^{-1})$ such that $|g_i{}^{m-1}(x) - g_j{}^{m-1}(x)| \leq 2^{-k}$ whenever $x \in B_k(i,j)$. We may assume that $N(1) < N(2) < \cdots$. Then $A_m \equiv \bigcap_{k=1}^{\infty} B_k(N(k), N(k+1))$ is integrable, $\mu(S_m - A_m) \leq m^{-1}$, and the subsequence $\{g_n{}^m\} \equiv \{g_{N(k)}^{m-1}\}_{k=1}^{\infty}$ of $\{g_n{}^{m-1}\}$ converges uniformly on A_m. This completes the induction. The sequence $\{g_n{}^n\}_{n=1}^{\infty}$ is, except for finitely many terms, a subsequence of each of the sequences $\{g_n{}^m\}_{n=1}^{\infty}$. It therefore converges uniformly on each of the sets A_m to a function $f: \bigcup_{m=1}^{\infty} A_m \to \mathbf{R}$. By Lemma 4, we see that $\{g_n{}^n\}$ converges almost uniformly to f. Since $\{g_n{}^n\}$ is a subsequence of $\{f_n\}$, and the latter sequence is Cauchy in measure, $\{f_n\}$ converges to f in measure.

As a corollary to Theorem 4, we see that almost-uniform convergence and convergence almost everywhere are equivalent.

We come now to some results concerning interchange of integrals and limits. The first is Lebesgue's famous *monotone-convergence theorem*.

Theorem 5 *Let $f_1 \leq f_2 \leq \cdots$ be a monotone sequence of integrable functions, such that $L \equiv \lim_{n \to \infty} \int f_n \, d\mu$ exists. Then $\{f_n\}$ converges almost uniformly to an integrable function f, and $\int f \, d\mu = L$. Conversely, if $\{f_n\}$*

converges almost uniformly to an integrable function f, then L exists and $L = \int f \, d\mu$.

Proof　Replacing $\{f_n\}$ with $\{f_n - f_1\}$, if necessary, we reduce to the case in which each f_i is nonnegative.

Assume first that L exists. Consider any integrable set K and any $\epsilon > 0$. Choose the positive integer N so large that

$$L - \int f_N \, d\mu \leq \frac{\epsilon^2}{8}$$

Consider integers $i \geq j \geq N$. Since $f_i - f_j$ is measurable, there exists an integrable set $A \subset K$ with $\mu(K - A) \leq \epsilon/2$ and a simple function χ such that

$$0 \leq \chi \leq \chi_A(f_i - f_j) \leq \chi_A \left(\chi + \frac{\epsilon}{2} \right) \text{ a.e.}$$

Then
$$\int \chi \, d\mu \leq \int (f_i - f_j) \, d\mu \leq L - \int f_N \, d\mu \leq \frac{\epsilon^2}{8}$$

By Theorem 1, there exists a representation

$$\chi = c_1\chi_1 + \cdots + c_n\chi_n \text{ a.e.}$$

where $\chi_1, \ldots, \chi_n$ are characteristic functions of disjoint integrable subsets $A_1, \ldots, A_n$ of A, and $c_1, \ldots, c_n$ are nonnegative constants. A rearrangement of the indices reduces the argument to the case in which there exists an integer k $(0 \leq k \leq n)$ such that $c_i \geq \epsilon/4$ for $1 \leq i \leq k$ and $c_i \leq \epsilon/2$ for $k + 1 \leq i \leq n$.

Then
$$\frac{\epsilon}{4} \left(\mu(A_1) + \cdots + \mu(A_k) \right) \leq \int \chi \, d\mu \leq \frac{\epsilon^2}{8}$$

Therefore $\mu(A_1) + \cdots + \mu(A_k) \leq \epsilon/2$. If we write

$$B \equiv A - A_1 - \cdots - A_k$$

it follows that $\mu(A - B) \leq \epsilon/2$, and consequently $\mu(K - B) \leq \epsilon$. Also, $\chi \leq \epsilon/2$ a.e. on B. Therefore $|f_i - f_j| = f_i - f_j \leq \chi + \epsilon/2 \leq \epsilon$ a.e. on B. It follows that $\{f_n\}$ is Cauchy in measure. By Theorem 4, some subsequence converges almost uniformly to a measurable function f. Since $\{f_n\}$ is monotone, $\{f_n\}$ itself converges almost uniformly to f.

To prove that $\int f \, d\mu = L$, we must show that L is the least upper bound of the set

$$S = \{\textstyle\int \chi \, d\mu : \chi \text{ is simple and } 0 \le \chi \le f \text{ a.e.}\}$$

Consider an element $\int \chi \, d\mu$ of S. Let K be an integrable set such that $\chi_K \chi = \chi$ a.e. For each $\epsilon > 0$ there exists an integrable set $A \subset K$, with $\mu(K - A) \le \epsilon$, and n in Z^+ such that $|f_n - f| \le \epsilon$ a.e. on A. Then $\chi_A(\chi - \epsilon) \le f_n$ a.e. Therefore

$$\begin{aligned}
L \ge \textstyle\int f_n \, d\mu &\ge \textstyle\int \chi_A(\chi - \epsilon) \, d\mu \\
&= \textstyle\int \chi_K(\chi - \epsilon) \, d\mu - \textstyle\int \chi_{K-A}(\chi - \epsilon) \, d\mu \\
&\ge \textstyle\int \chi \, d\mu - \epsilon\mu(K) - \epsilon c
\end{aligned}$$

where c is any bound for χ. Thus $L \ge \int \chi \, d\mu$, and thus L is an upper bound for S. To see that it is the least upper bound, we must find an element $\int \chi \, d\mu$ of S which is arbitrarily near to L. This is done by taking n so large that $\int f_n \, d\mu$ is arbitrarily near to L, and then taking χ with $0 \le \chi \le f_n$ a.e. such that $\int \chi \, d\mu$ is arbitrarily near to $\int f_n \, d\mu$.

Assume next that $\{f_n\}$ converges almost uniformly to the integrable function f. For each $\epsilon > 0$ there exists a simple function χ, with $0 \le \chi \le f$ a.e., such that $\int f \, d\mu - \int \chi \, d\mu \le \epsilon$. Since $\{f_n\}$ converges almost uniformly to f, we see that $\int \chi \, d\mu - \int f_n \, d\mu \le \epsilon$ if n is sufficiently large. Therefore $\int f \, d\mu - \int f_n \, d\mu \le 2\epsilon$ if n is sufficiently large. It follows that L exists and equals $\int f \, d\mu$.

A measurable function f is *integrable* on a measurable set A if $\chi_A f$ is integrable. The number $\displaystyle\int_A f \, d\mu \equiv \int \chi_A f \, d\mu$ is called the *integral* of f on A.

Proposition 8 *Let A_n be a sequence of disjoint integrable sets, such that $A \equiv \bigcup_{n=1}^{\infty} A_n$ is integrable, and $\mu(A) = \sum_{n=1}^{\infty} \mu(A_n)$. Let the measurable function $f \ge 0$ be integrable on each A_n. Then f is integrable on A if and only if $L = \sum_{n=1}^{\infty} \int_{A_n} f \, d\mu$ exists, in which case $\int_A f \, d\mu = L$.*

Proof The series $\sum_{n=1}^{\infty} \chi_{A_n} f$ converges uniformly to $\chi_A f$ on each of the sets $-A, A_1, A_2, \ldots$. By Lemma 4, the series converges almost uniformly to $\chi_A f$. By Theorem 5, the proposition follows.

The next lemma gives additional information about the dependence of $\int_A f \, d\mu$ on A.

Lemma 5 *For each integrable function f and each $\epsilon > 0$ there exists $\delta > 0$ such that $\left| \int_A f \, d\mu \right| \leq \epsilon$ whenever A is integrable and $\mu(A) \leq \delta$.*

Proof It is enough to consider the case $f \geq 0$. Choose n in Z^+ so that $\int (f - f^n) \, d\mu \leq \epsilon/2$, where $f^n \equiv \min \{f, n\}$. Write $\delta \equiv (2n)^{-1}\epsilon$. Then if A is integrable and $\mu(A) \leq \delta$, we have

$$\int_A f \, d\mu \leq \int (f - f^n) \, d\mu + \int_A f^n \, d\mu \leq \frac{\epsilon}{2} + n\delta = \epsilon$$

The next two results shed additional light on the interchange of integrals and limits. The first gives a good condition for the convergence of an integral to 0. The second is Lebesgue's famous *dominated-convergence theorem*.

Proposition 9 *Let the sequence $\{f_n\}$ of nonnegative integrable functions converge in measure to 0. Then $\lim\limits_{n \to \infty} \int f_n \, d\mu = 0$ if and only if for each $\epsilon > 0$ there exists an integrable set K and N in Z^+ such that $\int_A f_n \, d\mu \leq \epsilon$ for all $n \geq N$ and all measurable sets A with $\mu(A \cap K) \leq N^{-1}$.*

Proof The condition is clearly necessary. To show that it is sufficient, for each $\epsilon > 0$ choose K and N as described. Take $n_0 \geq N$ in Z^+ so that for each $n \geq n_0$ there exists an integrable set $B \subset K$ with $\mu(K - B) \leq N^{-1}$ and $|f_n| \leq \epsilon\tilde{\mu}(K)^{-1}$ a.e. on B. Then

$$\int f_n \, d\mu = \int_{-B} f_n \, d\mu + \int_B f_n \, d\mu \leq \epsilon + \epsilon = 2\epsilon$$

whenever $n \geq n_0$. Therefore $\int f_n \, d\mu \to 0$ as $n \to \infty$.

Theorem 6 *Let the sequence $\{f_n\}$ of integrable functions converge in measure to the measurable function f. Let there exist an integrable function g such that $|f_n| \leq g$ a.e. for all n in Z^+. Then f is integrable, and $\int f_n \, d\mu \to \int f \, d\mu$ as $n \to \infty$.*

Proof Since $|f| \leq g$ a.e., we see that f is integrable. To show that $\int f_n \, d\mu \to \int f \, d\mu$ as $n \to \infty$, it is enough to show that $\int |f_n - f| \, d\mu \to 0$

as $n \to \infty$. By Proposition 9, it is enough to find for each $\epsilon > 0$ an integrable set K and N in Z^+ such that $\int_A |f - f_n| \, d\mu \leq \epsilon$ whenever $n \geq N$ and $\mu(A \cap K) \leq N^{-1}$. Since $|f - f_n| \leq 2g$ a.e., we may set $N \equiv 1$ and conclude the existence of K from Lemma 5.

Our next lemma prepares the way for a study of the sets on which the value of a measurable function is less than a given real number.

Lemma 6　*If $f: X \to \mathbf{R}$ is measurable, and if $\varphi: \mathbf{R} \to \mathbf{R}$ is continuous, then $\varphi \circ f$ is measurable.*

Proof　Consider any integrable set K and any $\epsilon > 0$. Choose an integrable set $B \subset K$ with $\mu(K - B) \leq \epsilon/2$ so that $|f|$ is bounded on B by some constant $c > 0$. Let ω be the modulus of continuity of φ on the interval $[-c,c]$. Choose an integrable set $A \subset B$ with $\mu(B - A) \leq \epsilon/2$ and a simple function χ with $|\chi| \leq c$ a.e. and $|f - \chi| \leq \omega(\epsilon)$ a.e. on A. Then $\mu(K - A) \leq \epsilon$ and $\chi_A(\varphi \circ \chi)$ is a simple function, with

$$|\varphi \circ f - \chi_A(\varphi \circ \chi)| \leq \epsilon \text{ a.e. on } A$$

Thus $\varphi \circ f$ is measurable.

It is very convenient to be able to speak of the set A on which the value of a measurable function is less than a given real number, and to know that A is measurable. The theorem which asserts that this is actually the case bears a strong resemblance to Theorem 4 of Chap. 6.

Theorem 7　*Let $f: X \to \mathbf{R}$ be measurable. Then for all except countably many α in $\mathbf{R}$ there exists a measurable set A such that $f < \alpha$ a.e. on A and $f \geq \alpha$ a.e. on $-A$.*

Proof　It is enough to prove the assertion for all except countably many real numbers α in each interval of the form $I \equiv [-a,a]$, with $a > 0$. For this purpose we may replace f with

$$f^a \equiv \max \{\min \{f,a\}, -a\}$$

so that $|f| \leq a$ a.e. Assume first that $X_0 \in \mathfrak{M}$, and that f is therefore integrable. Choose a sequence

$$a \geq \chi^1 \geq \chi^2 \geq \cdots \geq \chi^n \geq \cdots \text{ a.e.}$$

of simple functions converging to f almost uniformly. Write χ^n in the form

$$\chi^n = c_1{}^n \chi_1{}^n + \cdots + c_N{}^n \chi_N{}^n$$

where $\chi_1{}^n, \ldots, \chi_N{}^n$ are the characteristic functions of disjoint integrable sets $A_1{}^n, \ldots, A_N{}^n$, respectively, whose union is a full set. (Of course, N depends on n.) Let ν be the positive measure on I defined by

$$\int h \, d\nu \equiv \int h \circ f \, d\mu \qquad (h \in C(I))$$

All except countably many α in I are steady relative to the measure ν, and satisfy the inequalities $\alpha \neq c_i{}^n$ for all n and i. Consider such a value of α. Since $\alpha \neq c_i{}^n$ for all i, there exists an integrable set A_n such that $\chi^n < \alpha$ a.e. on A_n and $\chi^n > \alpha$ a.e. on $-A_n$. Since $\chi^n \geq \chi^{n+1}$ a.e., $A_n - A_{n+1}$ is a null set for each n. We shall show that

$$(3.1) \qquad \lim_{n \to \infty} \mu(A_n) = \nu(I_\alpha)$$

where $I_\alpha \equiv \{t \in I : t \leq \alpha\}$. To this end, notice first that because α is steady, for each $\epsilon > 0$ there exists a monotone-nonincreasing function h in $C(I)$, with $0 \leq h \leq 1$, and $h(t) = 1$ whenever $t \leq \alpha$, such that $\int h \, d\nu - \nu(I_\alpha) \leq \epsilon$.

Then $\quad \nu(I_\alpha) \geq \int h \, d\nu - \epsilon = \int (h \circ f) \, d\mu - \epsilon \geq \int (h \circ \chi^n) \, d\mu - \epsilon$
$\qquad\qquad \geq \int \chi_{A_n} \, d\mu - \epsilon = \mu(A_n) - \epsilon \qquad (n \in Z^+)$

Therefore

$$(3.2) \qquad \mu(A_n) \leq \nu(I_\alpha) \qquad (n \text{ in } Z^+)$$

Notice next that for each $\epsilon > 0$ there exists h in $C(I)$ with $0 \leq h \leq 1$ and $h(t) = 0$ if $t \geq \alpha$, so that $\nu(I_\alpha) - \int h \, d\nu \leq \epsilon$. Now $\{h \circ \chi^n\}$ converges to $h \circ f$ almost uniformly. By Theorem 5, we have $\int (h \circ \chi^n) \, d\mu \to \int (h \circ f) \, d\mu$ as $n \to \infty$. Therefore we can choose N so large that $\int (h \circ \chi^n) \, d\mu \geq \int (h \circ f) \, d\mu - \epsilon$ for all $n \geq N$. Then

$$(3.3) \quad \nu(I_\alpha) \leq \int h \, d\nu + \epsilon = \int (h \circ f) \, d\mu + \epsilon \leq \int (h \circ \chi^n) \, d\mu + 2\epsilon$$
$$\leq \int \chi_{A_n} \, d\mu + 2\epsilon = \mu(A_n) + 2\epsilon \qquad (n \geq N)$$

Inequalities (3.2) and (3.3) imply (3.1). The set $A \equiv \bigcup_{n=1}^{\infty} A_n$ is integrable by (3.1) and the fact that $A_n - A_{n+1}$ is a null set for each n. Clearly $f < \alpha$ a.e. on A and $f \geq \alpha$ a.e. on $-A$.

Next consider the general case. Let the sequence $\{S_n\}$ of integrable sets have property (a) of Definition 1. Write $B_1 \equiv S_1$ and

$$B_n \equiv S_n - S_1 - S_2 - \cdots - S_{n-1}$$

for $n \geq 2$. By the case just considered, for each n there exists an integrable subset A_n of B_n such that $f < \alpha$ a.e. on A_n and $f \geq \alpha$ a.e. on $B_n - A_n$. Therefore the measurable set $A \equiv \bigcup_{n=1}^{\infty} A_n$ has the property that $f < \alpha$ a.e. on A and $f \geq \alpha$ a.e. on $-A$.

Corollary *A nonnegative integrable function f whose integral is 0 vanishes almost everywhere.*

Proof There exist arbitrarily small positive constants α such that a measurable set A exists with $f < \alpha$ a.e. on A and $f \geq \alpha$ a.e. on $-A$. Thus $f \geq \alpha \, \chi_{-A}$ a.e. It follows that $-A$ is a null set, and thus $f < \alpha$ a.e. Therefore $f = 0$ a.e.

4. THE L_p SPACES

Our next goal is to introduce certain basic metric spaces of measurable functions. This requires some preliminary inequalities.

Lemma 7 *Let a, b, α, β be real numbers with a, $b \geq 0$; α, $\beta > 0$; and $\alpha + \beta = 1$. Then*

$$(4.1) \qquad\qquad \alpha a + \beta b - a^\alpha b^\beta \geq 0$$

Proof Consider $\alpha a + \beta b - a^\alpha b^\beta$ as a function f of a for fixed values of b, α, and β. Then

$$f'(a) = \alpha - \alpha a^{\alpha-1} b^\beta = \alpha(1 - (ba^{-1})^\beta)$$

if $a \neq 0$. Thus $f'(a) \geq 0$ if $a > b$, and $f'(a) \leq 0$ if $0 < a \leq b$. Since $f(b) = 0$, it follows that $f(a) \geq 0$ whenever $a > 0$. Thus (4.1) is valid whenever $a > 0$. By continuity, (4.1) is valid for $a \geq 0$.

Successive applications of Lemma 7 show that if $\alpha_1, \ldots, \alpha_n$ are positive constants whose sum is 1 and if $a_1, \ldots, a_n \geq 0$, then

$$\alpha_1 a_1 + \cdots + \alpha_n a_n - a_1^{\alpha_1} \cdots a_n^{\alpha_n} \geq 0$$

Theorem 8　*Let $f_1, \ldots, f_n$ be nonnegative integrable functions and $\alpha_1, \ldots, \alpha_n$ positive constants whose sum is 1. Then $f_1^{\alpha_1} \cdots f_n^{\alpha_n}$ is integrable, and*

$$(4.2) \qquad \int f_1^{\alpha_1} \cdots f_n^{\alpha_n} \, d\mu \leq \left(\int f_1 \, d\mu\right)^{\alpha_1} \cdots \left(\int f_n \, d\mu\right)^{\alpha_n}$$

Proof　Let $c_1, \ldots, c_n$ be positive numbers with $c_i \geq \int f_i \, d\mu$ ($1 \leq i \leq n$). By (4.1),

$$\int (c_1^{-1} f_1)^{\alpha_1} \cdots (c_n^{-1} f_n)^{\alpha_n} \, d\mu \leq \int (\alpha_1 c_1^{-1} f_1 + \cdots + \alpha_n c_n^{-1} f_n) \, d\mu$$
$$= \alpha_1 c_1^{-1} \int f_1 \, d\mu + \cdots + \alpha_n c_n^{-1} \int f_n \, d\mu$$
$$\leq \alpha_1 + \cdots + \alpha_n = 1$$

or
$$\int f_1^{\alpha_1} \cdots f_n^{\alpha_n} \, d\mu \leq c_1^{\alpha_1} \cdots c_n^{\alpha_n}$$

Letting $c_i \to \int f_i \, d\mu$ ($1 \leq i \leq n$) gives (4.2).

Theorem 8 is called *Hölder's inequality*. It is basic in the theory of certain spaces—the L_p spaces—which we now define.

Definition 9　For each $p \geq 1$ the set L_p consists of all measurable functions f such that $|f|^p$ is integrable. The equality relation on L_p is equality almost everywhere. The *norm* $\|f\|_p$ of an element f of L_p is

$$\|f\|_p \equiv \left(\int |f|^p \, d\mu\right)^{1/p}$$

We now derive a remarkable inequality between certain L_p spaces, which is actually a reformulation of Hölder's inequality.

Proposition 10　*Let $p > 1$ and $q > 1$ satisfy the equation*

$$p^{-1} + q^{-1} = 1$$

Let $f \in L_p$ and $g \in L_q$. Then $fg \in L_1$ and

$$\int fg \, d\mu \leq \|f\|_p \|g\|_q$$

Proof　Application of Hölder's inequality, with $\alpha_1 \equiv p^{-1}$ and $\alpha_2 \equiv q^{-1}$, to the integrable functions $|f|^p$ and $|g|^q$, gives

$$\int fg \, d\mu \leq \int |f| \, |g| \, d\mu = \int |f|^{p\alpha_1} |g|^{q\alpha_2} \, d\mu$$
$$\leq \left(\int |f|^p \, d\mu\right)^{\alpha_1} \left(\int |g|^q \, d\mu\right)^{\alpha_2} = \|f\|_p \|g\|_q$$

Hölder's inequality leads to another remarkable inequality, known as *Minkowski's inequality*.

Theorem 9 *Let f and g be in L_p ($p \geq 1$). Then $f + g \in L_p$, and*

$$(4.3) \qquad \|f + g\|_p \leq \|f\|_p + \|g\|_p$$

Proof Since

$$|f + g|^p \leq (2 \max \{|f|, |g|\})^p \leq 2^p(|f|^p + |g|^p)$$

it follows that $f + g \in L_p$. Let α be a constant ($0 < \alpha < p^{-1}$). Then $p\alpha < 1$, and thus $|f + g|^{p\alpha} \leq |f|^{p\alpha} + |g|^{p\alpha}$. Using Hölder's inequality, we compute

$$(4.4) \quad \begin{aligned} \|f + g\|_p^p &= \int |f + g|^p \, d\mu = \int |f + g|^{p\alpha} |f + g|^{p(1-\alpha)} \, d\mu \\ &\leq \int |f|^{p\alpha} |f + g|^{p(1-\alpha)} \, d\mu + \int |g|^{p\alpha} |f + g|^{p(1-\alpha)} \, d\mu \\ &\leq \{(\int |f|^p \, d\mu)^\alpha + (\int |g|^p \, d\mu)^\alpha\} (\int |f + g|^p \, d\mu)^{1-\alpha} \\ &= (\|f\|_p^{p\alpha} + \|g\|_p^{p\alpha}) \|f + g\|_p^{p(1-\alpha)} \end{aligned}$$

Now for each $\epsilon > 0$ either $\|f + g\|_p \leq \epsilon$ or $\|f + g\|_p > 0$. In the first case,

$$(4.5) \qquad \|f + g\|_p \leq \epsilon + \|f\|_p + \|g\|_p$$

In the second case, dividing both sides of (4.4) by $\|f + g\|_p^{p(1-\alpha)}$ and then letting $\alpha \to p^{-1}$ gives $\|f + g\|_p \leq \|f\|_p + \|g\|_p$. Therefore (4.5) holds in all cases. Letting $\epsilon \to 0$ gives (4.3).

Minkowski's inequality shows that L_p is a metric space. In fact more is true.

Theorem 10 *Under the metric*

$$\rho(f,g) \equiv \|f - g\|_p$$

L_p is a complete metric space.

Proof By the Corollary to Theorem 7, we see that $\rho(f,g) = 0$ if and only if $f = g$ a.e. Clearly $\rho(f,g) = \rho(g,f)$. By Minkowski's inequality,

$$\rho(f,g) = \|f - g\|_p \leq \|f - h\|_p + \|g - h\|_p = \rho(f,h) + \rho(g,h)$$

Thus ρ is a metric on L_p. To show that L_p is complete, consider a Cauchy sequence $\{f_n\}$ in L_p. For each $\epsilon > 0$ there exists N in Z^+ such that $\int |f_n - f_m|^p \, d\mu \leq \epsilon^{1+p}$ whenever $m, n \geq N$. Consider integers $m \geq$

$n \geq N$. By Theorem 7, there exists α in $(\epsilon, 2\epsilon)$ and a measurable set A such that $|f_n - f_m| < \alpha$ a.e. on A and $|f_n - f_m| \geq \alpha$ a.e. on $-A$. Since $|f_n - f_m|^p$ is integrable, $-A$ is an integrable set, and

$$\epsilon^{1+p} \geq \int |f_n - f_m|^p \, d\mu \geq \epsilon^p \mu(-A)$$

or $\mu(-A) \leq \epsilon$. It follows that $\{f_n\}$ is Cauchy in measure. Thus, by passing to a subsequence, we may assume that $\{f_n\}$ converges almost uniformly to a measurable function f.

By passing to another subsequence, we may assume that $\|f_n - f_{n+1}\|_p \leq 2^{-n}$ for all n. Write

$$g_n \equiv |f_1| + |f_1 - f_2| + \cdots + |f_{n-1} - f_n| \qquad (n \in Z^+)$$

For $m \geq n$, we have

$$\|g_m - g_n\|_p \leq \|f_m - f_{m-1}\|_p + \cdots + \|f_{n+1} - f_n\|_p \leq 2^{-n+1}$$

Therefore $\{g_n\}$ is a Cauchy sequence in L_p. and thus some subsequence $\{h_n\}$ converges almost uniformly to a measurable function h. Hence $\{g_n{}^p\}$ converges monotonely almost uniformly to h^p. By Theorem 5, we see that h^p is integrable. Since $|f_n|^p \leq g_n{}^p \leq h^p$ for all n, we have $|f|^p \leq h^p$. Therefore $|f|^p$ is integrable. Since

$$|f - f_n|^p \leq (2 \max \{|f|, |f_n|\})^p \leq (2h)^p$$

we have $\|f - f_n\|_p \to 0$ as $n \to \infty$, by Theorem 6. In other words, $\{f_n\}$ converges to f in L_p. Thus L_p is complete.

5. PRODUCT MEASURES

We shall construct the product of two measure spaces and study the resulting product measure. The first step is to define the product of complemented sets.

Definition 10   Let X_1 and X_2 be sets, and F_1 and F_2 be families of real-valued functions on X_1 and X_2, respectively. Then the family F consists of all $f : X \equiv X_1 \times X_2 \to \mathbf{R}$ of the form $f \equiv f_\alpha \circ \pi_\alpha$, where $\alpha = 1$ or 2, where $f_\alpha \in F_\alpha$, and where π_α is the projection of X onto X_α. For each complemented set $A_1 \equiv (A_1{}^1, A_1{}^2)$ in X_1 and $A_2 \equiv (A_2{}^1, A_2{}^2)$ in X_2 we define the complemented set $A_1 \times A_2$ in $X_1 \times X_2$ by

$$A_1 \times A_2 \equiv (A_1{}^1 \times A_2{}^1, \ A_1{}^2 \times X_2 \cup X_1 \times A_2{}^2)$$

Such a complemented set is called a *rectangle*.

It is clear that the rectangles actually are complemented sets (relative to F).

Each complemented set A in X has *sections*, defined as follows. Let x_1 be a point of X_1. The complemented set $A(x_1,2)$ in X_2, called the x_1 *section of* A, is defined by

$$A(x_1,2) \equiv (\{x_2 \in X_2 : (x_1,x_2) \in A\}, \{x_2 \in X_2 : (x_1,x_2) \in -A\})$$

The following algebraic laws are satisfied.

$$(-A)(x_1,2) = -A(x_1,2)$$
$$(\bigcup_{n=1}^{\infty} A_n)(x_1,2) = \bigcup_{n=1}^{\infty} A_n(x_1,2)$$
$$(\bigcap_{n=1}^{\infty} A_n)(x_1,2) = \bigcap_{n=1}^{\infty} A_n(x_1,2)$$

Moreover, for each rectangle $A_1 \times A_2$ we have

$$(A_1 \times A_2)(x_1,2) = A_2 \qquad (x_1 \in A_1)$$

and
$$(A_1 \times A_2)(x_1,2) = \phi_0 \qquad (x_1 \in -A_1)$$

where ϕ_0 is the void complemented set in X_2.

For each x_2 in X_2 the section $A(x_2,1)$ is defined similarly. It is a complemented set in X_1, and the same algebraic laws are satisfied.

We now have enough machinery to begin defining the product of two measure spaces.

Definition 11 Let $(X_\alpha, F_\alpha, \mathfrak{I}_\alpha, \mathfrak{M}_\alpha, \mu_\alpha)$ be measure spaces, for $\alpha = 1$ and $\alpha = 2$. Define X and F as in Definition 10. Call a rectangle $A_1 \times A_2$ *integrable* if $A_1 \in \mathfrak{M}_1$ and $A_2 \in \mathfrak{M}_2$. Let $\mathfrak{F}$ consist of all complemented sets in X (relative to F). Let $\mathfrak{M}$ consist of all A in $\mathfrak{F}$ such that for almost all x in X_α ($\alpha = 1$ or 2) the section $A(x,\beta)$ belongs to $\mathfrak{M}_\beta$ (where $\beta = 2$ if $\alpha = 1$, and $\beta = 1$ if $\alpha = 2$), and such that for each k in Z^+ there exists a finite union B_k of integrable rectangles which *approximates* A *to within* k^{-1}. This means that there exist functions $\omega_1{}^k$ in $L_1(\mu_1)$ and $\omega_2{}^k$ in $L_1(\mu_2)$ such that

$$(5.1) \qquad \|\omega_1{}^k\|_1 \leq k^{-1} \qquad \|\omega_2{}^k\|_1 \leq k^{-1}$$

and

$$(5.2) \qquad \mu_\beta([A - B_k](x,\beta)) + \mu_\beta([B_k - A](x,\beta)) \leq \omega_\alpha{}^k(x)$$

for almost all x in X_α ($\alpha = 1$ or 2).

For each A in $\mathfrak{M}$ and each α ($\alpha = 1$ or 2) the function $\lambda(A,\alpha)$ whose value at x is

$$\lambda(A,x,\alpha) \equiv \mu_\beta(A(x,\beta))$$

is defined almost everywhere on X_α. To study these functions we need some lemmas.

Lemma 8 *Let*

$$(5.3) \qquad\qquad B \equiv A_{11} \times A_{21} \cup \cdots \cup A_{1n} \times A_{2n}$$

with $A_{1i} \in \mathfrak{M}_1$ and $A_{2i} \in \mathfrak{M}_2$ ($1 \le i \le n$) be a finite union of integrable rectangles. Then there exists a finite union B_0 of disjoint integrable rectangles such that $B(x,\beta) = B_0(x,\beta)$ a.e. for almost all x in $X_\alpha(\alpha = 1$ or $2)$.

Proof By Theorem 1, there exists a finite family $\mathfrak{N}_1$ of disjoint elements of $\mathfrak{M}_1$ and a finite family $\mathfrak{N}_2$ of disjoint elements of $\mathfrak{M}_2$ such that each of the sets A_{1i} (respectively, A_{2i}) is almost everywhere equal to the union of some finite subfamily of $\mathfrak{N}_1$ (respectively, $\mathfrak{N}_2$). Therefore the union B_0 of some finite subfamily of the finite family $\{C_1 \times C_2 : C_1 \in \mathfrak{N}_1, C_2 \in \mathfrak{N}_2\}$ of disjoint integrable rectangles has the desired property.

Lemma 9 *Each finite union B of integrable rectangles belongs to $\mathfrak{M}$. Moreover, the functions $\lambda(B,1)$ and $\lambda(B,2)$ are integrable, and*

$$(5.4) \qquad\qquad \textstyle\int \lambda(B,1)\, d\mu_1 = \int \lambda(B,2)\, d\mu_2$$

Proof That $B \in \mathfrak{M}$ is an immediate consequence of Definition 11. By Lemma 8, we may take B to have the form (5.3), where the rectangles $A_{1i} \times A_{2i}$ ($1 \le i \le n$) are mutually disjoint. Then

$$\lambda(B,\alpha) = \sum_{i=1}^{n} \mu_\beta(A_{\beta i})\chi_{A_{\alpha i}} \text{ a.e.} \qquad (\alpha = 1 \text{ or } 2)$$

Therefore $\lambda(B,\alpha)$ is integrable and

$$\int \lambda(B,\alpha)\, d\mu_\alpha = \sum_{i=1}^{n} \mu_\beta(A_{\beta i})\mu_\alpha(A_{\alpha i}) = \int \lambda(B,\beta)\, d\mu_\beta$$

Lemma 10 *For each A in $\mathfrak{M}$ the functions $\lambda(A,\alpha)$ ($\alpha = 1$ or 2) are integrable, and*

$$\textstyle\int \lambda(A,1)\, d\mu_1 = \int \lambda(A,2)\, d\mu_2$$

Proof For each k in Z^+ and each α ($\alpha = 1$ or 2) choose B_k and $\omega_\alpha{}^k$ as in Definition 11. For almost all x in X_α,

$$(5.5) \quad |\lambda(A,x,\alpha) - \lambda(B_k,x,\alpha)| = |\mu_\beta(A(x,\beta)) - \mu_\beta(B_k(x,\beta))| \leq \omega_\alpha{}^k(x)$$

Thus $\int |\lambda(B_j,\alpha) - \lambda(B_k,\alpha)| \, d\mu_\alpha \leq \int (\omega_\alpha{}^j + \omega_\alpha{}^k) \, d\mu_\alpha \leq j^{-1} + k^{-1}$

for all j and k, and thus $\{\lambda(B_k,\alpha)\}$ is a Cauchy sequence in $L_1(\mu_\alpha)$, whose limit we call λ^α. Since $\{\omega_\alpha{}^k\}$ converges to 0 in $L_1(\mu_\alpha)$, some subsequence converges to 0 almost everywhere. From (5.5) it follows that $\lambda^\alpha = \lambda(A,\alpha)$ a.e. Thus $\lambda(A,\alpha)$ is integrable and

$$\int \lambda(A,1) \, d\mu_1 = \lim_{k\to\infty} \int \lambda(B_k,1) \, d\mu_1 = \lim_{k\to\infty} \int \lambda(B_k,2) \, d\mu_2 = \int \lambda(A,2) \, d\mu_2$$

Theorem 11 *For each A in $\mathfrak{M}$ write*

$$\mu(A) \equiv \int \lambda(A,1) \, d\mu_1 = \int \lambda(A,2) \, d\mu_2$$

Then $(X,F,\mathfrak{F},\mathfrak{M},\mu)$ is a measure space, called the product *of $(X_1,F_1,\mathfrak{F}_1, \mathfrak{M}_1,\mu_1)$ and $(X_2,F_2,\mathfrak{F}_2,\mathfrak{M}_2,\mu_2)$.*

Proof If $A \equiv A_1 \times A_2$ and $B \equiv B_1 \times B_2$ are integrable rectangles, then simple computations prove that $A \cap B = (A_1 \cap B_1) \times (A_2 \cap B_2)$ and $A - B = \{(A_1 - B_1) \times A_2\} \cup \{A_1 \times (A_2 - B_2)\}$. Thus the class of all finite unions of integrable rectangles is closed under finite unions, intersections, and differences. By Definition 11, we see that $\mathfrak{M}$ is closed under finite unions, intersections, and differences. We must show that conditions (a) to (f) of Definition 1 are satisfied

To check (b), consider A in $\mathfrak{F}$, B in $\mathfrak{M}$, and N in $\mathfrak{M}$, as described. Then $\mu_\beta(N(x,\beta)) = \lambda(N,x,\alpha) = 0$ for almost all x in X_α. Applying (b) to the sets $A(x,\beta) \in \mathfrak{F}_\beta$, $B(x,\beta) \in \mathfrak{M}_\beta$, and $N(x,\beta) \in \mathfrak{M}_\beta$, we see that $A(x,\beta) \in \mathfrak{M}_\beta$ and $\mu(A(x,\beta)) = \mu(B(x,\beta))$, for almost all x in X_α. Moreover, if a finite union of integrable rectangles approximates B to within k^{-1}, it does the same for A. Thus $A \in \mathfrak{M}$ and $\mu(A) = \mu(B)$.

To check (c), consider sets A in $\mathfrak{M}$ and B in $\mathfrak{F}$, with $A \cap B \in \mathfrak{M}$. If B_1 and B_2 are finite unions of integrable rectangles, such that B_1 approximates A to within $(2k)^{-1}$ and B_2 approximates $A \cap B$ to within $(2k)^{-1}$, then $B_1 - B_2$ approximates $A - B$ to within k^{-1}. Thus $A - B$ is integrable. Moreover,

$$\begin{aligned}
\mu(A) &= \int \lambda(A,1) \, d\mu_1 = \int \mu_2(A(x,2)) \, d\mu_1(x) \\
&= \int \{\mu_2(A(x,2) - B(x,2)) + \mu_2(A(x,2) \cap B(x,2))\} \, d\mu_1(x) \\
&= \int \{\lambda(A - B, x, 1) + \lambda(A \cap B, x, 1)\} \, d\mu_1(x) \\
&= \mu(A - B) + \mu(A \cap B)
\end{aligned}$$

To check (d), consider A and B in $\mathfrak{M}$. We compute

$$\begin{aligned}
\mu(A) + \mu(B) &= \int \{\mu_2(A(x,2)) + \mu_2(B(x,2))\}\, d\mu_1(x) \\
&= \int \{\mu_2(A(x,2) \cup B(x,2)) + \mu_2(A(x,2) \cap B(x,2))\}\, d\mu_1(x) \\
&= \mu(A \cup B) + \mu(A \cap B)
\end{aligned}$$

To check (e), consider a sequence $\{A_n\}$ of sets in $\mathfrak{M}$ such that $c \equiv \lim_{n \to \infty} \mu(\bigcup_{k=1}^{n} A_k)$ exists. [The case $c \equiv \lim_{n \to \infty} \mu(\bigcap_{k=1}^{n} A_k)$ is similar.] Write $A \equiv \bigcup_{n=1}^{\infty} A_n$. We may assume $A_1 \subset A_2 \subset \cdots$. Then $\{\lambda(A_n,\alpha)\}$ is a nondecreasing sequence of integrable functions on X_α whose integrals converge to c. By Lebesgue's monotone-convergence theorem, this sequence converges almost everywhere to an integrable function λ^α. Since $\lambda(A_n,x,\alpha) = \mu_\beta(A_n(x,\beta))$, and since $A(x,\beta) = \bigcup_{n=1}^{\infty} A_n(x,\beta)$, it follows that $A(x,\beta)$ is integrable for almost all x in X_α, and

$$\lambda(A,x,\alpha) \equiv \mu_\beta(A(x,\beta)) = \lambda^\alpha(x)$$

Therefore

$$\int \lambda(A,\alpha)\, d\mu_\alpha = c$$

Thus to check (e) we need only show that for each k in Z^+ there exists a finite union B of integrable rectangles which approximates A to within k^{-1}.

To this end choose n so large that $c - \mu(A_n) \leq (2k)^{-1}$, and take a finite union B of integrable rectangles which approximates A_n to within $(2k)^{-1}$. Choose ω_α $(\alpha = 1$ or $2)$ to satisfy (5.1) and (5.2) relative to the sets A_n and B, and the integer $2k$. Then

$$\begin{aligned}
\mu_\beta(A(x,\beta) &- B(x,\beta)) + \mu_\beta(B(x,\beta) - A(x,\beta)) \\
&\leq \mu_\beta(A(x,\beta) - A_n(x,\beta)) + \mu_\beta(A_n(x,\beta) - B(x,\beta)) \\
&\qquad\qquad\qquad\qquad\qquad + \mu_\beta(B(x,\beta) - A_n(x,\beta)) \\
&\leq \mu_\beta(A(x,\beta)) - \mu_\beta(A_n(x,\beta)) + \omega_\alpha(x) \\
&= \lambda(A,x,\alpha) - \lambda(A_n,x,\alpha) + \omega_\alpha(x) \quad \text{a.e.}
\end{aligned}$$

Since

$$\int \{\lambda(A,x,\alpha) - \lambda(A_n,x,\alpha) + \omega_\alpha(x)\}\, d\mu_\alpha(x) \leq (2k)^{-1} + (2k)^{-1} = k^{-1}$$

it follows that B approximates A to within k^{-1}.

To check (f), consider any A in $\mathfrak{M}$ with $\mu(A) > 0$. By the definition of $\mu(A)$, there exists x_1 in X_1 such that $A(x_1,2)$ is in $\mathfrak{M}_2$ and $\mu_2(A(x_1,2))$

> 0. By condition (f) for the measure μ_2, there exists x_2 in $A(x_1,2)$. Hence $(x_1,x_2) \in A$, and therefore A is nonvoid.

To check (a), let $\{S_n{}^\alpha\}_{n=1}^\infty$ be a sequence of integrable sets in X_α $(\alpha = 1 \text{ or } 2)$ having property (a) for the measure space $(X_\alpha, F_\alpha, \mathfrak{I}_\alpha, \mathfrak{M}_\alpha, \mu_\alpha)$. For each n write $S_n \equiv S_n{}^1 \times S_n{}^2$. Clearly $X = \bigcup_{n=1}^\infty S_n$. Let A be any set in $\mathfrak{M}$. For each k in Z^+ let B be a finite union of integrable rectangles which approximates A to within k^{-1}. Then $\mu(A - B) \le k^{-1}$. There exists N in Z^+ such that $\mu(B - S_n) \le k^{-1}$ for $n \ge N$. Therefore

$$\mu(A - S_n) \le \mu(A - B) + \mu(B - S_n) \le 2k^{-1} \qquad (n \ge N)$$

and thus $\mu(A \cap S_n) \to \mu(A)$ as $n \to \infty$.

Our next result is the theorem of Fubini, which states that an integration with respect to a product measure is equivalent to an iterated integration.

Theorem 12 *Let f be integrable with respect to the product measure μ. Then for almost all x_1 in X_1 the function $x_2 \to f(x_1,x_2)$ is defined almost everywhere on X_2, and is integrable. Moreover, the function $x_1 \to \int f(x_1,x_2)\,d\mu_2(x_2)$ is integrable, and*

$$(5.6) \qquad \int \left(\int f(x_1,x_2)\,d\mu_2(x_2) \right) d\mu_1(x_1) = \int f\,d\mu$$

Proof There is no loss of generality in taking $f \ge 0$. In case f is the characteristic function of an integrable set, the result follows from the definition of the product measure μ. Hence the result holds whenever f is a simple function.

In the general case take a sequence

$$0 \le f_1 \le f_2 \le \cdots$$

of nonnegative simple functions converging almost uniformly to f, so that

$$\int \left(\int f_n(x_1,x_2)\,d\mu_2(x_2) \right) d\mu_1(x_1) = \int f_n\,d\mu \to \int f\,d\mu$$

as $n \to \infty$. By Lebesgue's monotone-convergence theorem, the sequence $\{ \int f_n(x_1,x_2)\,d\mu_2(x_2) \}$ of integrable functions on X_1 converges almost uniformly to an integrable function $h(x_1)$ on X_1, and $\int h\,d\mu_1 = \int f\,d\mu$. Hence

$$(5.7) \qquad \int f_n(x_1,x_2)\,d\mu_2(x_2) \to h(x_1) \qquad \text{as } n \to \infty$$

for almost all x_1 in X_1.

Now, $\{f_n(x_1,x_2)\}$ converges to $f(x_1,x_2)$ for all (x_1,x_2) in a full set $A \in \mathfrak{F}$. Since $-A$ is a null set, $-A(x_1,\beta)$ is a null set [that is, $A(x_1,\beta)$ is a full set] for almost all x_1 in X_1. Thus for almost all x_1 in X_1 the sequence $\{f_n(x_1,x_2)\}$ converges for almost every value of x_2 to $f(x_1,x_2)$. By Lebesgue's monotone-convergence theorem again, (5.7) implies that $f(x_1,x_2)$ is an integrable function of x_2, for almost all x_1 in X_1, with integral $h(x_1)$. Since $\int h \, d\mu_1 = \int\!\int f \, d\mu$, this is equivalent to (5.6).

PROBLEMS

1. Let A and C be integrable sets, with $\mu(A) = \mu(C)$, and B a Borel set with $A \subset B \subset C$. Show that B is integrable.

2. Show that equality almost everywhere is an equivalence relation on the class of measurable sets.

3. Let $(X,F,\mathfrak{F},\mathfrak{M},\mu)$ be a measure space. Let $\mathfrak{F}'$ consist of all complemented sets relative to F. Let $\mathfrak{M}'$ consist of all B in $\mathfrak{F}'$ such that there exist A and N in $\mathfrak{M}$ with $\mu(N) = 0$ such that $x \in B$ whenever $x \in A - N$ and $x \in -B$ whenever $x \in -A - N$. For each B in $\mathfrak{M}'$ write $\mu'(B) \equiv \mu(A)$. Show that $(X,F,\mathfrak{F}',\mathfrak{M}',\mu')$ is a measure space [called the *completion* of $(X,F,\mathfrak{F},\mathfrak{M},\mu)$].

4. A function f defined on a full subset S of a proper compact interval $[a,b]$ is *Riemann integrable* if for each $\epsilon > 0$ there exists $\delta > 0$ such that whenever

$$P \equiv \{a_0 = a, a_1, \ldots, a_n = b\}$$
$$\text{and} \qquad Q \equiv \{b_0 = a, b_1, \ldots, b_m = b\}$$

are partitions of $[a,b]$ of mesh less than δ, and $x_1, \ldots, x_n, y_1, \ldots, y_m$ are constants with $a_0 \leq x_1 \leq a_1 \leq \cdots \leq a_{n-1} \leq x_n \leq a_n$ and $b_0 \leq y_1 \leq b_1 \leq \cdots \leq b_{n-1} \leq y_n \leq b_n$, then

$$|S(f,P) - S(f,Q)| \leq \epsilon$$

$$\text{where} \qquad S(f,P) \equiv \sum_{i=1}^{n} f(x_i)(a_i - a_{i-1})$$

and $S(f,Q)$ is defined similarly. Define the *Riemann integral* $\int_a^b f(x) \, dx$ of f, and show that f is Riemann integrable if for each $\epsilon > 0$ there exists $\delta > 0$ and a compact integrable set $K \subset S$ with $\mu(-K) \leq \epsilon$ such that $|f(x) - f(y)| \leq \epsilon$ whenever $x \in K$, $y \in S$, and $|x - y| \leq \delta$. Show that the converse is *not* true.

5.   Show how to define a neighborhood structure on the set M of measurable functions corresponding to convergence in measure in such a way that a sequence $\{f_n\}$ converges in measure to a limit f if and only if all except finitely many terms of $\{f_n\}$ belong to any given neighborhood of f.

6.   Let p and q be real numbers, $p > 1$, $q > 1$, $p^{-1} + q^{-1} = 1$. Let $f \in L_p$ and $g \in L_q$. Show that $\int fg\, d\mu = \|f\|_p \|g\|_q$ if and only if $fg \geq 0$ a.e. and $|f|^p \|g\|_q^q = |g|^q \|f\|_p^p$ a.e., or equivalently,

$$f|f|^{p-1}\|g\|_q^q = g|g|^{q-1}\|f\|_p^p \text{ a.e.}$$

7.   Let f and g belong to L_p ($p \geq 1$). Show that $\|f + g\|_p = \|f\|_p + \|g\|_p$ if and only if $f\|g\|_p = g\|f\|_p$ a.e.

8.   Let f be an integrable function on the finite measure space $(X, F, \mathfrak{F}, \mathfrak{M}, \mu)$. Define the measure ν on $\mathbf{R}$ by $\int h\, d\nu \equiv \int h \circ f\, d\mu$, for all h in $C(\mathbf{R})$. Let the compact set S be weakly integrable with respect to ν. Show there exists an integrable set A with respect to μ such that $f(x) \in S$ for almost all x in A and $f(x) \in -S$ for almost all x in $-A$.

9.   Prove the formulas

$$(A_1 \times A_2) \cap (B_1 \times B_2) = (A_1 \cap B_1) \times (A_2 \cap B_2)$$

and

$$(A_1 \times A_2) - (B_1 \times B_2) = \{A_1 \times (A_2 - B_2)\} \cup \{(A_1 - B_1) \times A_2\}$$

10.   Let f be measurable with respect to the product measure μ, with $f \geq 0$ a.e., and let the iterated integral $c \equiv \int (\int f(x_1, x_2)\, d\mu_1(x_1))\, d\mu_2(x_2)$ exist. Show $\int f\, d\mu$ exists and equals c.

11.   If A and B are complemented sets in X, and x is a point of X, with $x \in B - B$, then $x \in A$. What is the justification for this assertion (implicit use of which was made in the proof of the first statement of Proposition 2)? How could we avoid its use?

NOTES

The definition of a measure space is much more complicated than the classical definition. In spite of this, the theory develops along very similar lines.

Note how the constructive definition (Definition 7) of convergence almost everywhere differs from the classical definition (pointwise convergence on a full set). The classical definition would be of no use constructively.

CHAPTER 8

LIMIT OPERATIONS IN MEASURE THEORY

Certain parts of measure theory are hard to develop constructively, because limits that are classically proved to exist simply do not exist constructively. To retrieve such results, we can either (i) strengthen the hypothesis, (ii) weaken the conclusion, or (iii) apply a combination of both. Section 1 concerns (a) the decomposition of a measure into positive and negative parts, and (b) the Radon-Nikodym theorem. Method (i) is used. The treatment follows the classical pattern, except that it is much messier. This material is included only for completeness. The reader is advised not to spend time reading the proofs. Section 2 treats the martingale theorem and Lebesgue's theorem that a function of bounded variation has a derivative almost everywhere, by method (i). To this end certain "norms" $\| \ \|_\lambda$ are introduced, which lack the usual properties of subadditivity and homogeneity, but are well suited to situations in which the linear combinations that occur are all convex. Section 3 presents a general ergodic theorem, in terms of upcrossing inequalities. Method (ii) is used. This treatment of ergodic theory, which was developed in order to constructivize the subject, seems more natural, even from the classical point of view, than the usual approach by means of a maximal ergodic lemma. Some corollaries are obtained: a constructive version of the Chacon-Ornstein theorem and a constructive substitute for the above-mentioned result of Lebesgue.

214

The derivative of a set function with respect to a measure is not easy to define constructively, because the classical results rely in an essential way on the limited principle of omniscience. Therefore these results are not valid constructively. Nevertheless it is possible to find satisfactory constructive substitutes. As we shall see, there are many possibilities. We begin with the study of measures on a locally compact space.

1. DECOMPOSITIONS OF MEASURES

Measures μ and ν on a locally compact space X are said to be mutually singular if, roughly speaking, their masses are distributed over disjoint sets. This concept is easily expressed in a precise way in terms of the values of the measures on test functions.

Definition 1   Two positive measures μ and ν on a locally compact space X are *mutually singular* if for each nonnegative test function f and each $\epsilon > 0$ there exists a test function g with $0 \leq g \leq f$ such that

$$(1.1) \qquad \int g \, d\mu + \int (f - g) \, d\nu \leq \epsilon$$

It is natural to ask whether an arbitrary measure μ on X is the difference of two positive measures. Even more natural is the question whether it is the difference of two mutually singular positive measures. Before giving necessary and sufficient conditions (Theorem 1 below), we prove a simple lemma.

Lemma 1   *Let μ be a measure on a locally compact space X, and let f and g be test functions with $0 \leq g \leq f$. Then*

$$(1.2) \qquad \int \varphi \, d\mu \leq \int g \, d\mu + \epsilon \qquad (\varphi \in C(X), 0 \leq \varphi \leq f)$$

if and only if

$$(1.3)$$
$$\int (\varphi_1 - \varphi_2) \, d\mu \leq \epsilon \qquad (\varphi_1, \varphi_2 \in C(X), 0 \leq \varphi_1 \leq f - g, 0 \leq \varphi_2 \leq g)$$

Proof   Assume that (1.2) is satisfied. Then for φ_1 and φ_2 as described we have $0 \leq g - \varphi_2 + \varphi_1 \leq f$, and thus $\int (g - \varphi_2 + \varphi_1) \, d\mu \leq \int g \, d\mu + \epsilon$, which is just (1.3).

Conversely, if (1.3) is satisfied, consider φ as described. Write $\varphi_1 \equiv \min \{\varphi, f - g\}$, $\varphi_2 \equiv g - (\varphi - \varphi_1)$. Then $0 \leq \varphi_1 \leq f - g$, and $0 \leq \varphi_2 \leq g$. Therefore $\int (\varphi_1 - \varphi_2) \, d\mu \leq \epsilon$. This is equivalent to (1.2).

Theorem 1 *Let μ be a measure on a locally compact space X. Let $\{f_n\}$ be a sequence of nonnegative test functions on X, such that for each f in $C(X)$ there exists n in Z^+ with $f = ff_n$. Then μ is the difference of two mutually singular positive measures μ_1 and μ_2 if the least upper bound*

$$\eta(f_n) \equiv \text{l.u.b.} \ \{\textstyle\int \varphi \, d\mu : \varphi \in C(X), 0 \le \varphi \le f_n\}$$

exists for all n in Z^+. Conversely, if μ is the difference of mutually singular positive measures μ_1 and μ_2, then $\eta(f)$ exists for all nonnegative test functions f and equals $\int f \, d\mu_1$.

Proof Assume first that μ is the difference of mutually singular positive measures μ_1 and μ_2. Consider any nonnegative test function f. For each $\epsilon > 0$ choose g in $C(X)$ with $0 \le g \le f$ and $\int (f - g) \, d\mu_1 + \int g \, d\mu_2 \le \epsilon$. If φ_1 and φ_2 are test functions, with $0 \le \varphi_1 \le f - g$ and $0 \le \varphi_2 \le g$, then

$$\int (\varphi_1 - \varphi_2) \, d\mu = \int (\varphi_1 - \varphi_2) \, d\mu_1 - \int (\varphi_1 - \varphi_2) \, d\mu_2$$
$$\le \int (f - g) \, d\mu_1 + \int g \, d\mu_2 \le \epsilon$$

By Lemma 1, we have $\int \varphi \, d\mu \le \int g \, d\mu + \epsilon$ whenever $\varphi \in C(X)$ and $0 \le \varphi \le f$. Therefore $\eta(f)$ exists.

Assume conversely that $\eta(f_n)$ exists for all n. To show that $\eta(f)$ exists for all nonnegative test functions f, we may restrict consideration to those f for which $0 \le f \le 1$. Choose n in Z^+ with $f = ff_n \le f_n$. For a given $\epsilon > 0$ choose g in $C(X)$ with $0 \le g \le f_n$ and $\eta(f_n) \le \int g \, d\mu + \epsilon$. By Lemma 1, we have $\int \varphi_1 \, d\mu - \int \varphi_2 \, d\mu \le \epsilon$ whenever φ_1 and φ_2 are test functions with $0 \le \varphi_1 \le f_n - g$ and $0 \le \varphi_2 \le g$. Therefore $\int \varphi_1 \, d\mu - \int \varphi_2 \, d\mu \le \epsilon$ whenever $0 \le \varphi_1 \le f - fg$ and $0 \le \varphi_2 \le fg$. Lemma 1 gives $\int \varphi \, d\mu \le \int fg \, d\mu + \epsilon$ whenever $0 \le \varphi \le f$. It follows that $\eta(f)$ exists.

For arbitrary f_1 and f_2 in $C(X)$, with $0 \le f_1, 0 \le f_2$, we clearly have $\eta(f_1) + \eta(f_2) \le \eta(f_1 + f_2)$. For all φ in $C(X)$ with $0 \le \varphi \le f_1 + f_2$ we have $\varphi = \varphi_1 + \varphi_2$, where $0 \le \varphi_1 \equiv \min \{\varphi, f_1\} \le f_1$ and $0 \le \varphi_2 \equiv \varphi - \varphi_1 \le f_2$. It follows that

$$\int \varphi \, d\mu = \int \varphi_1 \, d\mu + \int \varphi_2 \, d\mu \le \eta(f_1) + \eta(f_2)$$

Therefore $\eta(f_1 + f_2) \le \eta(f_1) + \eta(f_2)$. Thus $\eta(f_1 + f_2) = \eta(f_1) + \eta(f_2)$.

Consider any f in $C(X)$. Choose f_1 and f_2 in $C(X)$ with $f_1 \ge 0, f_2 \ge 0$, and $f_1 - f_2 = f$. Write

$$\eta(f) \equiv \eta(f_1) - \eta(f_2)$$

The number $\eta(f)$ does not depend on the choice of f_1 and f_2, since for a different choice f_1' and f_2' we have $f_1' - f_2' = f = f_1 - f_2$, and thus $f_1' + f_2 = f_1 + f_2'$, and consequently $\eta(f_1') + \eta(f_2) = \eta(f_1) + \eta(f_2')$.

Thus η is a linear functional from $C(X)$ to $\mathbf{R}$, with $\eta(f) \geq 0$ whenever $f \geq 0$. In other words, η is a positive measure μ_1 on X. Write $\mu_2 \equiv \mu_1 - \mu$. Then μ_2 is a measure, since μ and μ_1 are measures. If $f \in C(X)$ and $f \geq 0$, then

$$\int f \, d\mu_2 = \int f \, d\mu_1 - \int f \, d\mu = \eta(f) - \int f \, d\mu \geq 0$$

Thus μ_2 is also a positive measure.

It remains to show that μ_1 and μ_2 are mutually singular. Consider any $f \geq 0$ in $C(X)$, and any $\epsilon > 0$. Choose g in $C(X)$ with $0 \leq g \leq f$ and

$$\int f \, d\mu_1 = \eta(f) \leq \int g \, d\mu + \epsilon$$

We have

$$\int g \, d\mu_2 + \int (f - g) \, d\mu_1 = \int g \, d\mu_1 - \int g \, d\mu + \int (f - g) \, d\mu_1$$
$$= \int f \, d\mu_1 - \int g \, d\mu \leq \epsilon$$

Therefore μ_1 and μ_2 are mutually singular.

The problem solved by Theorem 1 is related to the problem of representing a measure as an integral with respect to another measure.

Theorem 2 *Let μ and ν be measures on a locally compact space X, with μ positive. Then there exists a measurable function $h: X \to \mathbf{R}$ such that (i) $fh \in L_1(\mu)$ for all f in $C(X)$ and (ii) $\nu = h\mu$, in the sense that $\int f \, d\nu = \int fh \, d\mu$ for all f in $C(X)$, if and only if*

(a) There exists a dense subset S of $\mathbf{R}$ such that for each t in S the measure $\nu - t\mu$ is the difference of two mutually singular positive measures, and

(b) For each $f \geq 0$ in $C(X)$ and each $\epsilon > 0$ there exists $c > 0$ such that

$$\left| \int g \, d\nu \right| \leq \epsilon + c \int g \, d\mu$$

whenever $g \in C(X)$ and $0 \leq g \leq f$.

Proof Assume first that the function h exists. Let f be any nonnegative test function and ϵ any positive constant. Choose $c > 0$ so that $h = h_1 + h_2$, where h_1, h_2 are measurable functions with $|h_1| \leq c$ and $\int f|h_2| \, d\mu \leq \epsilon$. Then whenever $g \in C(X)$ and $0 \leq g \leq f$ we have

$$\left| \int g \, d\nu \right| = \left| \int gh \, d\mu \right| \leq \int g|h_1| \, d\mu + \int g|h_2| \, d\mu \leq c \int g \, d\mu + \epsilon$$

Thus condition (b) is satisfied.

To see that condition (a) is satisfied, consider any t in $\mathbf{R}$. Then $\nu - t\mu = (h - t)\mu$. Let f be a test function, with $0 \le f \le 1$. Let the compact μ-integrable set K be a support for f. By Theorem 7 of Chap. 7, for each $\epsilon > 0$ there exists an integrable subset A of K such that $h - t \ge 0$ on A and $h - t \le \epsilon$ on $K - A$. Choose a test function g to approximate A so closely that

$$\int_A f(h - t)\, d\mu \le \int_A fg(h - t)\, d\mu + \epsilon$$

and
$$\left| \int_{K-A} fg(h - t)\, d\mu \right| \le \epsilon$$

Then for each test function f_0 with $0 \le f_0 \le f$ we have

$$
\begin{aligned}
\int f_0\, d(\nu - t\mu) &= \int f_0(h - t)\, d\mu \\
&= \int_A f_0(h - t)\, d\mu + \int_{K-A} f_0(h - t)\, d\mu \\
&\le \int_A f(h - t)\, d\mu + \epsilon\mu(K - A) \\
&\le \int_A fg(h - t)\, d\mu + \epsilon + \epsilon\mu(K - A) \\
&= \int fg(h - t)\, d\mu - \int_{K-A} fg(h - t)\, d\mu \\
&\qquad\qquad\qquad\qquad\qquad + \epsilon + \epsilon\mu(K - A) \\
&\le \int fg\, d(\nu - t\mu) + 2\epsilon + \epsilon\mu(K)
\end{aligned}
$$

Since ϵ is arbitrary, it follows from Theorem 1 that $\nu - t\mu$ is the difference of two mutually singular positive measures. Thus (a) is satisfied.

Assume conversely that conditions (a) and (b) are satisfied. First assume that X is compact. Let k be any positive integer. Choose c so large that $|\int g\, d\nu| \le k^{-1} + c\int g\, d\mu$ whenever $g \in C(X)$ and $0 \le g \le 1$. Choose α and β in S, with $\alpha > c$ and $\beta < -c$. Our first task will be to approximate ν by a measure ν' with $\beta\mu \le \nu' \le \alpha\mu$. Since $\alpha \in S$, we have

$$\nu - \alpha\mu = \nu_1 - \nu_2$$

where ν_1 and ν_2 are mutually singular positive measures. For each ϵ with $0 < \epsilon < k^{-1}$ there exists g in $C(X)$ with $0 \le g \le 1$ and

$$\int d\nu_1 \le \int g\, d(\nu - \alpha\mu) + \epsilon \le \int g\, d(\nu - c\mu) + k^{-1} \le 2k^{-1}$$

Similarly, $\nu - \beta\mu = \nu'_1 - \nu'_2$, where ν'_1 and ν'_2 are mutually singular positive measures with $\int d\nu'_2 \le 2k^{-1}$. Then

$$\nu_1 - \nu_2 = \nu - \alpha\mu \le \nu - \beta\mu = \nu'_1 - \nu'_2$$

With g as above, for all f in $C(X)$ with $0 \leq f \leq 1$ we get

$$
\begin{aligned}
\int f \, d\nu_1 &= \int d\nu_1 - \int (1 - f) \, d\nu_1 \\
&\leq \int g \, d(\nu_1 - \nu_2) + \epsilon - \int (1 - f) \, g \, d\nu_1 \\
&\leq \int g \, d(\nu_1 - \nu_2) + \epsilon - \int (1 - f) g \, d(\nu_1 - \nu_2) \\
&= \int fg \, d(\nu_1 - \nu_2) + \epsilon \\
&\leq \int fg \, d(\nu_1' - \nu_2') + \epsilon \leq \int f \, d\nu_1' + \epsilon
\end{aligned}
$$

Since ϵ is arbitrary, it follows that $\nu_1 \leq \nu_1'$. Similarly, $\nu_2' \leq \nu_2$. We define

$$
\nu' \equiv \nu - \nu_1 + \nu_2'
$$

Then

$$
(1.4) \qquad \left| \int f \, d(\nu' - \nu) \right| \leq 4k^{-1} \qquad (f \in C(X), 0 \leq f \leq 1)
$$

Also,

$$
(1.5) \qquad\qquad \nu' \leq \nu - \nu_1 + \nu_2 = \alpha\mu
$$

and

$$
(1.6) \qquad\qquad \nu' \geq \nu - \nu_1' + \nu_2' = \beta\mu
$$

Our next task is to find a function h_k such that $h_k\mu$ is a sufficiently close approximation to ν'. Let $\alpha_1 = \beta < \alpha_2 < \cdots < \alpha_n = \alpha$ be elements of S with $\alpha_{i+1} - \alpha_i \leq k^{-1}$ $(1 \leq k \leq n - 1)$. Then

$$
\nu - \alpha_i\mu = \nu_1{}^i - \nu_2{}^i
$$

where $\nu_1{}^i$ and $\nu_2{}^i$ are mutually singular positive measures. As above, we have $\nu_1 \leq \nu_1{}^i$ and $\nu_2' \leq \nu_2{}^i$. Now, $\nu' - \alpha_i\mu = \nu_1{}^i - \nu_1 - (\nu_2{}^i - \nu_2')$, and thus $\nu' - \alpha_i\mu$ is the difference of mutually singular positive measures. By Theorem 1, there exists f_i in $C(X)$, with $0 \leq f_i \leq 1$, such that

$$
(1.7) \qquad \int g \, d(\nu' - \alpha_i\mu) \leq \int f_i \, d(\nu' - \alpha_i\mu) + (n^2 k)^{-1}
$$

whenever $g \in C(X)$ and $0 \leq g \leq 1$. By (1.5) and (1.6), we may take $f_1 \equiv 1$ and $f_n \equiv 0$. For $1 \leq i \leq n$, write

$$
f_i' \equiv \max \{f_i, \ldots, f_n\}
$$

Then
$$
1 = f_1' \geq f_2' \geq \cdots \geq f_n' = 0
$$

Also,
$$
f_i' = f_i + g_{i+1}^i + \cdots + g_n{}^i
$$

where $g_j{}^i \in C(X)$ and $0 \leq g_j{}^i \leq f_j$ $(i + 1 \leq j \leq n)$. By Lemma 1 and (1.7),

$$
\int g_j{}^i \, d(\nu' - \alpha_i\mu) \geq -(n^2 k)^{-1}
$$

By (1.7) again,

$$(1.8) \quad \int g \, d(\nu' - \alpha_i\mu) \leq \int f_i' \, d(\nu' - \alpha_i\mu)$$
$$- \sum_{j=i+1}^{n} \int g_j{}^i \, d(\nu' - \alpha_i\mu) + (n^2k)^{-1}$$
$$\leq \int f_i' \, d(\nu' - \alpha_i\mu) + (n-1)(n^2k)^{-1} + (n^2k)^{-1}$$
$$= \int f_i' \, d(\nu' - \alpha_i\mu) + (nk)^{-1}$$

whenever $g \in C(X)$ and $0 \leq g \leq 1$.

We define

$$h_k \equiv \alpha_2(f_1' - f_2') + \cdots + \alpha_n(f_{n-1}' - f_n')$$

Let g be any function in $C(X)$ with $0 \leq g \leq 1$. Then for $1 \leq i \leq n-1$ we have

$$(1.9) \qquad \int g(f_i' - f_{i+1}') \, d(\nu' - \alpha_{i+1}\mu) \leq (nk)^{-1}$$

by (1.8) and Lemma 1, since $g(f_i' - f_{i+1}') \leq (f_i' - f_{i+1}') \leq 1 - f_{i+1}'$. Similarly,

$$(1.10) \quad \int g(f_i' - f_{i+1}') \, d(\nu' - \alpha_{i+1}\mu)$$
$$= \int g(f_i' - f_{i+1}') \, d(\nu' - \alpha_i\mu) - (\alpha_{i+1} - \alpha_i)\int g(f_i' - f_{i+1}') \, d\mu$$
$$\geq -(nk)^{-1} - k^{-1}\int (f_i' - f_{i+1}') \, d\mu$$

by (1.8) and Lemma 1, since $g(f_i' - f_{i+1}') \leq f_i'$. From (1.9) and (1.10) we get

$$\left| \int g(f_i' - f_{i+1}') \, d(\nu' - \alpha_{i+1}\mu) \right| \leq (nk)^{-1} + k^{-1}\int (f_i' - f_{i+1}') \, d\mu$$

Summation from $i = 1$ to $i = n - 1$ gives

$$\left| \int g \, d\nu' - \int gh_k \, d\mu \right| < k^{-1} + k^{-1}\int (f_1 - f_n) \, d\mu = k^{-1} + k^{-1}\int d\mu$$

Together with (1.4), this gives

$$(1.11) \quad \left| \int g \, d\nu - \int gh_k \, d\mu \right| \leq 5k^{-1} + k^{-1}\int d\mu$$
$$(g \in C(X), 0 \leq g \leq 1)$$

In particular, $\{h_k\}$ is a Cauchy sequence in $L_1(\mu)$, whose limit we denote by h. From (1.11) it follows that $\int g \, d\nu = \int gh \, d\mu$ for all g in $C(X)$. Thus $\nu = h\mu$, and the theorem is proved in case X is compact.

Now consider the general case. Let ρ be a compactifier for X. Choose the sequence $t_1 < t_2 < \cdots$ of real numbers so that $\lim_{n \to \infty} t_n = \infty$ and so that

$$X_n \equiv \{x \in X \colon \rho(x) \leq t_n\}$$

is compact and integrable with respect to μ for each n. Let f_n be a test function with $0 \leq f_n \leq 1$ such that $f_n(x) = 1$ for all x in X_{n-1} and $f_n(x) = 0$ for all x in $-X_n$. Define measures μ_n and ν_n on X_n by $\int f \, d\mu_n \equiv \int f f_n \, d\mu$ and $\int f \, d\nu_n \equiv \int f f_n \, d\nu$ for all $f \in C(X_n)$, where, of course, $f f_n$ is to be interpreted as that function in $C(X)$ whose value at a point $x \in X_n$ is $f(x) f_n(x)$ and whose value at a point of $-X_n$ is 0. Then the measures μ_n and ν_n on X_n satisfy conditions (a) and (b). By the case already considered, there exists h_n in $L_1(\mu_n)$ such that $\nu_n = h_n \mu_n$. For any f in $C(X)$ which is supported by X_{n-1} we have

$$(1.12) \qquad \int f \, d\nu = \int f \, d\nu_n = \int f h_n \, d\mu_n = \int f h'_n \, d\mu$$

where h'_n is that integrable function on X whose value at a point x in X_n is $h_n(x)$ and whose value at a point x in $-X_n$ is 0. Similarly, $\int f \, d\nu = \int f h'_{n+1} \, d\mu$, and thus $\int f(h'_{n+1} - h'_n) \, d\mu = 0$. Since this is true for all f in $C(X)$ which are supported by X_{n-1}, it follows that $h'_{n+1}(x) = h'_n(x)$ for almost all x in X_{n-1}. Thus there exists a measurable function h on X such that for each n we have $h(x) = h'_n(x)$ for almost all x in X_{n-1}. By (1.12), we see that h has the desired property.

The function h constructed in the proof of Theorem 2 can be thought of as the derivative of the measure ν with respect to the measure μ. It is called the *Radon-Nikodym derivative*.

2. DERIVATIVES OF SET FUNCTIONS

It is plausible that at least in certain cases the Radon-Nikodym derivative h constructed in Theorem 2 can be obtained from the measures μ and ν by a limiting process involving quotients of the form $\nu(A)\mu(A)^{-1}$, where A is a set integrable with respect to both μ and ν. More generally, we wish to find conditions under which limits of quotients of the form $\nu(A)\mu(A)^{-1}$ exist. This circle of ideas includes the theory of martingales (a concept which will presently be defined). In the usual presentations of martingale theory a distinction is made between the L_p case $(p > 1)$ and the L_1 case; the L_p case is treated separately, in order to take advantage of the uniform convexity. The treatment we shall give applies simultaneously to all $p \geq 1$. This is made possible by introducing "norms" with respect to fairly general weight functions λ. We lose the homogeneity of the norm but gain a weak form of uniform convexity, which is more important.

A twice-differentiable nonnegative function $\lambda: \mathbf{R} \to \mathbf{R}$ is *admissible* if $\lambda(0) = 0$, if $\lambda(x) = \lambda(-x)$ for all x, and if for each $x > 0$ the number $\theta(x) \equiv \inf \{\lambda''(y): |y| \leq x\}$ is positive. Not only is such a function convex, but we have the following critical estimate.

Lemma 2 *If λ is admissible, then for all $x_1, \ldots, x_n$ in $\mathbf{R}$ and $\alpha_1, \ldots, \alpha_n$ in $\mathbf{R}^{0+}$, with $\Sigma \alpha_i = 1$, we have*

$$(2.1) \qquad \Sigma \alpha_i \lambda(x_i) - \lambda(\bar{x}) \geq \tfrac{1}{2} \Sigma \alpha_i (x_i - \bar{x})^2 \theta(\max \{|x_i|, |\bar{x}|\})$$

where $\bar{x} \equiv \Sigma \alpha_i x_i$.

Proof We compute

$$
\begin{aligned}
\sum \alpha_i \lambda(x_i) - \lambda(\bar{x}) &= \sum \alpha_i \int_{\bar{x}}^{x_i} \lambda'(t) \, dt \\
&= \sum \alpha_i \int_{\bar{x}}^{x_i} (\lambda'(t) - \lambda'(\bar{x})) \, dt \\
&= \sum \alpha_i \int_{\bar{x}}^{x_i} \int_{\bar{x}}^{t} \lambda''(u) \, du \, dt \\
&\geq \sum \alpha_i \int_{\bar{x}}^{x_i} \int_{\bar{x}}^{t} \theta(\max \{|x_i|, |\bar{x}|\}) \, du \, dt \\
&= \frac{1}{2} \sum \alpha_i (x_i - \bar{x})^2 \theta(\max \{|x_i|, |\bar{x}|\})
\end{aligned}
$$

A finite martingale is a sequence $f_1, \ldots, f_n$ of integrable functions (on some measure space) such that the expected value of one of the functions, say f_k, given that the preceding functions $f_1, \ldots, f_{k-1}$ have certain values $a_1, \ldots, a_{k-1}$, is equal to a_{k-1}. For instance, let a gambler bet n times, each bet being fair in the sense that his total expected gain is 0. Let the functions $f_1, \ldots, f_n$ represent his winnings at stages $1, \ldots, n$, respectively (so that $f_1, \ldots, f_n$ are integrable functions, on the space of all possible outcomes of the n bets weighted with the appropriate weights). Then $f_1, \ldots, f_n$ is a finite martingale.

Here is the precise definition.

Definition 2 Consider integrable functions $f_1, \ldots, f_k$ (on a certain measure space), and a measurable set A, such that there exists a Borel set S in $\mathbf{R}^k$ with $(f_1(x), \ldots, f_k(x)) \in S$ for almost all x in A and $(f_1(x), \ldots, f_k(x)) \in -S$ for almost all x in $-A$. We say that A is *determined* by the functions $f_1, \ldots, f_k$, and loosely write

$$A = \{x\colon (f_1(x), \ldots, f_k(x)) \in S\}$$

A *martingale* is a finite or infinite sequence $f_1, f_2, \ldots$ of integrable functions, such that for each function f_k of the sequence and each measurable set A determined by the functions $f_1, \ldots, f_k$ we have

$$(2.2) \qquad \int_A f_{k+1}\, d\mu = \int_A f_k\, d\mu$$

The fact that a sequence $f_1, f_2, \ldots$ of integrable functions is a martingale will be expressed by the notation

$$f_1 \prec f_2 \prec \cdots$$

A fundamental property of martingales is the following.

Lemma 3 *Let $f_1 \prec \cdots \prec f_n$ be a finite martingale. For $1 \leq k \leq n - 1$ let A_k be an integrable set determined by $f_1, \ldots, f_k$. Write $B_1 \equiv A_1$, $B_k \equiv A_k - \bigcup\limits_{i=1}^{k-1} A_i \ (2 \leq k \leq n - 1)$, and $B_n \equiv X - \bigcup\limits_{i=1}^{n-1} A_i$. Let h be that integrable function which equals f_k on $B_k \ (1 \leq k \leq n)$. Then*

$$f_1 \prec h \prec f_n$$

Proof Let C be an integrable set determined by f_1 and h. We see that the set $C \cap B_k \ (1 \leq k \leq n)$ is determined by $f_1, \ldots, f_k$. Hence

$$\int_C h\, d\mu = \sum_{k=1}^{n} \int_{C \cap B_k} h\, d\mu = \sum_{k=1}^{n} \int_{C \cap B_k} f_k\, d\mu$$

$$= \sum_{k=1}^{n} \int_{C \cap B_k} f_n\, d\mu = \int_C f_n\, d\mu$$

In case C is determined by f_1 alone, we have in addition

$$\int_C f_n\, d\mu = \int_C f_1\, d\mu$$

and thus $\int_C h\, d\mu = \int_C f_1\, d\mu$. Thus $f_1 \prec h \prec f_n$.

Now consider any measurable function f for which $\lambda(f) \equiv \lambda \circ f$ is integrable, and write

$$\|f\|_\lambda \equiv \int \lambda(f)\,d\mu$$

As remarked above, this "norm" is not homogeneous. It is, however, almost uniformly convex. For us the most useful manifestation of this almost-uniform convexity is the following *martingale inequality.*

Lemma 4   *If* $f \prec g$, *and* $\|f\|_\lambda$ *and* $\|g\|_\lambda$ *exist, then*

$$(2.3) \qquad \tfrac{1}{2}\int |f - g|^2 \theta(\max\{|f|, |g|\})\,d\mu \le \|g\|_\lambda - \|f\|_\lambda$$

Proof   We can approximate f arbitrarily closely by a simple function $\tilde{f} = \sum_{A \in \alpha} a_A \chi_A$, where α is a finite family of disjoint integrable sets of positive measures determined by f. At the same time we approximate g arbitrarily closely by a simple function $\tilde{g} \equiv \sum_{B \in \beta} b_B \chi_B$, where β is a finite family of disjoint integrable sets, which is a *refinement* of α in the sense that for each B in β there exists A in α with $B \subset A$. We can do this so that (i) each A in α is determined by $\tilde{f}$, and (ii) $\tilde{f} \prec \tilde{g}$. Then for each A in α we have

$$a_A = \mu(A)^{-1} \int_A \tilde{f}\,d\mu = \mu(A)^{-1}\int_A \tilde{g}\,d\mu$$
$$= \mu(A)^{-1} \sum_{B \subset A} \mu(B) b_B$$

It is enough to prove (2.3) with f replaced by $\tilde{f}$ and g by $\tilde{g}$. We compute, using (2.1),

$$\|\tilde{g}\|_\lambda - \|\tilde{f}\|_\lambda = \sum_A \mu(A)\left\{ \sum_{B \subset A} \mu(A)^{-1}\mu(B)\lambda(b_B) - \lambda(a_A) \right\}$$
$$= \sum_A \mu(A)\left\{ \sum_{B \subset A} \mu(A)^{-1}\mu(B)\lambda(b_B) \right.$$
$$\left. - \lambda\left(\sum_{B \subset A} \mu(A)^{-1}\mu(B) b_B \right) \right\}$$
$$\ge \sum_A \mu(A)\tfrac{1}{2} \sum_{B \subset A} \mu(A)^{-1}\mu(B)|b_B - a_A|^2 \theta(\max\{|b_B|, |a_A|\})$$
$$= \tfrac{1}{2}\int |\tilde{f} - \tilde{g}|^2 \theta(\max\{|\tilde{f}|, |\tilde{g}|\})\,d\mu$$

We can now prove the *martingale theorem.*

Theorem 3   *For arbitrary positive constants K and ϵ there exists a positive constant δ such that for every finite martingale $f_1 \prec \cdots \prec f_n$,* *with $\|f_n\|_\lambda \leq K$ and $\|f_n\|_\lambda - \|f_1\|_\lambda \leq \delta$, we have $\mu(A_2 \cup \cdots \cup A_n) \leq \epsilon$ for each constant t, with $t \geq \epsilon$, for which the sets*

$$A_i \equiv \{x \colon |f_1(x) - f_i(x)| \geq t\} \qquad (1 \leq i \leq n)$$

are integrable.

Proof Take the positive constant r so large that $K\lambda(r)^{-1} \leq \frac{1}{3}\epsilon$. Write $\delta \equiv \frac{1}{6}\epsilon^3\theta(r)$. Consider a finite martingale $f_1 < \cdots < f_n$ satisfying the prescribed conditions. Define the integrable function h as in Lemma 3. Then $|f_1 - h| \geq t$ a.e. on $A_2 \cup \cdots \cup A_n$. There is no loss of generality in taking $C_1 \equiv \{x \colon |f_1(x)| \geq r\}$ and $C_2 \equiv \{x \colon |h(x)| \geq r\}$ to be integrable sets. We have

$$\mu(C_1) \leq \lambda(r)^{-1}\|f_1\|_\lambda \leq \lambda(r)^{-1}\|f_n\|_\lambda \leq \lambda(r)^{-1}K \leq \frac{1}{3}\epsilon$$

Since $h < f_n$, by Lemma 3, and thus $\|h\|_\lambda \leq \|f_n\|_\lambda$, we also have $\mu(C_2) \leq \frac{1}{3}\epsilon$. We compute

$$\begin{aligned}
\mu(A_2 \cup \cdots \cup A_n) &\leq t^{-2} \int_{-C_1 \cap -C_2} (f_1 - h)^2 \, d\mu + \mu(C_1) + \mu(C_2) \\
&\leq \epsilon^{-2}\theta(r)^{-1} \int (f_1 - h)^2 \theta(\max\{|f_1|, |h|\}) \, d\mu + \tfrac{2}{3}\epsilon \\
&\leq \epsilon^{-2}\theta(r)^{-1}2(\|h\|_\lambda - \|f_1\|_\lambda) + \tfrac{2}{3}\epsilon \\
&\leq \epsilon^{-2}\theta(r)^{-1}2\delta + \tfrac{2}{3}\epsilon = \epsilon
\end{aligned}$$

as was to be proved.

It is customary to state the martingale theorem for an infinite martingale, as follows.

Corollary *Let $f_1 < f_2 < \cdots$ be an infinite martingale, such that the monotone-nondecreasing sequence $\{\|f_n\|_\lambda\}$ converges. Then $\{f_n\}$ converges almost everywhere.*

Proof Fix any $\epsilon > 0$. By Theorem 3, there exists a sequence $n_1 < n_2 < \cdots$ of positive integers such that the sets

$$A_k \equiv \{x \colon |f_{n_k}(x) - f_j(x)| \geq 2^{-k}\epsilon \text{ for some } j \ (n_k \leq j \leq n_{k+1})\}$$

(which may be assumed to be integrable) satisfy the conditions $\mu(A_k) \leq 2^{-k}\epsilon$ ($k \in Z^+$). Thus the set $A \equiv \cup A_k$ is integrable, and $\mu(A) \leq \epsilon$. Consider any $j \geq n_1$. Then $n_k \leq j \leq n_{k+1}$ for some k, and for almost all x in $-A$ we have

$$|f_{n_1}(x) - f_J(x)| \le \sum_{i=1}^{k-1} |f_{n_i}(x) - f_{n_{i+1}}(x)| + |f_{n_k}(x) - f_J(x)| \le \epsilon$$

Hence $\{f_n\}$ converges almost everywhere.

Next we develop a constructive substitute for the theorem that a function of bounded variation has a derivative almost everywhere. Fix a family ζ of integrable sets of positive measure in a finite measure space $(X,F,\mathfrak{F},\mathfrak{M},\mu)$. Take Γ to consist of all finite partitions α of X, using sets in ζ. (In other words, α is a finite set of disjoint elements of ζ whose union equals X almost everywhere.) We define a partial order $\le$ on Γ by taking $\alpha \le \beta$ to mean that β is a *refinement* of α, in the sense that to each set I of the partition β corresponds a set $J \equiv J(I,\alpha)$ of the partition α such that $I \subset J$. Of special importance is the case in which ζ satisfies the following *condition* (a).

Definition 3 For each α in Γ let ζ_α consist of all I in ζ such that there exists $J \equiv J(I,\alpha)$ in α with $I \subset J$. The class ζ is said to satisfy *condition* (a) if there exists $\rho > 0$ having the property that for each α in Γ and each finite subset ζ_0 of ζ_α there exists a finite subset ζ_1 of ζ_0 and an element β of Γ, with $\beta \ge \alpha$ and $\zeta_1 \subset \beta$, such that

$$(2.4) \qquad \mu(\bigcup_{I \in \zeta_1} I) > \rho\mu(\bigcup_{I \in \zeta_0} I)$$

The most important case in which condition (a) is satisfied is obtained by taking $X \equiv [0,1]$, letting μ be Lebesgue measure, and taking ζ to consist of all proper subintervals, or of all proper subintervals whose end points lie in a given dense set $S \subset [0,1]$, with $0 \in S$ and $1 \in S$. In fact, condition (a) is satisfied for all $\rho < \frac{1}{2}$. To prove this, it is enough to show that corresponding to every finite subset ζ_0 of ζ there exists a disjoint subset ζ_1 of ζ_0 satisfying (2.4). It is sufficient to consider the case in which all the end points of the intervals in ζ_0 are distinct. Because of this condition, there exists a minimal collection $I_1, \ldots, I_n$ of intervals in ζ_0 whose union equals almost everywhere the union of ζ_0. We may assume these intervals to be ordered according to the values of their left end points. Let ζ_1' be the set of those I_k ($1 \le k \le n$) with odd index k, and ζ_1'' the set of those with even index. Each of these sets is disjoint. Therefore one of them will serve for ζ_1.

This example generalizes to higher dimensions. Take $X \equiv [0,1]^n$ and let μ be Lebesgue measure. Let ζ consist of all "rectangles"

$$I \equiv [a_1,b_1] \times \cdots \times [a_n,b_n]$$

where $0 \leq a_i < b_i \leq 1$ for each i, where we restrict the a_i and the b_i to some dense $S \subset [0,1]$, with $0 \in S$ and $1 \in S$, and where

$$\min \{b_1 - a_1, \ldots, b_n - a_n\} \geq \tau \max \{b_1 - a_1, \ldots, b_n - a_n\}$$

for some fixed $\tau > 0$. Then ζ can be shown to satisfy condition (a), with some constant $\rho > 0$ depending on τ and n.

Suppose $v: \zeta \to \mathbf{R}$ is *finitely additive*. This means that whenever $I_1, \ldots, I_n$ are disjoint elements of ζ such that $I_1 \cup \cdots \cup I_n$ is equal almost everywhere to some element I of ζ, then

$$v(I) = \Sigma v(I_i)$$

For each α in ζ we define the integrable function f_α on X by

$$f_\alpha(x) \equiv v(I)\mu(I)^{-1} \qquad (x \in I \in \alpha)$$

It is clear that $f_\alpha \prec f_\beta$ whenever $\alpha \leq \beta$.

We are interested in convergence properties of the family $\{f_\alpha\}$. Here is the basic estimate.

Proposition 1 *Suppose that ζ satisfies condition (a), and that the map v: $\zeta \to \mathbf{R}$ is finitely additive. For each α in Γ and each $\epsilon > 0$ let $\zeta_\alpha(\epsilon)$ consist of all I in ζ_α with $\|[J(I,\alpha),I]\| > \epsilon$, where*

$$[J(I,\alpha),I] \equiv v(J(I,\alpha))\mu(J(I,\alpha))^{-1} - v(I)\mu(I)^{-1}$$

Let K and ϵ be positive constants. Then there exists a positive constant δ, depending only on K, ϵ, and ρ, such that for each α in Γ having the property that $\|f_\beta\|_\lambda \leq K$ and $\|f_\beta\|_\lambda - \|f_\alpha\|_\lambda \leq \delta$ whenever $\beta \in \Gamma$ and $\alpha \leq \beta$, we have $\mu(A) \leq \epsilon$ for all finite unions A of elements of $\zeta_\alpha(\epsilon)$.

Proof We imitate the proof of Theorem 3. Take the positive constant r so large that $K\lambda(r)^{-1} \leq \frac{1}{3}\rho\epsilon$. Write $\delta \equiv \frac{1}{6}\epsilon^3\theta(r)\rho$. Consider an integrable set A of the form described. By condition (a), there exists β in Γ, with $\alpha \leq \beta$, and an integrable set $B \subset \{x: |f_\alpha(x) - f_\beta(x)| \geq \epsilon\}$ with $\mu(B) > \rho\mu(A)$. There is no loss of generality in taking the sets $B_1 \equiv \{x: |f_\alpha(x)| \geq r\}$ and $B_2 \equiv \{x: |f_\beta(x)| \geq r\}$ to be integrable. Since

$\|f_\alpha\|_\lambda \leq K$ and $\|f_\beta\|_\lambda \leq K$, we get $\mu(B_1) \leq \frac{1}{3}\rho\epsilon$ and $\mu(B_2) \leq \frac{1}{3}\rho\epsilon$, as before. We compute

$$\rho\mu(A) \leq \mu(B) \leq \epsilon^{-2} \int_{-B_1\cap -B_2} (f_\alpha - f_\beta)^2 \, d\mu + \mu(B_1) + \mu(B_2)$$

$$\leq \epsilon^{-2}\theta(r)^{-1} \int (f_\alpha - f_\beta)^2 \theta(\max\{|f_\alpha|, |f_\beta|\}) \, d\mu + \tfrac{2}{3}\rho\epsilon$$

$$\leq \epsilon^{-2}\theta(r)^{-1}2(\|f_\beta\|_\lambda - \|f_\alpha\|_\lambda) + \tfrac{2}{3}\rho\epsilon \leq \epsilon^{-2}\theta(r)^{-1}2\delta + \tfrac{2}{3}\rho\epsilon = \rho\epsilon$$

as was to be proved.

Proposition 1 can be used to give a simple *nonconstructive* proof that a function $f: S \to \mathbf{R}$ of bounded variation on a dense set $S \subset [0,1]$, with $0 \in S$ and $1 \in S$, has a derivative almost everywhere, as follows. Let ζ consist of all proper intervals with end points in S. For each interval I in ζ, with end points a and b, write $\nu(I) \equiv f(b) - f(a)$. Choose λ to have the special property that $\lambda(x) \leq c|x|$ for all x in $\mathbf{R}$ and some $c > 0$. Then for each α in Γ we have

$$\|f_\alpha\|_\lambda = \sum_{A\in\alpha} \mu(A)\lambda(\nu(A)\mu(A)^{-1}) \leq c \sum_{A\in\alpha} |\nu(A)| \leq cC$$

where C is a bound for the variation of f. Hence, *by the principle of omniscience*, the quantities $\|f_\alpha\|_\lambda$ have a least upper bound K. Given any $\epsilon > 0$, let δ be given by Proposition 1, and choose α as described in Proposition 1. Each finite union A of elements of $\zeta_\alpha(\epsilon)$ has measure at most ϵ. By omniscience again, it follows that there exists an integrable set B, with $\mu(B) \leq \epsilon$, such that $I \subset B$ a.e. for all I in $\zeta_\alpha(\epsilon)$. This implies that f has a derivative almost everywhere.

Constructively, things are not so simple. In fact there are two difficulties with the above proof. First, the least upper bound K may not exist. In general there is no possibility of proving constructively that K exists. We therefore assume that it exists. Second, the integrable set B may not exist. In case f is monotone-nondecreasing (or equivalently, ν is nonnegative), this second difficulty can be overcome, by means of the following lemma.

Lemma 5 *Let f be a monotone-nondecreasing function defined on a dense set $S \subset [0,1]$, with $0 \in S$ and $1 \in S$. Let ζ consist of all proper intervals with end points in S. Then for given constants $K > 0$ and $\epsilon > 0$ there exists $\delta > 0$, depending only on K and ϵ, such that for all α in Γ having the property that $\|f_\beta\|_\lambda \leq K$ and $\|f_\beta\|_\lambda - \|f_\alpha\|_\lambda \leq \delta$ whenever $\beta \geq \alpha$, and all $\gamma \geq \alpha$, there exists an integrable set B, with $\mu(B) \leq \epsilon$, such that for each I in $\zeta_\alpha(\epsilon)$ either $I \subset B$ or $I \in \zeta_\gamma(\epsilon/4)$.*

Proof By a slight modification of the proof of Proposition 1, we can find $\delta > 0$ so small that for all α as described and all finite subsets ζ_0 of $\zeta_\alpha(\epsilon/5)$ we have $\mu(\bigcup_{I \in \zeta_0} I') \leq \epsilon/2$, where I' is any interval containing I whose length is less than twice the length of I. Consider α as described, and consider any $\gamma \geq \alpha$. Let B_0 be a finite union of open intervals that contains the end points of each of the intervals in γ, with $\mu(B_0) \leq \epsilon/2$. Choose $t > 0$ so small that (i) $\mu(I) \geq t/2$ for all I in γ and (ii) for each I in ζ, with $\mu(I) \leq t$, either $I \subset B_0$ or $I \in \zeta_\gamma$.

Take finitely many intervals $I_1, \ldots, I_n, I'_1, \ldots, I'_n$ from ζ_α, with $I_k \subset I'_k$ and $\mu(I'_k) < 2\mu(I_k)$ for $1 \leq k \leq n$, such that for each I in ζ_α, with $\mu(I) \geq t/2$, there exists k ($1 \leq k \leq n$) for which

$$(2.5) \qquad I_k \subset I \subset I'_k \text{ and}$$

$$\nu(I_k)\{\mu(I_k)^{-1} - \mu(I)^{-1}\} < \frac{\epsilon}{4}$$

$$\nu(I'_k)\{\mu(I)^{-1} - \mu(I'_k)^{-1}\} < \frac{\epsilon}{4}$$

For each k either (a) one of the intervals I_k, I'_k belongs to $\zeta_\alpha(\epsilon/5)$ or (b) the numbers $|[J(I_k,\alpha),I_k]|$ and $|[J(I_k,\alpha),I'_k]|$ are both less than $\epsilon/4$. Let B_1 be the union of those I'_k corresponding to case (a). By the choice of δ, we see that $\mu(B_1) \leq \epsilon/2$. Write $B \equiv B_0 \cup B_1$. Then $\mu(B) \leq \epsilon$.

Consider any I in $\zeta_\alpha(\epsilon/2)$, for which $\mu(I) \geq t/2$. We shall prove that $I \subset B$. Now, (2.5) holds for some k, and either (i) $[J(I,\alpha),I] > \epsilon/2$ or (ii) $[J(I,\alpha),I] < -\epsilon/2$. If (i) holds, then

$$\nu(J(I,\alpha))\mu(J(I,\alpha))^{-1} - \nu(I_k)\mu(I)^{-1} \geq [J(I,\alpha),I] > \frac{\epsilon}{2}$$

By (2.5), it follows that $[J(I,\alpha),I_k] > \epsilon/4$. Hence $I'_k \in B_1$. Similarly, if (ii) holds, then $I'_k \in B_1$. Thus $I \subset I'_k \subset B_1 \subset B$ whenever $I \in \zeta_\alpha(\epsilon/2)$ and $\mu(I) \geq t/2$.

Now consider any I in $\zeta_\alpha(\epsilon)$. To prove Lemma 5, we must show that either $I \subset B$ or $I \in \zeta_\gamma(\epsilon/4)$. Either $\mu(I) \geq t/2$ or $\mu(I) \leq t$. If $\mu(I) \geq t/2$, the above proof gives $I \subset B$. Consider the case $\mu(I) \leq t$. By the construction of B_0, either $I \subset B_0$ or $I \in \zeta_\gamma$. We need only consider the alternative $I \in \zeta_\gamma$. Since

$$\epsilon < |[J(I,\alpha),I]| \leq |[J(I,\alpha),J(I,\gamma)]| + |[J(I,\gamma),I]|$$

either $\qquad |[J(I,\gamma),I]| > \dfrac{\epsilon}{4} \qquad$ or $\qquad |[J(I,\alpha),J(I,\gamma)]| > \dfrac{\epsilon}{2}$

In the former case, $I \in \zeta_\gamma(\epsilon/4)$. In the latter case, $J(I,\gamma) \in \zeta_\alpha(\epsilon/2)$.

Recall that $\mu(J) \geq t/2$ for all J in γ, by the choice of t. In particular, $\mu(J(I,\gamma)) \geq t/2$. By the above, it follows that $J(I,\gamma) \subset B$. Hence $I \subset B$.

Here is our substitute for Lebesgue's result that a monotone function has a derivative almost everywhere.

Theorem 4 *Take* $X \equiv [0,1]$, *let* μ *be Lebesgue measure, let* S *be dense in* $[0,1]$, *with* $0 \in S$ *and* $1 \in S$, *and let* ζ *be all proper intervals with end points in* S. *Let* ν *be a nonnegative finitely additive function on* ζ. *Let the set* $\{\|f_\alpha\|_\lambda : \alpha \in \Gamma\}$ *be bounded. Let* $\alpha_1 < \alpha_2 < \cdots$ *be elements of* Γ, *whose meshes approach* 0, *such that for each* $\epsilon > 0$ *there exists* n *in* Z^+ *such that* $\|f_\alpha\|_\lambda - \|f_{\alpha_n}\|_\lambda \leq \epsilon$ *for all* $\alpha \geq \alpha_n$. *Then for each* $\epsilon > 0$ *there exists* n *in* Z^+ *and an integrable set* B, *with* $\mu(B) \leq \epsilon$, *such that* $I \subset B$ *for all* I *in* $\zeta_{\alpha_n}(\epsilon)$.

Proof Using Lemma 5, we construct by induction a subsequence $\beta_0 < \beta_1 < \cdots$ of $\alpha_1 < \alpha_2 < \cdots$ and a sequence $B_0, B_1, \ldots$ of integrable sets, with $\mu(B_k) \leq \epsilon 2^{-k-1}$, such that for each I in $\zeta_{\beta_k}(\epsilon 4^{-k})$ either $I \subset B_k$ or $I \in \zeta_{\beta_{k+1}}(\epsilon 4^{-k-1})$. Since the mesh of the coverings $\alpha_1 < \alpha_2 < \cdots$ approaches 0, it follows that if $I \in \zeta_{\beta_0}(\epsilon)$, then $I \subset B_k$ for some k. The integrable set $B \equiv \cup B_k$ and the covering $\alpha_n \equiv \beta_0$ therefore satisfy the prescribed conditions.

The principal case in which the hypotheses of Theorem 4 can be verified is the following.

Theorem 5 *Let* g *be an integrable function on* $[0,1]$ *(relative to Lebesgue measure). For each* x *in* $[0,1]$ *write*

$$f(x) \equiv \int_{[0,x]} g \, d\mu$$

Then $f' = g$ *almost everywhere, in the sense that for each* $\epsilon > 0$ *there exists* α *in* Γ *and an integrable set* B, *with* $\mu(B) \leq \epsilon$, *such that for each* $I \equiv [a,b]$ *in* ζ_α *and each* x *in* $I - B$ *we have*

$$|g(x) - (f(b) - f(a))(b - a)^{-1}| \leq \epsilon$$

Proof We may assume that $g \geq 0$ a.e. Choose λ so that $\lambda(x) \leq c|x|$ for all x in $\mathbf{R}$. We first show that if the mesh of the covering α is sufficiently small, then $\|f_\alpha - g\|_1$, and hence $\|f_\alpha - g\|_\lambda$, will be arbitrarily small. This is certainly true if g is continuous. The general case then follows from the fact that the continuous functions are dense in L_1. Thus any sequence $\alpha_1 < \alpha_2 < \cdots$ of coverings whose meshes converge to 0 satisfies the hypotheses of Theorem 4. We can therefore choose n so large that there exists an integrable set B_0, with $\mu(B_0) \leq \epsilon/2$, such that $I \subset B_0$ for all I in $\zeta_\alpha(\epsilon/4)$, where $\alpha \equiv \alpha_n$. In addition, we can take n so large that $\|f_\alpha - g\|_1 < \epsilon^2/4$. From this we see that there exists an integrable set B_1, with $\mu(B_1) \leq \epsilon/2$, such that g is defined at all points of $-B_1$ and

$$|f_\alpha(x) - g(x)| < \frac{\epsilon}{2} \qquad (x \in -B_1)$$

Write $B \equiv B_0 \cup B_1$. Consider $I \equiv [a,b]$ in ζ_α and x in $I - B$. Either $I \in \zeta_\alpha(\epsilon/4)$ or $|f_\alpha(x) - (f(b) - f(a))(b - a)^{-1}| \leq \epsilon/2$. The former case is ruled out, since this would imply that $I \subset B_0 \subset B$. Hence

$$
\begin{aligned}
|g(x) &- (f(b) - f(a))(b - a)^{-1}| \\
&\leq |g(x) - f_\alpha(x)| + |f_\alpha(x) - (f(b) - f(a))(b - a)^{-1}| \\
&\leq \frac{\epsilon}{2} + \frac{\epsilon}{2} = \epsilon
\end{aligned}
$$

as was to be proved.

3. UPCROSSING INEQUALITIES

Usually many different roads can be followed to constructivize the same theorem of classical mathematics. To take a simple example, let P be the classical result that if $f: [0,1] \to \mathbf{R}$ is a continuous function with $f(0) < 0$ and $f(1) > 0$, then $f(x) = 0$ for some x in $[0,1]$. Here are three constructive substitutes for P. The first, Q_1, states that if f is a polynomial function on $[0,1]$ with $f(0) < 0$ and $f(1) > 0$, then $f(x) = 0$ for some x in $[0,1]$. The second, Q_2, states that if $f: [0,1] \to \mathbf{R}$ is a continuous function with $f(0) < 0$ and $f(1) > 0$, then for each $\epsilon > 0$ there exists $x \in [0,1]$ with $|f(x)| < \epsilon$. The third, Q_3, states that if $f: [0,1] \to \mathbf{R}$ is continuous, with $f(0) < 0$ and $f(1) > 0$, such that for each x in $[0,1]$ either $f(x) = 0$ or $f(x) \neq 0$, then $f(x) = 0$ for some x in $[0,1]$.

It is clear that Q_3 is a bad result, in that its constructive content is almost nil. Nevertheless, it does imply P with the aid of the principle of omniscience. Both Q_1 and Q_2 are good results. Result Q_1 is not a complete constructive substitute for P, because there is no simple argument based on the principle of omniscience by which P can be derived from Q_1. For that matter, the derivation of P from Q_2, using the principle of omniscience, is not so simple either. Nevertheless, Q_2

comes very near to capturing the full flavor of the classical theorem P, and in most situations it affords a workable substitute. We might say that Q_2 is an *equal-hypothesis* substitute for P: The hypotheses are the same, although the conclusion of Q_2 is weaker. An equal-hypothesis substitute for a classical result is of special value. It gives information about the *general* case, ideally as much information as can be obtained without appeal to the principle of omniscience. A constructive substitute such as Q_1, in which the hypothesis is changed, gives no information at all (constructively) for those cases in which the hypothesis of P is satisfied but the hypothesis of Q is not. For instance, Theorem 4 of this chapter gives no information unless the integer n exists for each $\epsilon > 0$. A similar remark applies to Theorem 2.

Our goal is to find good equal-hypothesis substitutes for (i) the ergodic theorem and (ii) Lebesgue's theorem that a function of bounded variation has a derivative almost everywhere.

We first look at the ergodic theorem. Consider a finite measure space $(X,F,\mathfrak{F},\mathfrak{M},\mu)$ and a function T from a full subset D_T of X to X, such that (i) if A is any full subset of X, then $\{x \in D_T : Tx \in A\}$ is also a full subset of X, and (ii) if f is any integrable function on X, then $f \circ T$ [which is defined almost everywhere by (i)] is also integrable, and

$$\int f \circ T \, d\mu = \int f \, d\mu$$

Such a function T is called a measure-preserving transformation. For each nonnegative integer n let T^n be the nth iterate of T: $T^0 x \equiv x$, $T^1 x \equiv Tx$, $T^2 x \equiv TTx$, and so on. Then T^n is also a measure-preserving transformation on X. Examples indicate that the transforms x, Tx, $T^2 x$, . . . of a general point x in X display some sort of recurrent behavior. One might hope, for instance, that for each integrable set A there exists for almost all x in X a constant $\alpha \geq 0$, depending on x and A, such that the transforms x, Tx, . . . lie in A with probability α; in other words,

$$\frac{1}{n+1} \left(\chi_A(x) + \chi_A(Tx) + \cdots + \chi_A(T^n x) \right) \to \alpha$$

as $n \to \infty$. This turns out to be the case (according to Birkhoff). Birkhoff even proved the stronger statement that for each f in L_1 the limit

$$\lim_{n \to \infty} \frac{1}{n+1} \left(f(x) + f(Tx) + \cdots + f(T^n x) \right)$$

exists for almost all x.

Birkhoff's ergodic theorem has been generalized in many ways by many authors. Here we consider one of the most recent and beautiful of these generalizations, the Chacon-Ornstein ergodic theorem, which concerns a bounded linear operator T from L_1 to L_1, assumed to be positive and bounded by 1. (Assuming T to be positive means that $Tf \leq Tg$ a.e. whenever f and g are elements of L_1 with $f \leq g$ a.e.; assuming T to be bounded by 1 means that $\|Tf\|_1 \leq \|f\|_1$ for all f in L_1.) The Chacon-Ornstein theorem states that if f and p are measurable functions, with $f \in L_1$ and $p \geq 0$ a.e., then the limit

$$\lim_{n \to \infty} \frac{f(x) + Tf(x) + \cdots + T^n f(x)}{p(x) + Tp(x) + \cdots + T^n p(x)} \equiv L(x)$$

exists for almost all x in the set $\{x : p(x) > 0\}$. Birkhoff's ergodic theorem is a special case. To see this, define the operator T on L_1 by $Tf(x) \equiv f(Tx)$. Then T satisfies the hypotheses of the Chacon-Ornstein theorem, and $T1 = 1$. Birkhoff's theorem is then obtained from the Chacon-Ornstein theorem by taking p to be 1.

Unfortunately even Birkhoff's theorem is *not* constructively valid, as the following heuristic argument shows. Think of X as the union of two equal tanks of fluid, and T as a motion of the fluid, which is supposed to keep the fluid confined to the tank in which it has been placed. Imagine that there may be a small leak, which would in fact allow the fluid in the two tanks to mix, but that we are not able to decide whether a leak actually exists. Since the leak if it exists, is small, there will be little mixing between the tanks after unit time (that is, under the transformation T), but after a long time (that is, under the transformation T^n for some large n) the mixing may be substantial. Let χ_A be the characteristic function of the first tank. If there is no leak, the limit α exists for almost all x in X, and equals $\chi_A(x)$. On the other hand, if there is a leak, it is more likely that $\alpha = \frac{1}{2}$. Since we cannot test for a leak, there is no reason to expect the limit α to exist. This heuristic argument can be made precise. The upshot is that Birkhoff's theorem is nonconstructive.

Birkhoff's theorem is an instance in which a sequence $\{a_n\}$ of real numbers that converges classically fails to converge constructively. A simpler instance is obtained as follows. Let $\{r_n\}$ be a monotone-non-decreasing sequence of real numbers, each of which is either 0 or 1, for which we are unable to rule out either the possibility that all r_n are 0 or the possibility that some r_n is 1. This sequence converges classically but *not* constructively. However, constructively it does have the following property: it upcrosses from an arbitrary real number α to an arbitrary real number $\beta > \alpha$, in the sense of the following definition, at most once.

Definition 4   A (finite or infinite) sequence a_1, a_2, . . . of real numbers *upcrosses* from a real number α to a real number $\beta > \alpha$ *at most n times* if $N \leq n$ whenever N is a positive integer such that there exists a subsequence b_1, b_2, . . . , b_{2N} such that $b_i \leq \alpha$ for $i = 1, 3, \ldots,$ $2N - 1$ and $b_i \geq \beta$ for $i = 2, 4, \ldots, 2N$. The least such n (alternatively, the greatest such N), if it exists, is called the *number of upcrossings* of the sequence from α to β.

It is clear that a sequence $\{a_n\}$ of real numbers converges classically if and only if it is bounded, and for all real numbers α and β with $\alpha < \beta$, upcrosses from α to β at most a certain finite number $N(\alpha,\beta)$ of times. Constructively, however, convergence is much the stronger property. Thus we have found a possible constructive substitute for the convergence of a sequence of real numbers.

In the light of these remarks, the following developments will be seen to constitute a constructive substitute for the Chacon-Ornstein ergodic theorem.

We first prove a combinatorial lemma.

Lemma 6   *Let $a(-1)$, $a(0)$, . . . , $a(n)$ and $b(-1)$, $b(0)$, . . . , $b(n)$ be real numbers. Let N be any positive integer such that there exist integers*

$$(3.1) \qquad -1 \leq u_1 < v_1 < u_2 < v_2 < \cdots < u_N < v_N \leq n$$

with

$$(3.2) \qquad\qquad a(u_i) \leq b(v_i) \qquad (1 \leq i \leq N)$$

and

$$(3.3) \qquad\qquad a(u_{i+1}) \leq b(v_i) \qquad (1 \leq i \leq N - 1)$$

Call a finite or void sequence $P \equiv \{s_1, t_1, \ldots, s_m, t_m\}$ of integers (where $m \equiv m_P$ is a nonnegative integer depending on P) admissible if

$$-1 \leq s_1 < t_1 \leq s_2 < t_2 \leq \cdots \leq s_m < t_m \leq n$$

For each admissible P write

$$S(P) \equiv \sum_{i=1}^{m} b(t_i) - a(s_i)$$

[Of course, $S(P) \equiv 0$, if $m_P = 0$.] Then for each admissible P there exists an admissible Q with $S(Q) \geq S(P)$ and $m_Q \geq N$.

Proof Consider any admissible P. It will be enough to show that if $m_P < N$, there exists an admissible Q with $S(Q) \geq S(P)$ and

$$m_Q = m_P + 1$$

since by repeating this construction enough times we can boost the value of m_Q to be N. Assume therefore that $m \equiv m_P < N$.

For convenience, we supplement our notation by setting $s_{m+1} \equiv n$. Choose integers (3.1) satisfying (3.2) and (3.3). Then v_{m+1} is defined, since $m + 1 \leq N$, and $v_{m+1} \leq n \equiv s_{m+1}$. Hence there exists a least positive integer i with $v_i \leq s_i$. If $i = 1$, let Q be the finite sequence obtained from P by adjoining the points u_1 and v_1, so that

$$Q \equiv \{u_1, v_1, s_1, t_1, \ldots, t_m\}$$

Consider next the case $i > 1$. Then $v_{i-1} > s_{i-1}$, because $v_{i-1} \leq s_{i-1}$ would contradict the choice of i. Hence $s_{i-1} < v_{i-1} < u_i < v_i \leq s_i$. If $u_i < t_{i-1}$, let Q be the finite sequence obtained by adjoining the points v_{i-1} and u_i to P so that

$$Q \equiv \{s_1, t_1, \ldots, s_{i-1}, v_{i-1}, u_i, t_{i-1}, \ldots, t_m\}$$

If $u_i \geq t_{i-1}$, let Q be the finite sequence obtained by adjoining the points u_i and v_i to P so that

$$Q \equiv \{s_1, t_1, \ldots, t_{i-1}, u_i, v_i, s_i, \ldots, t_m\}$$

Then Q is admissible in every case, and $m_Q = m_P + 1$. Also, in every case, either

$$S(Q) = S(P) + b(v_i) - a(u_i)$$
or
$$S(Q) = S(P) + b(v_{i-1}) - a(u_i)$$

Hence $S(Q) \geq S(P)$.

Definition 5 Let T be a positive operator from L_1 to L_1 that is bounded by 1. Let n be a positive integer. Let $f \equiv \{f_0, f_1, \ldots, f_n\}$ be a sequence of measurable functions such that f_j^+ is integrable for each j and

$$(3.4) \qquad T \Big(\sum_{j \in \Omega} f_j \Big)^+ \geq \sum_{j \in \Omega} f_{j+1} \text{ a.e.}$$

for each finite subset Ω of $\{0, 1, \ldots, n - 1\}$. Let $p \equiv \{p_0, \ldots, p_n\}$ be a sequence of nonnegative measurable functions such that for each j with $0 \leq j \leq n - 1$ and each integrable function h with $h \leq p$, a.e. we have $Th \leq p_{j+1}$ a.e. For each integer $u \leq n$ write

$$\sum (f,u) \equiv \sum_{j=0}^{u} f_j \qquad \sum (f - p,u) \equiv \sum_{j=0}^{u} (f_j - p_j)$$

(Of course, these sums are taken to be 0 if $u \leq -1$.) Let $\Sigma(x,f,u)$ be the value of $\Sigma(f,u)$ at the point x, and define $\Sigma(x, f - p, u)$ similarly. For each pair u, v of integers with $-1 \leq u, v \leq n$ let $A(u,v)$ be a measurable set such that

$$(3.5) \qquad \Sigma(x,f,u) \leq \Sigma(x, f - p, v) \qquad (x \in A(u,v))$$

Then for almost all x in X the maximum integer N such that there exist integers (3.1) with

$$(3.6) \qquad x \in \bigcap_{j=1}^{N} A(u_j,v_j) \cap \bigcap_{j=1}^{N-1} A(u_{j+1},v_j)$$

is well defined. We let $\omega_n(x)$ be this maximum integer N.

Here is our ergodic theorem.

Theorem 6 *Under the conditions of Definition 5, the function $\omega_n p_0$ is integrable, and*

$$(3.7) \qquad \int \omega_n p_0 \, d\mu \leq \int f_0{}^+ \, d\mu$$

Proof For each choice F of integers $-1 \leq u_1 < v_1 < \cdots < u_k < v_k \leq n$, let Φ_F be the characteristic function of the intersection of the sets $A(u_j,v_j)$ $(1 \leq j \leq k)$ and $A(u_{j+1},v_j)$ $(1 \leq j \leq k - 1)$. Then ω_n is the maximum of the measurable functions $k\Phi_F$, taken over the finitely many possible choices of F, and is therefore measurable.

Consider any admissible sequence $P \equiv \{s_1, t_1, \ldots, s_m, t_m\}$ of integers, so that $-1 \leq s_1 < t_1 \leq \cdots \leq s_m < t_m \leq n$. Write

$$S(f,p,P) \equiv \sum_{i=1}^{m} \left\{ \sum (f - p, t_i) - \sum (f,s_i) \right\}$$

Since there are only finitely many admissible P, the maximum $\lambda(f,p)$ of the functions $S(f,p,P)$ exists. Since each of the functions $S(f,p,P)^+$ is integrable, and $S(f,p,P) \equiv 0$ in case $m_P = 0$, it follows that $\lambda(f,p)$ is a nonnegative integrable function. For each x for which $\lambda(f,p)$ is defined let $\lambda(x,f,p)$ be the value of $\lambda(f,p)$ at x. Define $S(x,f,p,P)$ similarly.

Let f' and p' be the finite sequences $f' \equiv \{f'_0, \ldots, f'_{n-1}\}$ and $p' \equiv \{p'_0, \ldots, p'_{n-1}\}$, respectively, where $f'_i \equiv f_{i+1}$ and $p'_i \equiv p_{i+1}$ $(0 \le i \le n - 1)$. We shall prove that

$$(3.8) \qquad f_0(x)^+ - \omega_n(x)p_0(x) \ge \lambda(x,f,p) - \lambda(x,f',p')$$

for all x such that the quantities involved are all defined, and thus for almost all x. We compute

$$
\begin{aligned}
S(x,f,p,P) &= \sum_{i=1}^{m} \left\{ \sum (x, f - p, t_i) - \sum (x,f,s_i) \right\} \\
&= \sum (x, f' - p', t_1 - 1) - \sum (x, f', s_1 - 1) + a - p_0(x) \\
&\quad + \sum_{i=2}^{m} \left\{ \sum (x, f' - p', t_i - 1) - \sum (x, f', s_i - 1) - p_0(x) \right\}
\end{aligned}
$$

where $a = 0$ if $s_1 \ge 0$ and $a = f_0(x)$ if $s_1 = -1$. In either case

$$S(x,f,p,P) \le \lambda(x,f',p') + f_0(x)^+ - mp_0(x)$$

In case

$$(3.9) \qquad\qquad\qquad m \equiv m_P \ge \omega_n(x)$$

this gives

$$
(3.10) \qquad
\begin{aligned}
f_0(x)^+ - \omega_n(x)p_0(x) &\ge f_0(x)^+ - mp_0(x) \\
&\ge S(x,f,p,P) - \lambda(x,f',p')
\end{aligned}
$$

By Lemma 6, for each admissible Q there exists an admissible P satisfying (3.9) with $S(x,f,p,P) \ge S(x,f,p,Q)$. Therefore (3.10) gives

$$f_0(x)^+ - \omega_n(x)p_0(x) \ge S(x,f,p,Q) - \lambda(x,f',p')$$

for all admissible Q. Taking maxima with respect to Q gives (3.8).

Consider any admissible P with $t_m \le n - 1$. Then

$$S(f,p,P) = \sum_{j \in \Omega} f_j - \sum_{j=0}^{n-1} v_j p_j$$

where $\Omega \subset \{0, 1, \ldots, n - 1\}$, and $\nu_0, \ldots, \nu_{n-1}$ are nonnegative integers. For each j let h_j be the integrable function

$$h_j \equiv \min \left\{ p_j, \left(\sum_{k \in \Omega} f_k \right)^+ \right\}$$

Then
$$S(f,p,P)^+ \geq \left(\sum_{j \in \Omega} f_j \right)^+ - \sum_{j=0}^{n-1} \nu_j h_j$$

and $Th_j \leq p_{j+1}$ $(0 \leq j \leq n - 1)$. Hence

$$TS(f,p,P)^+ \geq \sum_{j \in \Omega} f_{j+1} - \sum_{j=0}^{n-1} \nu_j p_{j+1} = S(f',p',P)$$

Since T is a positive operator and $\lambda(f,p) \geq S(f,p,P)^+$, this gives $T\lambda(f,p) \geq S(f',p',P)$. Taking maxima with respect to P, we get $T\lambda(f,p) \geq \lambda(f',p')$. In view of (3.8), this gives

$$f_0^+ - \omega_n p_0 \geq \lambda(f,p) - T\lambda(f,p) \text{ a.e.}$$

Integrating this inequality, we get

$$\begin{aligned} \int (f_0^+ - \omega_n p_0) \, d\mu &\geq \int \lambda(f,p) \, d\mu - \int T\lambda(f,p) \, d\mu \\ &= \|\lambda(f,p)\|_1 - \|T\lambda(f,p)\|_1 \geq 0 \end{aligned}$$

which is equivalent to (3.7).

We now specialize Theorem 6 to obtain a constructive version of the Chacon-Ornstein theorem. Let g and q be integrable functions, with $q > 0$ a.e. Let α and β be real numbers, with $\alpha < \beta$, and n a positive integer. For arbitrary integers u and v with $-1 \leq u, v \leq n$ let $A(u,v)$ be a measurable set, with

$$\sum_{i=0}^{u} T^i(g - \alpha q) \leq \sum_{i=0}^{v} T^i(g - \beta q) \qquad \text{on } A(u,v)$$

Then for almost all x in X the maximum integer $N \equiv \sigma_n(x)$ such that there exist integers (3.1) satisfying (3.6) is well defined, and we have the following corollary.

Corollary   *The function $\sigma_n q$ is integrable, with*

$$(3.11) \qquad \int \sigma_n q \, d\mu \leq (\beta - \alpha)^{-1} \int (g - \alpha q)^+ \, d\mu$$

Proof For each j $(0 \leq j \leq n)$ write $f_j \equiv T^j(g - \alpha q)$ and $p_j \equiv (\beta - \alpha)T^j q$. The sequences $f \equiv \{f_0, \ldots, f_n\}$ and $p \equiv \{p_0, \ldots, p_n\}$ and the sets $A(u,v)$ satisfy the requirements of Definition 5. Hence (3.7) holds. Since $\sigma_n = \omega_n$, $f_0 = g - \alpha q$, and $p_0 = (\beta - \alpha)q$, inequality (3.7) reduces to (3.11).

For the exact relation of our results to the Chacon-Ornstein theorem, see Prob. 12 below.

Lebesgue's theorem (that a function of bounded variation has a derivative almost everywhere) and the martingale theorem can also be treated in terms of upcrossing inequalities. In fact the relevant inequalities can be obtained as special cases of Theorem 6. We do this for Lebesgue's theorem only.

Theorem 7 *Let f be a function of bounded variation defined on a full subset S of $\mathbf{R}$ that vanishes outside some finite interval. Let $0 < t_0 < t_1 < t_2 < \cdots < t_m$ be real numbers. Let α and β be real numbers, with $0 \leq \alpha < \beta$. For arbitrary integers i and j with $0 \leq i, j \leq m$ let $B(i,j)$ be a measurable set, with*

$$f(x + t_i) - f(x) - \alpha t_i \leq f(x + t_j) - f(x) - \beta t_j \qquad (x \in B(i,j))$$

Then for almost all x in X the maximum integer $N \equiv \rho(x)$ such that there exist integers $0 \leq i_1 < j_1 < \cdots < i_N < j_N \leq m$ with

$$x \in \bigcap_{k-1}^{N} B(i_k, j_k) \cap \bigcap_{k-1}^{N-1} B(i_{k+1}, j_k)$$

is well defined, the function ρ is integrable, and

$$\int \rho \, d\mu \leq (\beta - \alpha)^{-1} V$$

where V is any positive constant such that

$$\sum_{i=1}^{k-1} \{f(x_{i+1}) - f(x_i) - \alpha(x_{i+1} - x_i)\} \leq V$$

whenever $x_1 < x_2 < \cdots < x_k$ are points of S.

Proof By an approximation argument, which is left to the reader, we reduce to the case in which the numbers $t_0, \ldots, t_m$ are all rational. Let t be any positive constant such that $t_0, t_1, \ldots, t_m$ are all integral multiples of t: $t_j \equiv n_j t$ $(0 \leq j \leq m)$. Let T be the measure-preserving transformation on $\mathbf{R}$ defined by $Tx \equiv x + t$. Write

$$f_j(x) \equiv t^{-1}\{f(x + jt + t) - f(x + jt)\} - \alpha$$

and $p_j(x) \equiv (\beta - \alpha)$ for $0 \leq j \leq n$, where $n \equiv n_m$. Since each of the functions $f_j{}^+$ is bounded and measurable, and vanishes outside a finite interval, it is integrable. The sequences $f \equiv \{f_0, \ldots, f_n\}$ and $p \equiv \{p_0, \ldots, p_n\}$ satisfy the conditions of Definition 5. Also, for each integer u, with $-1 \leq u \leq n$, we have

$$\Sigma(x,f,u) = t^{-1}\{f(x + ut + t) - f(x)\} - (u + 1)\alpha$$
$$\text{and} \quad \Sigma(x, f - p, u) = t^{-1}\{f(x + ut + t) - f(x)\} - (u + 1)\beta$$

Let u and v be arbitrary integers, with $-1 \leq u, v \leq n$. Write $A(u,v) \equiv B(i,j)$, in case $u = n_i - 1$ and $v = n_j - 1$ for certain integers i and j with $0 \leq i, j \leq m$. Otherwise, write $A(u,v) \equiv \phi_0$. To see that the sets $A(u,v)$ satisfy the conditions of Definition 5, it is enough to consider the case $u = n_i - 1$ and $v = n_j - 1$. Consider x in $A(u,v) \equiv B(i,j)$. Then

$$\begin{aligned}
t\{\Sigma(x, f - p, v) - \Sigma(x,f,u)\} &= f(x + vt + t) - f(x) - t(v + 1)\beta \\
&\quad - \{f(x + ut + t) - f(x) - t(u + 1)\alpha\} \\
&= f(x + t_j) - f(x) - t_j\beta \\
&\quad - \{f(x + t_i) - f(x) - t_i\alpha\} \geq 0
\end{aligned}$$

Hence the sets $A(u,v)$ satisfy the conditions of Definition 5. Moreover, $\omega_n = \rho$ a.e. From Theorem 6 it follows that ρ is integrable and $\int\rho \, d\mu \leq (\beta - \alpha)^{-1}\int f_0{}^+ \, d\mu$. To complete the proof, we must show that $\int f_0{}^+ \, d\mu \leq V$. The summation in the following computation is actually finite.

$$\begin{aligned}
\int f_0{}^+ \, d\mu &= \int_0^t \left(\sum_{k=-\infty}^{\infty} f_0(x + kt)^+ \right) dx \\
&= t^{-1} \int_0^t \left(\sum_{k=-\infty}^{\infty} \{f(x + kt + t) - f(x + kt) - \alpha t\}^+ \right) dx \\
&\leq t^{-1} \int_0^t V \, dx = V
\end{aligned}$$

For the exact relation of Theorem 7 to Lebesgue's theorem, see Prob. 13 below.

PROBLEMS

1. Construct a measure on $[0,1]$ that is *not* the difference of two mutually singular positive measures. (*Hint:* See Prob. 1 of Chap. 6.)

2. Let μ and ν be mutually singular positive measures on a locally compact space X. Show that there exists a complemented set A, measurable with respect to $\mu + \nu$, with $\mu(A) = \nu(-A) = 0$.

3. Let $\{a_n^1\}$, $\{a_n^2\}$, $\ldots$ be monotone-nondecreasing bounded sequences of real numbers. Show that there exists a monotone-nondecreasing bounded sequence $\{a_n\}$ of real numbers that converges if and only if each of the sequences $\{a_n^i\}$ converges.

4. Let μ be a measure on a locally compact space. Show that there exists a monotone-nondecreasing bounded sequence $\{a_n\}$ of real numbers, which converges if and only if μ is the difference of two mutually singular positive measures.

5. Let μ and ν be measures on a locally compact space X, with μ positive, satisfying (b) of Theorem 2. Show that there exists a measurable function $h\colon X \to \mathbf{R}$, satisfying (i) and (ii) of Theorem 2, if and only if a certain monotone-nondecreasing bounded sequence of real numbers converges.

6. Let μ be a measure on a compact space X, with $|\int f\,d\mu| \leq 1$ whenever $\|f\| \leq 1$. Let ϵ be any positive constant. Say that μ *splits* to within ϵ if there exists f_0 in $C(X)$, with $0 \leq f_0 \leq 1$, so that $\int f\,d\mu \leq \int f_0\,d\mu + \epsilon$ for all f in $C(X)$ with $0 \leq f \leq 1$. Use the limited principle of omniscience n times, where n is any integer with $2^{-n} < \epsilon$, to show that μ splits to within ϵ. Show that μ splits to within ϵ, for all $\epsilon > 0$, if and only if μ is the difference of two mutually singular positive measures.

7. Let μ be Lebesgue measure on $[0,1]$ and ν any positive measure on $[0,1]$. Show that there exists a bounded double sequence $\{a_n^i\}_{n,i=1}^{\infty}$ of real numbers, such that (i) $a_1^i \leq a_2^i \leq \cdots$ for each i, and (ii) $a_n^1 \geq a_n^2 \geq \cdots$ for each n, having the following property: The iterated limit $\lim\limits_{i\to\infty} \lim\limits_{n\to\infty} a_n^i$ exists if and only if ν has a representation

$$(*) \qquad\qquad \nu = h\,d\mu + \nu_1$$

where h is an integrable function (with respect to μ), and ν_1 and μ are mutually singular.

8. Construct a positive measure ν on $[0,1]$ so that there does *not* exist any monotone-nondecreasing bounded sequence $\{a_n\}$ whose convergence implies that ν has a representation $(*)$ (this is an unsolved problem).

9. Construct an infinite martingale $f_1 < f_2 < \cdots$ such that the sequence $\{\|f_n\|_1\}$ converges, but $\{f_n\}$ does *not* converge almost everywhere.

10. Let g be an integrable function on $[0,1]^n$, relative to Lebesgue measure μ. Let τ be a positive constant, and define ζ as in the text. For each I in ζ write

$$\nu(I) \equiv \int_I g \, d\mu$$

Show that the derivative of ν with respect to μ exists and equals g almost everywhere, in the sense that for each $\epsilon > 0$ there exists α in Γ and an integrable set B, with $\mu(B) < \epsilon$, such that for each I in ζ_α and each x in $I - B$ we have

$$|g(x) - \nu(I)\mu(I)^{-1}| \le \epsilon$$

11. Make precise the heuristic argument, given in the text, that Birkhoff's ergodic theorem is *not* constructively valid.

12. Under the hypotheses of Theorem 6, use the principle of omniscience to show that the limit $L(x)$ exists almost everywhere on any integrable set on which $p > 0$ a.e.

13. Use the principle of omniscience and Theorem 7 to show that a function of bounded variation has a derivative almost everywhere from the right.

NOTES

The theorems of Sec. 1 are unsatisfying, because the principle of omniscience is avoided by blatantly building it into the hypotheses. This material is presented primarily for completeness, but also as a point of departure for some interesting questions (raised in Probs. 3 to 8).

The norms $\|\ \|_\lambda$ seem not to have been used classically, perhaps because they are not norms in the classical sense. They merit further investigation.

Definition 2 can possibly be put in a better way. As given, it is adequate for getting the theorems we want, but it may possibly be too restrictive or too hard to verify in actual cases. More work is needed to determine the exact class of complemented sets A for which (2.2) should be assumed to hold.

Note how use of the norms $\|\ \|_\lambda$ makes possible estimates of the type given in Theorem 3. Such estimates do not hold in the norm $\|\ \|_1$.

Since Theorem 6 is new, from both the classical and the constructive points of view, we have given it in a fair degree of generality. This makes the definitions and proof somewhat more complicated than they would be if used just to state and prove the corollary. One should compare Theorem 6 with recent work of Chacon [4].

CHAPTER 9

NORMED LINEAR SPACES

In Sec. 1 we define a normed linear space and prove some simple results. Section 2 is concerned with the L_p spaces. Clarkson's inequalities are used to determine the form of a normable linear functional on L_p, for $p > 1$. (In contrast to the classical theory, a bounded linear functional need not have a norm.) In Sec. 3 we prove the separation theorem (and its corollary, the Hahn-Banach theorem) under the assumption that a certain convex set is located. In Sec. 4 we make some elementary remarks about operators, define Hilbert space, and prove the spectral theorem. For algebras of normable operators, the spectral theorem admits a significant strengthening (see the supplement to Theorem 8). Next we prove the compactness of the unit sphere of the dual, and a result (Theorem 10) which makes it plausible that every constructively defined linear functional on the dual B^* is induced by an element of B—a curious form of reflexivity! The last section contains a proof of the Krein-Milman theorem. In contrast to the classical treatment, detailed estimates are needed.

243

Most classes of functions that arise in analysis are endowed with both a linear structure and a norm. This makes it natural to introduce the concept of a normed linear space.

1. DEFINITIONS AND EXAMPLES

To be able to discuss real and complex linear spaces simultaneously, we let $\mathbf{F}$ denote either the real or the complex numbers. The elements of $\mathbf{F}$ are called *scalars*.

Definition 1 A *linear space* (or *vector space*) over $\mathbf{F}$ is an abelian group V with an operation $(a,v) \to av$ from $\mathbf{F} \times V$ to V such that

$$(a_1 + a_2)v = a_1 v + a_2 v \qquad (a_1, a_2 \in \mathbf{F}, v \in V)$$
$$a(v_1 + v_2) = av_1 + av_2 \qquad (a \in \mathbf{F}, v_1, v_2 \in V)$$
$$\text{and} \qquad a_1(a_2 v) = (a_1 a_2)v, \ 1v = v \qquad (a_1, a_2 \in \mathbf{F}, v \in V)$$

The elements of V are called *vectors*. We assume the reader to be familiar with the elementary theory of linear spaces.

Definition 2 A *seminorm* $\| \ \|$ on a linear space V is a function $v \to \|v\|$ from V to $\mathbf{R}^{0+}$ such that

$$\|av\| = |a| \, \|v\| \qquad \text{and} \qquad \|v_1 + v_2\| \leq \|v_1\| + \|v_2\|$$
$$(a \in \mathbf{F}, v, v_1, v_2 \in V)$$

In case $v = 0$ whenever $\|v\| = 0$, the seminorm $\| \ \|$ is called a *norm*. A linear space with a norm is called a *normed linear space*.

There is no loss of generality in confining our attention to norms, since a seminorm $\| \ \|$ on a linear space V becomes a norm if the equality relation on V is changed to mean that $\|v_1 - v_2\| = 0$.

A seminorm $\| \ \|$ on a linear space V determines a pseudometric ρ on V:

$$\rho(v_1,v_2) \equiv \|v_1 - v_2\| \qquad (v_1, v_2 \in V)$$

which is a metric when $\| \ \|$ is a norm. A normed linear space V is thus a metric space. It is easily seen that addition and subtraction are continuous functions from $V \times V$ to V, and that multiplication of a vector by a scalar is a continuous function from $\mathbf{F} \times V$ to V.

The norm can be described geometrically in terms of the closed unit sphere. The proof is left to the reader.

Proposition 1 *The closed unit sphere*

$$S \equiv \{v \in V : \|v\| \le 1\}$$

in a seminormed linear space V is convex *[meaning that if v_1 and v_2 are any vectors in S, and t is any real number with $0 \le t \le 1$, then the vector $tv_1 + (1 - t)v_2$ is in S] and* circular *[which means that for each v in V the set $S_v \equiv \{a \in \mathbf{F} : av \in S\}$ (i) is locally compact, (ii) contains some non-zero scalar, and (iii) contains all a in $\mathbf{F}$ such that there exists b in $\mathbf{F}$ with $|a| \le |b|$]. Conversely, if S is a convex circular subset of a linear space V, there exists a unique seminorm $\|\ \ \|$ on V defined by*

$$\|v\| \equiv \text{g.l.b.} \ \{|a|^{-1} : a \in S_v,\ a \ne 0\}$$

such that S is the closed unit sphere of V relative to the seminorm $\|\ \ \|$. The seminorm $\|\ \ \|$ is a norm if and only if S_v is bounded for all $v \ne 0$.

Definition 3 A *linear functional* on a vector space V is a function $\lambda : V \to \mathbf{F}$ such that

$$\lambda(av) = a\lambda(v) \qquad \text{and} \qquad \lambda(v_1 + v_2) = \lambda(v_1) + \lambda(v_2)$$

whenever $a \in \mathbf{F}$ and $v, v_1, v_2 \in V$.

In case V has a norm and therefore a metric, we can say what it means for a linear functional λ on V to be continuous: it means that λ is uniformly continuous on bounded subsets of V. Therefore λ is bounded on some sphere about the origin of V, and hence it is bounded on the unit sphere S of V. Conversely, if λ is any linear functional on V which is bounded on S, then λ is continuous, in fact uniformly continuous. To see this, assume that $|\lambda(v)| \le c$ for all v in S. Consider any vector v. For each $a > 0$ with $a\|v\| \le 1$ we have $av \in S$, and therefore $a|\lambda(v)| = |\lambda(av)| \le c$. It follows that $|\lambda(v)| \le c\|v\|$. Putting $v \equiv v_1 - v_2$ gives

$$|\lambda(v_1) - \lambda(v_2)| = |\lambda(v_1 - v_2)| \le c\|v_1 - v_2\|$$

for all vectors v_1 and v_2. Hence λ is uniformly continuous.

We have thus proved the following proposition.

Proposition 2 *The following are equivalent conditions for a linear functional λ on a normed linear space V:*

(a) λ is continuous
(b) λ is uniformly continuous

 (c) *λ is bounded on S*
 (d) *λ is bounded on each bounded subset of V*
 (e) *There exists $c > 0$ such that*

$$|\lambda(v)| \le c\|v\| \qquad (v \in V)$$

A linear functional λ with property (e) is called *bounded*, and c is called a *bound* for λ. The set of all bounded linear functionals on V is denoted by V^*.

The *null space* $N(\lambda)$ of a linear functional λ on a linear space V is the set of all vectors v for which $\lambda(v) = 0$. We can characterize a nonzero bounded linear functional λ in terms of its null space. [Calling λ nonzero means that $\lambda(v) \ne 0$ for some v in V.] A linear subset L of a normed linear space V is a *hyperplane* if there exists u in V and $c > 0$ such that (i) $\|u - w\| > c$ for all w in L, and (ii) every vector v can be written in the form $v = au + w$, with $a \in \mathbf{F}$ and $w \in L$. We have the following result.

Proposition 3 *If λ is a nonzero bounded linear functional on a normed linear space V, then $N(\lambda)$ is a hyperplane. Conversely, if L is a hyperplane with associated vector u, there exists a unique bounded linear functional λ with $N(\lambda) = L$ and $\lambda(u) = 1$.*

Proof Let λ be a nonzero bounded linear functional, and u any vector with $\lambda(u) = 1$. By Proposition 2, there exists $c > 0$ such that $c|\lambda(v)| \le \|v\|$ for all vectors v. If w is any vector in $N(\lambda)$, this gives

$$\|u - w\| \ge c|\lambda(u - w)| = c$$

For each vector v we have $v = au + w$, where $a \equiv \lambda(v)$ and $w \equiv v - \lambda(v)u$; clearly $w \in N(\lambda)$.

Consider, conversely, a hyperplane L with associated vector u satisfying (i) and (ii). The representation $v = au + w$ is unique, since if $v = a_1 u + w_1$ is another representation of the same vector v, we have $(a - a_1)u = w_1 - w$, and thus $a = a_1$ and $w = w_1$ by (i). Define the linear functional λ on V by $\lambda(v) \equiv a$. Clearly $\lambda(u) = 1$ and $L = N(\lambda)$. For each vector v for which $\lambda(v) \equiv a > 0$ we have, by (i),

$$\|v\| = |a|\,\|u - (-a^{-1}w)\| \ge |a|c = |\lambda(v)|c$$

or $|\lambda(v)| \le c^{-1}\|v\|$. Thus λ is bounded.

The most useful normed linear spaces are those which are both separable and complete. Such a space is called a *Banach space*.

Recall that a metric space is separable if it has a countable dense set. There are no important constructively defined normed linear spaces that are not separable. Thus, as far as existing mathematics is concerned, there is no loss of generality in postulating separability, and there is a great gain in power and convenience.

The fact that every metric space can be completed leads us to hope that the same is true of every normed linear space V. This turns out to be the case, and the completion of V as a normed linear space can be identified with its completion as a metric space. Here is an outline of the construction. We take $\tilde{V}$ to consist of all Cauchy sequences from V, with termwise addition and termwise scalar multiplication. Two elements $\{x_n\}$ and $\{y_n\}$ of $\tilde{V}$ are *equal* if $\lim_{n\to\infty} \|x_n - y_n\| = 0$. Clearly $\tilde{V}$ is a linear space. Under the norm

$$\|\{x_n\}\| \equiv \lim_{n\to\infty} \|x_n\|$$

it is a complete normed linear space, called the *completion* of V. The *inclusion map* $i: V \to \tilde{V}$ defined by $i(v) \equiv \{v_n\}$, where $v_n \equiv v$ for each n, preserves norms and realizes V as a dense linear subset of $\tilde{V}$. We have thus proved the following result.

Proposition 4 *To each normed linear space V is associated a complete normed linear space $\tilde{V}$, and a norm-preserving linear inclusion map i from V into $\tilde{V}$. Moreover, V is dense in $\tilde{V}$.*

Remarks (a) In case V is separable, $\tilde{V}$ is also separable and therefore a Banach space.

(b) The bounded linear functionals on V are in one-one correspondence with the bounded linear functionals on $\tilde{V}$: If λ is any bounded linear functional on V and $\{x_n\}$ any point in $\tilde{V}$, then $\{\lambda(x_n)\}$ is a Cauchy sequence in $\mathbf{F}$ whose limit $\lambda(\{x_n\}) \equiv \lim_{n\to\infty} \lambda(x_n)$ defines an extension of λ to $\tilde{V}$.

The simplest instances of Banach spaces are the euclidean spaces $\mathbf{F}^n$, with norm

$$\|(a_1, \ldots, a_n)\| \equiv \Big(\sum_{i=1}^{n} |a_i|^2 \Big)^{1/2}$$

These spaces are *finite-dimensional*, in the following sense.

Definition 4 A normed linear space V is *finite-dimensional* if it has a finite basis, that is, if there exist nonzero vectors $v_1, \ldots, v_n$ and bounded linear functionals $\lambda_1, \ldots, \lambda_n$ such that

$$(a) \quad v = \sum_{i=1}^{n} \lambda_i(v) v_i \qquad (v \in V)$$

and

$$(b) \quad \lambda_i(v_j) = 0 \qquad (1 \leq i, j \leq n, i \neq j)$$

The vectors $v_1, \ldots, v_n$ are the *basis vectors*, $\lambda_1, \ldots, \lambda_n$ are the *coordinate functions*, and the numbers $\lambda_1(v), \ldots, \lambda_n(v)$ are the *coordinates* of v.

The integer n is called the *dimension* of V. It does not depend on the choice of basis.

The mapping σ defined by

$$\sigma(v) \equiv (\lambda_1(v), \ldots, \lambda_n(v))$$

is a linear-space isomorphism of V with the linear space $\mathbf{F}^n$. It is uniformly continuous because each λ_i is uniformly continuous. The inverse mapping takes the vector $(a_1, \ldots, a_n) \in \mathbf{F}^n$ into the vector $a_1 v_1 + \cdots + a_n v_n \in V$. It also is uniformly continuous. Hence σ is a metric equivalence. Since $\mathbf{F}^n$ is separable and complete, so is V. Thus every finite-dimensional normed linear space is a Banach space.

Proposition 5   *The closed unit sphere S of a finite-dimensional Banach space B is compact.*

Proof   Since B is metrically equivalent to some $\mathbf{F}^n$, it is locally compact. Since S is a bounded subset of B, to show that S is compact it is enough, by Proposition 13 of Chap. 4, to show that S is located. This follows when we take $K \equiv \{0\}$ in the following lemma.

Lemma 1   *Let K be a located subset of a normed linear space V. Then for each $r \geq 0$ the set*

$$K_r \equiv \{y + z : y \in K, \|z\| \leq r\}$$

is located, and

$$(1.1) \qquad \rho(x, K_r) = \max\{0, \rho(x, K) - r\}$$

for all x in V.

                                                249

Proof For each y in K and z in V, with $\|z\| \leq r$, we have

$$\|x - (y + z)\| \geq \|x - y\| - \|z\| \geq \rho(x,K) - r$$

Therefore $$\|x - (y + z)\| \geq \alpha(x)$$

for all vectors $y + z$ in K_r, where $\alpha(x)$ is the right side of (1.1). Let y be any point of K. Let z be that nonnegative multiple of $x - y$ for which $\|z\| = \min\{r, \|x - y\|\}$. Then

$$\|x - (y + z)\| = \max\{\|x - y\| - r, 0\}$$

It follows that there exist points $y + z$ in K_r for which $\|x - (y + z)\|$ is arbitrarily near to $\alpha(x)$. Therefore $\rho(x,K_r)$ exists and equals $\alpha(x)$.

The *norm* of a bounded linear functional λ on a normed linear space V is the least upper bound

$$\|\lambda\| \equiv \text{l.u.b.}\ \{|\lambda(x)| : x \in S\}$$

if it exists. A bounded linear function λ for which $\|\lambda\|$ exists is called *normable*. Proposition 5 implies that every bounded linear functional on a finite dimensional normed linear space is normable.

Important examples of Banach spaces attach themselves to a locally compact space X. For each locally compact space X we define $C_\infty(X,\mathbf{F})$ to be all continuous functions $f : X \to \mathbf{F}$ such that for each $r > 0$ there exists a compact subset K of X with $|f(x)| < r$ for all x in $-K$. [For $C_\infty(X,\mathbf{R})$ we write simply $C_\infty(X)$.] The *norm*

$$\|f\| \equiv \sup\{|f(x)| : x \in X\}$$

is well defined, and makes $C_\infty(X,\mathbf{F})$ into a normed linear space, which can be identified with the set G of all continuous functions f on the one-point compactification X_0 of X for which $f(\infty) = 0$. Since $C(X_0,\mathbf{F})$ is complete and G is a closed subset of $C(X_0,\mathbf{F})$, it follows that G is complete. Hence $C_\infty(X,\mathbf{F})$ is complete. It is also separable, and therefore a Banach space. To see this, it is enough to consider the case $\mathbf{F} = \mathbf{R}$. Let $\{x_n\}$ be a dense sequence of points in X_0. The set of all polynomial functions with rational coefficients in the functions $x \to \rho(x,x_n)$ $(1 \leq n < \infty)$ is a countable subset Γ of $C(X_0)$, which by the Stone-Weierstrass theorem is dense in $C(X_0)$. For each function f in Γ, we see that $f - f(\infty)$ vanishes at ∞, and these functions are dense in G. Therefore G is separable.

As with every normed linear space, we are interested in the form of the bounded linear functionals of $C_\infty(X,\mathbf{F})$. It is enough to consider the case $\mathbf{F} = \mathbf{R}$, since a bounded linear functional λ on $C_\infty(X,\mathbf{C})$ can be written uniquely in the form

$$\lambda(f_1 + if_2) = (\lambda_1(f_1) + i\lambda_1(f_2)) + i(\lambda_2(f_1) + i\lambda_2(f_2))$$

where λ_1 and λ_2 are bounded linear functionals on $C_\infty(X)$. Consider therefore a bounded linear functional λ on $C_\infty(X)$. For each f in $C(X_0)$ write $Tf \equiv f - f(\infty)$. Then Tf belongs to G, and so can be identified with an element of $C_\infty(X)$. The map $\lambda \circ T$ is thus a bounded linear functional on $C(X_0)$, that is, a measure μ on X_0, that vanishes on the constant functions. Conversely, every measure μ on X_0 that vanishes on the constant functions defines a bounded linear functional λ on $C_\infty(X)$, by

$$\lambda(f) \equiv \int f \, d\mu$$

An interesting special case arises when X is taken to be the set Z^+ of positive integers. In this case we can describe the bounded linear functional λ explicitly. For each n in Z^+ let δ_n be the element of $C_\infty(Z^+)$ defined by $\delta_n(n) = 1$ and $\delta_n(m) = 0$ if $m \neq n$, and write

$$a_n \equiv \lambda(\delta_n)$$

For each pair of positive integers m and N there exists h in $C_\infty(Z^+)$ such that $\|h\| = 1$, $h(n) = 1$ whenever $a_n > m^{-1}$ and $n \leq N$, $h(n) = -1$ whenever $a_n < -m^{-1}$ and $n \leq N$, and $h(n) = 0$ whenever $n > N$. Then

$$|\lambda(h)| = \left| \sum_{n=1}^{N} a_n h(n) \right| \leq c$$

where c is any bound for λ. Holding N fixed and letting $m \to \infty$ gives

$$(1.2) \qquad \sum_{n=1}^{N} |a_n| \leq c \qquad (N \in Z^+)$$

It follows that

$$(1.3) \qquad \lambda(f) = \sum_{n=1}^{\infty} a_n f(n)$$

for all f in $C_\infty(Z^+)$.

Conversely, if $\{a_n\}$ is any sequence of real numbers satisfying (1.2), then (1.3) defines a bounded linear functional λ on $C_\infty(Z^+)$.

2. THE L_p SPACES

A measure space $(X,F,\mathfrak{F},\mathfrak{M},\mu)$ gives rise to a whole family of normed linear spaces, the L_p spaces of Chap. 7.

Definition 5 If $(X,F,\mathfrak{F},\mathfrak{M},\mu)$ is a measure space and $p \geq 1$ is a real number, then $L_p(\mu)$ is the set of all measurable functions f from X to **F** such that $|f|^p$ is integrable. (A complex-valued function defined almost everywhere on X is *measurable* if its real and imaginary parts are measurable.) *Equality* of functions in $L_p(\mu)$ means equality almost everywhere. The *norm* of a function f in $L_p(\mu)$ is

$$\|f\|_p \equiv (\textstyle\int |f|^p \, d\mu)^{1/p}$$

By Theorem 10 of Chap. 7, we see that $L_p(\mu)$ is a complete normed linear space.

Lemma 2 *The simple functions are dense in $L_p(\mu)$.*

Proof For each $f \geq 0$ in $L_p(\mu)$ and each $\epsilon > 0$ there exists an integrable set A and a positive integer N, with $|f| \leq N$ a.e. on A, such that $\|f - \chi_A f\|_p \leq \epsilon$. Moreover, there exists a simple function g with $0 \leq g \leq \chi_A f \leq g + \epsilon\tilde{\mu}(A)^{-1/p}$ a.e. Then

$$\|f - g\|_p \leq \|\chi_A f - g\|_p + \epsilon \leq \left(\int_A (\epsilon\tilde{\mu}(A)^{-1/p})^p \, d\mu \right)^{1/p} + \epsilon < 2\epsilon$$

Thus the simple functions are dense in $L_p(\mu)$.

To show that $L_p(\mu)$ is separable we must impose a restriction on the measure.

Definition 6 The measure space $(X,F,\mathfrak{F},\mathfrak{M},\mu)$ is *separable* if there exists a sequence $\{A_n\}$ of integrable sets which is *dense* in $\mathfrak{M}$, in the sense that for each A in $\mathfrak{M}$ and each $\epsilon > 0$ there exists n in Z^+ with

$$\textstyle\int |\chi_A - \chi_{A_n}| \, d\mu = \mu(A - A_n) + \mu(A_n - A) < \epsilon$$

We then have the following result.

Proposition 6 *If $(X,F,\mathfrak{F},\mathfrak{M},\mu)$ is a separable measure space, then $L_p(\mu)$ is a Banach space for each $p \geq 1$.*

Proof We must show that $L_p(\mu)$ is separable. It is enough to consider the case $\mathbf{F} = \mathbf{R}$. The set Γ of all linear combinations with rational coefficients of characteristic functions of the sets A_n is countable. Every simple function is in the closure of Γ in $L_p(\mu)$. Since the simple functions are dense, Γ is dense.

Of particular interest is a measure defined on a locally compact space. The following lemma will be useful later.

Lemma 3 *If μ is a positive measure on a locally compact space X, then the set $C(X,\mathbf{F})$ of all f in $C_\infty(X,\mathbf{F})$ with compact support is dense in $L_p(\mu)$.*

Proof It is enough to consider the case $\mathbf{F} = \mathbf{R}$. Since the simple functions are dense in $L_p(\mu)$, it is enough to show that if A is any integrable set, there are test functions f such that $\|\chi_A - f\|_p$ is arbitrarily small. This is done by means of an estimate which is interesting in its own right. Consider any $\epsilon > 0$, and let the test function f approximate A to within ϵ, with error sequence $\{f_j\}$. By Theorem 5 of Chap. 7, the series $\sum_{j=1}^{\infty} f_j$ converges at every point of some full set B to an integrable function h with $\int h\, d\mu < \epsilon$. Consider any x in the full set $C \equiv (A \cup -A) \cap B$. If $f(x) + h(x) < 1$, then $x \in -A$. Hence $f(x) \geq 1 - h(x)$ for all x in A. Similarly, $f(x) \leq h(x)$ for all x in $-A$. It follows that $|\chi_A(x) - f(x)| \leq h(x)$. Thus we have

$$\|\chi_A - f\|_p^p = \int |\chi_A - f|^p\, d\mu \leq \int |\chi_A - f|\, d\mu \leq \int h\, d\mu < \epsilon$$

Proposition 7 *If μ is a positive measure on a locally compact space X, then $(X,F,\mathfrak{F},\mathfrak{M},\mu)$ is separable.*

Proof Again, we may restrict our attention to the case $\mathbf{F} = \mathbf{R}$. As shown above, there exists a dense sequence $\{f_n\}$ in $C_\infty(X)$. For each n choose α_n, with $\frac{1}{2} < \alpha_n < \frac{3}{4}$, so that

$$A_n \equiv \{x : f_n(x) \geq \alpha_n\}$$

is compact and integrable. We shall show that the sequence $\{A_n\}$ satisfies the requirements of Definition 6. To this end consider an integrable set A and a constant ϵ with $0 < \epsilon < \frac{1}{4}$. Choose a compact integrable set $K \subset A$ with $\mu(A) - \mu(K) < \epsilon$. Choose a test function f, $0 \leq f \leq 1$, such that $f(x) = 1$ for all x in K and $\int f\, d\mu - \mu(K) < \epsilon$. Choose n in Z^+ with $\|f - f_n\| < \epsilon$. Then $K \subset A_n$, since $\epsilon < \frac{1}{4}$ and

$\alpha_n < \frac{3}{4}$. Also,

$$\mu(A_n - K) \leq 4 \int_{A_n - K} (f_n - \epsilon)\, d\mu \leq 4 \int_{A_n - K} f\, d\mu$$
$$\leq 4 \left(\int f\, d\mu - \mu(K) \right) \leq 4\epsilon$$

Therefore

$$\mu(A - A_n) + \mu(A_n - A) \leq \mu(A - K) + \mu(A_n - K) < 5\epsilon$$

To smooth the way to the incisive geometrical inequalities of Clarkson for L_p spaces, we prove some lemmas. The first is an inequality which replaces Hölder's inequality for $0 < p < 1$.

Lemma 4 *Let p and q be real numbers with $0 < p < 1$ and $p^{-1} + q^{-1} = 1$. Let $f \geq 0$ and $g \geq 0$ be measurable functions with $g > 0$ a.e. such that fg and g^q are integrable. Then*

$$(2.1) \qquad \int fg\, d\mu \geq (\int f^p\, d\mu)^{1/p} (\int g^q\, d\mu)^{1/q}$$

Proof Write (2.1) in the form

$$(2.2) \qquad \int f^p\, d\mu \leq (\int fg\, d\mu)^p (\int g^q\, d\mu)^{-p/q}$$

Define $u \equiv (fg)^p$, $v \equiv g^{-p}$, $r = p^{-1}$, and $s = -qp^{-1}$. Then $r > 1$, $r^{-1} + s^{-1} = 1$, and (2.2) becomes

$$\int uv\, d\mu \leq (\int u^r\, d\mu)^{1/r} (\int v^s\, d\mu)^{1/s}$$

or Hölder's inequality.

Our next lemma is a replacement for Minkowski's inequality for $0 < p < 1$.

Lemma 5 *If $0 < p < 1$, and if f and g are nonnegative measurable functions such that f^p and g^p are integrable, then $(f + g)^p$ is integrable, and*

$$(2.3) \qquad (\int (f + g)^p\, d\mu)^{1/p} \geq (\int f^p\, d\mu)^{1/p} + (\int g^p\, d\mu)^{1/p}$$

Proof The inequality $(f + g)^p \leq f^p + g^p$ implies that $(f + g)^p$ is integrable. Using an approximation argument, we see that it is enough to consider the case in which $f + g > 0$ a.e. Write $q \equiv p/(p - 1)$. By

Lemma 4,

$$\int (f + g)^p \, d\mu = \int f(f + g)^{p-1} \, d\mu + \int g(f + g)^{p-1} \, d\mu$$
$$\geq \left(\int f^p \, d\mu\right)^{1/p} \left(\int (f + g)^p \, d\mu\right)^{(p-1)/p}$$
$$+ \left(\int g^p \, d\mu\right)^{1/p} \left(\int (f + g)^p \, d\mu\right)^{(p-1)/p}$$

This is equivalent to (2.3).

Lemma 6 *Let x, y, p, and q be real numbers, with $p^{-1} + q^{-1} = 1$. Then*

$$(2.4) \qquad |x + y|^q + |x - y|^q \geq 2(|x|^p + |y|^p)^{q-1}$$

for $p \geq 2$, and

$$(2.5) \qquad |x + y|^q + |x - y|^q \leq 2(|x|^p + |y|^p)^{q-1}$$

for $1 < p \leq 2$.

Proof Assume first that $p \geq 2$, so that $1 < q \leq 2$. By continuity and symmetry we may take $|x| > |y| > 0$. Dividing by $|x|^q$ and setting $c \equiv |yx^{-1}|$ (so that $0 < c < 1$), we reduce (2.4) to

$$(2.6) \qquad (1 + c)^q + (1 - c)^q - 2(1 + c^p)^{q-1} \geq 0$$

Expanding the left side of (2.6) in powers of c, we get

$$(2.7) \quad 2 \sum_{k=1}^{\infty} ((2k)!)^{-1} q(q - 1) \cdots (q - 2k + 1) c^{2k}$$

$$- 2 \sum_{k=1}^{\infty} ((2k)!)^{-1}(q - 1) \cdots (q - 2k) c^{2kp}$$

$$- 2 \sum_{k=1}^{\infty} ((2k - 1)!)^{-1}(q - 1) \cdots (q - 2k + 1) c^{(2k-1)p}$$

$$= 2 \sum_{k=1}^{\infty} ((2k)!)^{-1}(q - 1) \cdots (q - 2k + 1)$$

$$\times \{ qc^{2k} - (q - 2k)c^{2kp} - 2kc^{(2k-1)p} \}$$

Each of the products $(q - 1) \cdots (q - 2k + 1)$ is nonnegative, because $1 < q \leq 2$. Consider k in Z^+, and write $\alpha \equiv 2k(p - 1)$. Then

$$qc^{2k} - (q - 2k)c^{2kp} - 2kc^{(2k-1)p}$$
$$= 2kpc^{2kp}\{\alpha^{-1}c^{-\alpha} - (\alpha^{-1} - p^{-1}) - p^{-1}c^{-p}\}$$
$$= 2kpc^{2kp}\{\alpha^{-1}(c^{-\alpha} - 1) - p^{-1}(c^{-p} - 1)\} \geq 0$$

for all k in Z^+, since $t^{-1}(c^{-t} - 1)$ is an increasing function of t for $t \geq 0$, and $\alpha \equiv 2k(p-1) \geq 2(p-1) = p + p - 2 \geq p$. Thus each term in the expansion (2.7) is nonnegative. Hence (2.4) is valid.

Assume next that $1 < p \leq 2$. Then $2 \leq q < \infty$, and thus, by the case already considered, we have

$$|x + y|^p + |x - y|^p \geq 2(|x|^q + |y|^q)^{p-1}$$

Substituting $(x + y)/2$ for x and $(x - y)/2$ for y gives

$$|x|^p + |y|^p \geq 2(\tfrac{1}{2})^{q(p-1)}(|x + y|^q + |x - y|^q)^{p-1}$$
$$= 2^{1-p}(|x + y|^q + |x - y|^q)^{p-1}$$

which is equivalent to (2.5).

We now prove Clarkson's basic inequalities for the L_p norms.

Theorem 1 *Take* **F** *to be* **R***. Let f and g be elements of L_p. For $p \geq 2$ we have*

$$(a) \quad 2(\|f\|_p^p + \|g\|_p^p)^{q-1} \leq \|f + g\|_p^q + \|f - g\|_p^q$$

and

$$(b) \quad \|f + g\|_p^p + \|f - g\|_p^p \leq 2(\|f\|_p^q + \|g\|_p^q)^{p-1}$$

while for $1 < p \leq 2$ the reverse inequalities hold.

Proof The inequality (b) arises from (a) when we replace f with $(f + g)/2$ and g with $(f - g)/2$. It is therefore enough to prove (a). Consider first the case $p \geq 2$. Write $s \equiv p/q$. From Minkowski's inequality and Lemma 6 we get

$$
\begin{aligned}
\|f + g\|_p^q + \|f - g\|_p^q &= \| \, |f + g|^q \|_s + \| \, |f - g|^q \|_s \\
&\geq \| \, |f + g|^q + |f - g|^q \|_s \\
&\geq 2\|(|f|^p + |g|^p)^{q-1}\|_s \\
&= 2(\|f\|_p^p + \|g\|_p^p)^{q/p} \\
&= 2(\|f\|_p^p + \|g\|_p^p)^{q-1}
\end{aligned}
$$

as desired. In case $1 < p \leq 2$, using first Lemma 5 with $s \equiv p/q$ and then Lemma 6, we get

$$
\begin{aligned}
\|f + g\|_p^q + \|f - g\|_p^q &= \| \, |f + g|^q \|_s + \| \, |f - g|^q \|_s \\
&\leq \| \, |f + g|^q + |f - g|^q \|_s \\
&\leq 2\|(|f|^p + |g|^p)^{q-1}\|_s \\
&= 2(\|f\|_p^p + \|g\|_p^p)^{q-1}
\end{aligned}
$$

as desired.

Corollary   *Take* **F** *to be* **R**. *Then for each* $p > 1$ *the space* L_p *is uniformly convex, in the sense that to each* $\epsilon > 0$ *corresponds a positive constant* $r < 1$ *such that* $\|\tfrac{1}{2}(f + g)\|_p \le r$ *whenever* $\|f\|_p = \|g\|_p = 1$ *and* $\|f - g\|_p \ge \epsilon$.

Proof   For $p \ge 2$ we have, by (*b*) of Theorem 1,

$$\|\tfrac{1}{2}(f + g)\|_p = \tfrac{1}{2}\|f + g\|_p \le \tfrac{1}{2}(2^p - \epsilon^p)^{1/p} = (1 - (\tfrac{1}{2}\epsilon)^p)^{1/p}$$

For $1 < p \le 2$ we have, by Theorem 1,

$$\|\tfrac{1}{2}(f + g)\|_p = \tfrac{1}{2}\|f + g\|_p \le \tfrac{1}{2}(2^q - \epsilon^q)^{1/q} = (1 - (\tfrac{1}{2}\epsilon)^q)^{1/q}$$

Thus the desired inequality holds for all $p > 1$, with

$$r \equiv \max \left\{ (1 - (\tfrac{1}{2}\epsilon)^p)^{1/p}, \; (1 - (\tfrac{1}{2}\epsilon)^q)^{1/q} \right\}$$

We can use Theorem 1 to determine all normable linear functionals on L_p, for $p > 1$.

Theorem 2   *Take* **F** *to be* **R**. *Then, for each* $p > 1$, *and each* g *in* L_q, *the linear functional* λ_g *on* L_p *defined by*

$$(2.8) \qquad \lambda_g(f) = \int fg \, d\mu$$

is normable, and $\|\lambda_g\| = \|g\|_q$. *Conversely, if* λ *is any normable linear functional on* L_p, *then* $\lambda = \lambda_g$ *for some* g *in* L_q.

Proof   By Hölder's inequality,

$$(2.9) \qquad |\lambda_g(f)| \le \|g\|_q \|f\|_p$$

There exists a unique element f_0 of L_p such that $f_0 g \ge 0$ a.e. and $|f_0|^p = |g|^q$ a.e. Then

$$(2.10) \quad \lambda_g(f_0) = \int f_0 g \, d\mu = \int |g|^{1+(q/p)} \, d\mu = \int |g|^q \, d\mu = \|g\|_q \|f_0\|_p$$

Together with (2.9), this implies that λ_g is normable with norm $\|g\|_q$.

Consider, conversely, a normable linear functional λ on L_p. Let ϵ be any constant, $0 < \epsilon < 1$. Choose f in L_p with $\|f\|_p = 1$ and $\lambda(f) \ge \|\lambda\| - \epsilon^2$. Let g be the unique element of L_q with $fg \ge 0$ and $|f|^p = |g|^q$ a.e. Then $\|g\|_q = 1$. Let h be an arbitrary vector in L_p with $\|h\| = 1$ and $\lambda_g(h) = 0$. Then

$$\lambda_g(f + \epsilon h) = \lambda_g(f - \epsilon h) = \lambda_g(f) = \int fg \, d\mu = \int |f|^p \, d\mu = 1$$

Therefore $\|f + \epsilon h\|_p \geq 1$ and $\|f - \epsilon h\|_p \geq 1$. In case $p \geq 2$, Theorem 1 gives

$$\|f + \epsilon h\|_p^p \leq 2(1 + \epsilon^q)^{p-1} - 1 \leq 1 + c_1\epsilon^q$$

where c_1 is some positive constant. Therefore

$$\|f + \epsilon h\|_p \leq 1 + c_2\epsilon^q$$

where c_2 is some positive constant. Similarly for $1 < p \leq 2$, Theorem 1 gives

$$\|f + \epsilon h\|_p^q \leq 2(1 + \epsilon^p)^{q-1} - 1$$

and thus

$$\|f + \epsilon h\|_p \leq 1 + c_3\epsilon^p$$

Thus in case $1 < p \leq 2$ or $2 \leq p$ we have

$$(2.11) \qquad\qquad \|f + \epsilon h\|_p \leq 1 + c_4\epsilon^t$$

where $t = \min\{p,q\}$. Hence (2.11) holds for all $p > 1$. Therefore

$$|\lambda(f + \epsilon h)| \leq (1 + c_4\epsilon^t)\|\lambda\|$$

and

$$\begin{aligned}
\lambda(h) &\leq \epsilon^{-1}(|\lambda(f + \epsilon h)| - \lambda(f)) \\
&\leq \epsilon^{-1}((1 + c_4\epsilon^t)\|\lambda\| - \|\lambda\| + \epsilon^2) \\
&\leq c_4\epsilon^{t-1}\|\lambda\| + \epsilon
\end{aligned}$$

It follows that for all h in L_p with $\lambda_g(h) = 0$ we have

$$(2.12) \qquad\qquad |\lambda(h)| \leq (c_4\epsilon^{t-1}\|\lambda\| + \epsilon)\|h\|$$

For any h in L_p,

$$\lambda_g(h - \lambda_g(h)f) = \lambda_g(h)(1 - \lambda_g(f)) = 0$$

If we set $c \equiv \lambda(f)$, inequality (2.12) therefore gives

$$\begin{aligned}
|\lambda(h) - \lambda_{cg}(h)| &= |\lambda(h - \lambda_g(h)f)| \\
&\leq (c_4\epsilon^{t-1}\|\lambda\| + \epsilon)(\|h\| + |\lambda_g(h)|\,\|f\|) \\
&\leq (c_4\epsilon^{t-1}\|\lambda\| + \epsilon)2\|h\|
\end{aligned}$$

Thus for each positive integer k there exists g_k in L_q such that

$$(2.13) \qquad\qquad \|\lambda(h) - \lambda_{g_k}(h)\| \leq k^{-1}\|h\|$$

for all h in L_p. By (2.13), we have $\|g_j - g_k\|_q \leq j^{-1} + k^{-1}$ for all j and k, and thus $\{g_k\}$ is a Cauchy sequence in L_q. Let g be the limit. By (2.13), we have $\lambda = \lambda_g$, as was to be proved.

Certain bounded linear functionals on L_p, as we shall see later, are *not* normable, and therefore do *not* come from elements of L_q. Nevertheless, we shall prove that the normable linear functionals are dense in the natural metric (to be described later) of the space of linear functionals.

3. THE EXTENSION OF LINEAR FUNCTIONALS

The normability of a linear functional can be simply characterized in terms of its null set.

Proposition 8   *A bounded nonzero linear functional λ on a normed linear space B is normable if and only if $N(\lambda)$ is located.*

Proof   Assume that λ is normable. Since λ is nonzero, $\|\lambda\| > 0$. For each k in Z^+ let x_k be a vector with $\|x_k\| = 1$ and $\lambda(x_k) > \|\lambda\| - k^{-1}$. For each x in B and y in $N(\lambda)$ we have

$$\|x - y\| \geq \|\lambda\|^{-1}|\lambda(x)|$$

On the other hand, the vector $z_k \equiv x - \lambda(x)\lambda(x_k)^{-1}x_k$ belongs to $N(\lambda)$, and

$$\|x - z_k\| = |\lambda(x)|\lambda(x_k)^{-1} < |\lambda(x)|(\|\lambda\| - k^{-1})^{-1}$$

Thus $\rho(x,N(\lambda))$ exists and equals $\|\lambda\|^{-1}|\lambda(x)|$.

Assume, conversely, that $\rho(x,N(\lambda))$ exists for all x in V. Since λ is nonzero, there exists a vector x_0 with $\lambda(x_0) = 1$. Hence

$$\inf \{\|x\| : \lambda(x) = 1\} = \inf \{\|x_0 - y\| : y \in N(\lambda)\} = \rho(x_0,N(\lambda))$$

exists. Therefore $\|\lambda\|$ exists and equals $\rho(x_0,N(\lambda))^{-1}$.

Lemma 7   *Let K be a bounded located convex subset of a normed linear space B, whose distance to 0 is positive. Then the cone*

$$c(K) \equiv \{tx : t > 0, x \in K\}$$

generated by K is located.

Proof For each $t \geq 0$ write

$$tK \equiv \{tx : x \in K\}$$

Then tK is located. Since $\rho(0,K) > 0$, for each x in B we have $\lim_{t \to \infty} \rho(x,tK) = \infty$. Therefore $\rho(x,c(K))$ exists and is the infimum of the continuous function $t \to \rho(x,tK)$ of t.

The fundamental geometric fact about normed linear spaces states that under certain conditions two convex sets can be separated by a normable linear functional. This result, which we now prove, is called the *separation theorem*.

Theorem 3 *Let F and G be bounded convex subsets of a separable normed linear space V, whose algebraic difference*

$$\{y - x : x \in F, y \in G\}$$

is located, and whose mutual distance

$$d \equiv \inf \{\|y - x\| : x \in F, y \in G\}$$

is positive. Then for each $\epsilon > 0$ there exists a normable linear functional λ on V of norm 1 such that

$$(3.1) \qquad\qquad \lambda_r(y) > \lambda_r(x) + d - \epsilon$$

for all x in F and y in G, where λ_r is the real part of λ.

Proof Consider first the case $\mathbf{F} = \mathbf{R}$. We may assume that $\epsilon < d$. By Lemma 1, the bounded open convex set

$$K \equiv \{y - x - z : x \in F, y \in G, \|z\| < d - \epsilon/2\}$$

is located, and the distance of K to 0 is $\epsilon/2$. As we shall see, to make (3.1) hold, it will be sufficient to make λ be positive on K. We shall get λ from successive enlargements of K, whose union will be the complete set of points where λ is positive.

Fix any vector x_0 with $-x_0 \in K$. Then

$$\begin{aligned}
\rho(x_0,c(K)) &= \inf \{\|tx - x_0\| : t > 0, x \in K\} \\
&= \inf \left\{ (t+1) \left\| \frac{t}{t+1} x + \frac{1}{t+1}(-x_0) \right\| : t > 0, x \in K \right\} \\
&\geq \rho(0,K) = \frac{\epsilon}{2}
\end{aligned}$$

Let $\{x_n\}$ be a sequence of vectors which is dense in V. We construct inductively a sequence $K \equiv K_0 \subset K_1 \subset \cdots$ of bounded located open convex sets such that the following properties hold for all $n \geq 1$.

(a) $\rho(x_0,c(K_n)) > (1 - 2^{-n})\rho(x_0,c(K_{n-1}))$

(b) $\rho(0,K_n) > 0$

(c) Either $\rho(x_n,c(K_n)) < n^{-1}$ or $\rho(-x_n,c(K_n)) < n^{-1}$

Assume that $K_0, \ldots, K_{n-1}$ have been constructed. If either $\rho(x_n,c(K_{n-1})) < n^{-1}$ or $\rho(-x_n,c(K_{n-1})) < n^{-1}$, we may take $K_n \equiv K_{n-1}$. Assume therefore that both these distances are at least $(2n)^{-1}$. Write

$$K_n^+ \equiv \{tx_n + (1 - t)k: 0 < t < 1, k \in K_{n-1}\}$$

and
$$K_n^- \equiv \{-tx_n + (1 - t)k: 0 < t < 1, k \in K_{n-1}\}$$

Since both K_n^+ and K_n^- satisfy (c), if one of these sets satisfies (a) and (b), we may take K_n to be that set. We first show that both K_n^+ and K_n^- satisfy (b). By symmetry, it is sufficient to check K_n^+. For each point $x \equiv tx_n + (1 - t)k$ in K_n^+ we have

$$\|x\| \geq (1 - t)\rho(0,K_{n-1}) - t\|x_n\|$$

and
$$\|x\| = t\|x_n + t^{-1}(1 - t)k\| \geq t\rho(-x_n,c(K_{n-1})) \geq (2n)^{-1}t$$

These two inequalities show that $\|x\|$ is bounded below by some positive constant. In other words, $\rho(0,K_n^+)$ is positive. Thus K_n^+, and similarly K_n^-, satisfies (b).

To verify (a), consider arbitrary elements $j_1 \in c(K_n^+)$ and $j_2 \in c(K_n^-)$. Then $j_1 = t_1x_n + k_1$ and $j_2 = -t_2x_n + k_2$, where $t_1 > 0$, $t_2 > 0$, $k_1 \in c(K_{n-1})$, and $k_2 \in c(K_{n-1})$. We compute

$$
\begin{aligned}
t_2\|x_0 - j_1\| + t_1\|x_0 - j_2\| \\
= t_2\|x_0 - (t_1x_n + k_1)\| + t_1\|x_0 - (-t_2x_n + k_2)\| \\
\geq \|(t_1 + t_2)x_0 - (t_2k_1 + t_1k_2)\| \\
\geq (t_1 + t_2)\rho(x_0,c(K_{n-1}))
\end{aligned}
$$

Thus either $\|x_0 - j_1\|$ or $\|x_0 - j_2\|$ is larger than

$$(1 - 2^{-(n+1)})\rho(x_0,c(K_{n-1}))$$

Therefore either K_n^+ or K_n^- satisfies (a). In the former case define $K_n \equiv K_n^+$, and in the latter define $K_n \equiv K_n^-$. This completes the construction of the sequence $\{K_n\}$.

The set

$$K_\infty \equiv \bigcup_{n=1}^{\infty} c(K_n)$$

is open, convex, and invariant under multiplication by positive scalars. By (a), we see that $\rho(x_0, K_\infty)$ exists and is positive.

Let $K_{-\infty}$ be the negative $\{x: -x \in K_\infty\}$ of K_∞. By (b), we see that K_∞ and $K_{-\infty}$ are disjoint, in the sense that if $x \in K_\infty$ and $y \in K_{-\infty}$, then $\|x - y\| > 0$. By (c), we see that $K_\infty \cup K_{-\infty}$ is dense in V. Since $-x_0 \in K \subset K_\infty$, it follows that $x_0 \in K_{-\infty}$.

Consider an arbitrary vector x in K_∞. Let L be the line segment

$$L \equiv \{tx + (1 - t)x_0 : 0 \leq t \leq 1\}$$

joining x to x_0. Then $K_\infty \cap L$ and $K_{-\infty} \cap L$ are convex nonvoid disjoint subsets of L. Since K_∞ and $K_{-\infty}$ are open, there exists $\delta > 0$ such that if $\|z\| < \delta$, then $x + z \in K_\infty$ and $x_0 + z \in K_{-\infty}$. Since $K_\infty \cup -K_\infty$ is dense in V, for each y in L there exists y_0 in $K_\infty \cup K_{-\infty}$ with $\|y - y_0\| < \delta^2$. Assume first that $y_0 \in K_\infty$. Then $x + \delta^{-1}(y - y_0) \in K_\infty$, by the choice of δ. Since K_∞ is convex,

$$(1 + \delta)^{-1}y_0 + \delta(1 + \delta)^{-1}(x + \delta^{-1}(y - y_0))$$
$$= \delta(1 + \delta)^{-1}x + (1 + \delta)^{-1}y \in L \cap K_\infty$$

Similarly,
$$\delta(1 + \delta)^{-1}x_0 + (1 + \delta)^{-1}y \in L \cap K_{-\infty}$$

if $y_0 \in -K_\infty$. Since δ can be made arbitrarily small, $(K_\infty \cup K_{-\infty}) \cap L$ is dense in L. Hence there exists a unique vector z in $L \cap N$, where N is the intersection of the closure of K_∞ and the closure of $K_{-\infty}$. The vector x is a linear combination of x_0 and z. Thus every vector x in K_∞ is a linear combination of the vector x_0 and a vector z in N.

Since both K_∞ and $K_{-\infty}$ are cones, N is a cone, and $x \in N$ if and only if $-x \in N$. Therefore N is a closed linear subspace of V. The distance of x_0 to N exists, and equals the distance r of x_0 to K_∞ [which is positive, by (a)], because we have seen that if x is any point in K_∞, there exists a point z of N on the line segment joining x_0 to x. Since any point x in K_∞ can be written in the form $x = ax_0 + z$, with $z \in N$, we see that $\rho(x, N)$ exists and equals $\rho(ax_0, N) = r|a|$. Since $K_\infty \cup K_{-\infty}$ is dense in V, it follows that N is located, and that any x in V can be written in the form $ax_0 + z$, with $z \in N$. Thus N is a hyperplane with associated vector x_0. From Propositions 3 and 8, we see that there exists a normable linear functional λ on V, with $\|\lambda\| = 1$ and $\lambda(x_0) < 0$, such that $N = N(\lambda)$. For each x in K_∞ we have seen that there exists a point z in $N \cap L$, where L is the line segment joining x to x_0. Since $\lambda(x_0) < 0$ and $\lambda(z) = 0$, it follows that $\lambda(x) > 0$. Thus λ is positive on K_∞, and therefore on K.

Consider arbitrary vectors $x \in F$ and $y \in G$. Since $\|\lambda\| = 1$, there exists a vector z with $\|z\| < d - \epsilon/2$ and $\lambda(z) > d - \epsilon$. Thus $y - x - z \in K$. Since λ is positive on K, this gives

$$\lambda(y) \geq \lambda(x + z) > \lambda(x) + d - \epsilon$$

This completes the proof of Theorem 3, for the case $\mathbf{F} = \mathbf{R}$.

Now consider the case $\mathbf{F} = \mathbf{C}$. By the case $\mathbf{F} = \mathbf{R}$ already considered, there exists a normable *real* linear functional μ on V (i.e., a linear functional relative to the real linear structure on V) such that (3.1) is satisfied, with λ_r replaced by μ. For each x in V write

$$\lambda(x) \equiv \mu(x) - i\mu(ix)$$

Since λ is real linear, to show that λ is complex linear we need only note that

$$\lambda(ix) = \mu(ix) - i\mu(i^2x) = \mu(ix) + i\mu(x) = i\lambda(x) \qquad (x \in V)$$

Clearly any bound for λ is a bound for μ. Conversely, for each x in V and each $\epsilon > 0$ we can choose a complex constant α with $|\alpha| = 1$ and $|\mu(\alpha y)| < \epsilon$. Then

$$|\lambda(y)| = |\lambda(\alpha y)| = |\mu(\alpha y) - i\mu(i\alpha y)| \leq (\epsilon^2 + \|\mu\|^2\|y\|^2)^{1/2}$$

Letting $\epsilon \to 0$ gives $|\lambda(y)| \leq \|\mu\| \, \|y\|$. Thus $\|\mu\| = 1$ is a bound for λ, and thus λ is normable and $\|\lambda\| = 1$. Since $\lambda_r = \mu$, we see that (3.1) is satisfied.

Corollary *Let x_0 be a vector in a separable normed linear space V, and ϵ a positive constant. Then there exists a normable linear functional λ on V of norm 1 with $\lambda(x_0) \geq \|x_0\| - \epsilon$.*

Proof Apply Theorem 3 to the sets $G \equiv \{x_0\}$ and $F \equiv \{0\}$.

We can now prove the famous Hahn-Banach theorem on the extension of linear functionals. To do this we make use of the notion of a quotient space. Let V be a located linear subset of a normed linear space B. We define a seminorm $\| \quad \|_0$ on B by

$$\|x\|_0 = \rho(x, V)$$

Relative to the equality relation induced by this seminorm, B is a normed linear space, with norm $\| \quad \|_0$, called the *quotient space* of B by V and written B/V.

Theorem 4 *Let λ be a nonzero linear functional on a linear subset V of a separable normed linear space B, whose null space $N(\lambda)$ is a located subset of B. Then for each $\epsilon > 0$ there exists a normable linear functional ν on B with $\nu(x) = \lambda(x)$ for all x in V, and $\|\nu\| \leq \|\lambda\| + \epsilon$.*

Proof Let B_0 be the quotient space $B/N(\lambda)$ of B by $N(\lambda)$, with norm $\|x\|_0 \equiv \rho(x, N(\lambda))$. Let x_0 be any vector of V with $\lambda(x_0) = 1$. For each z in $N(\lambda)$ we have

$$\|x_0 - z\| \geq \|\lambda\|^{-1}\lambda(x_0 - z) = \|\lambda\|^{-1}$$

Therefore

$$\|x_0\|_0 = \rho(x_0, N(\lambda)) \geq \|\lambda\|^{-1}$$

By the corollary to the separation theorem, there exists a normable linear functional ν_0 on B_0, with $\|\nu_0\| \leq \|\lambda\| + \epsilon$ and $\nu_0(x_0) = 1$. When considered as a linear functional on B, ν_0 is a normable linear functional ν with

$$\|\nu\| = \|\nu_0\| \leq \|\lambda\| + \epsilon$$

Since $\nu(x_0) = 1$ and $\nu(z) = 0$ for all z in $N(\lambda)$, it follows that $\nu(x) = \lambda(x)$ for all x in V.

4. HILBERT SPACE AND THE SPECTRAL THEOREM

One of the central problems of functional analysis is to analyze the structure of a linear transformation of a Banach space into itself.

Definition 7 A *bounded linear transformation* from a normed linear space B_1 to a normed linear space B_2 is a function T from B_1 to B_2 such that

$$T(ax + by) = aTx + bTy \qquad (a, b \in \mathbf{F}, \, x, y \in B_1)$$

and

$$\|Tx\| \leq c\|x\| \qquad (x \in B_1)$$

for some positive constant c, called a *bound* for T. (In case $B_1 = B_2$, a bounded linear transformation is called an *operator*.)

The set of all bounded linear transformations from B_1 to B_2 will be denoted by Hom (B_1, B_2). An element T of Hom (B_1, B_2) is called *normable* if

$$\|T\| \equiv \sup \{\|Tx\| : x \in B_1, \|x\| \leq 1\}$$

exists, in which case $\|T\|$ is called the *norm* of T. If T is normable, $\|T\|$ is the smallest bound for T.

Important special cases of Definition 1 occur when $B_1 = \mathbf{F}$ or $B_2 = \mathbf{F}$.

In case $B_1 = \mathbf{F}$, we see that Hom (B_1,B_2) can be identified with B_2. In case $B_2 = \mathbf{F}$, we see that Hom (B_1,B_2) is the set B_1^* of bounded linear functionals on B_1. Proposition 2 extends to the case of a linear transformation T from a normed linear space B_1 into a normed linear space B_2.

In case $B_1 = B_2$, the *identity operator* in Hom (B_1,B_2) which takes each x in B_1 into itself is denoted by I.

When B_1 is separable and $\{x_n\}$ is a countable dense subset, the *double norm*

$$|||T||| \equiv \sum_{n=1}^{\infty} 2^{-n}(1 + \|x_n\|)^{-1}\|Tx_n\|$$

makes Hom (B_1,B_2) into a normed linear space. In particular, the double norm makes B_1^* into a normed linear space. Although the double norm depends on the choice of the dense sequence $\{x_n\}$, we have the following important proposition.

Proposition 9 *If $\{x_n'\}$ is another dense sequence in B_1, and if the norm $||| \quad |||'$ on* Hom (B_1,B_2) *is defined by*

$$|||T|||' \equiv \sum_{n=1}^{\infty} 2^{-n}(1 + \|x_n'\|)^{-1}\|Tx_n'\|$$

the norms $||| \quad |||$ and $||| \quad |||'$ induce equivalent metrics on bounded subsets K of Hom (B_1,B_2). *[A set $K \subset$ Hom (B_1,B_2) is bounded if there exists $c > 0$ such that $\|Tx\| \leq c\|x\|$ for all x in B_1 and T in K.]*

Proof Let $\epsilon > 0$ be given. Choose N so large that $2^{-N}c \leq \epsilon/4$. For $1 \leq n \leq N$ choose an integer k_n for which $2c\|x_n' - x_{k_n}\| < \epsilon/4$. Choose $\delta > 0$ so small that $\delta < (\epsilon/4)\, 2^{-k_n}(1 + \|x_{k_n}\|)^{-1}$ for $1 \leq n \leq N$. Let T_1 and T_2 be elements of K with $|||T_1 - T_2||| < \delta$. Then $\|(T_1 - T_2)x_{k_n}\| < \delta 2^{k_n}(1 + \|x_{k_n}\|)$ for all n. Hence

$$|||T_1 - T_2|||' \leq \sum_{n=1}^{N} 2^{-n}\|(T_1 - T_2)x_n'\| + 2c2^{-N}$$

$$\leq \sum_{n=1}^{N} 2^{-n}\{\|(T_1 - T_2)(x_n' - x_{k_n})\| + \|(T_1 - T_2)x_{k_n}\|\} + \frac{\epsilon}{2}$$

$$\leq \sum_{n=1}^{N} 2^{-n}\{2c\|x_n' - x_{k_n}\| + \delta 2^{k_n}(1 + \|x_{k_n}\|)\} + \frac{\epsilon}{2}$$

$$< \sum_{n=1}^{N} 2^{-n}\left\{\frac{\epsilon}{4} + \frac{\epsilon}{4}\right\} + \frac{\epsilon}{2} < \epsilon$$

Thus the metric induced by $||| \quad |||'$ is uniformly continuous on K with respect to the metric induced by $||| \quad |||$, and conversely. In other words, the two metrics are equivalent on K.

The structure of a bounded linear transformation is best understood for a *Hilbert space*, which is a special case of an inner product space.

Definition 8   An *inner product space* (also called a *pre-Hilbert space*) is a linear space H and a function from $H \times H$ to $\mathbf{F}$, whose value on the pair of elements x and y is written (x,y), such that

(a) $(x,y) = (y,x)^*$
(b) $(a_1 x_1 + a_2 x_2, y) = a_1(x_1,y) + a_2(x_2,y)$
(c) $(x,x) \geq 0$ for all x in H, and $(x,x) = 0$ if and only if $x = 0$

The simplest example of an inner product space is euclidean space $\mathbf{F}^n$, with the inner product of vectors $x \equiv (x_1, \ldots, x_n)$ and $y \equiv (y_1, \ldots, y_n)$ defined by

$$(x,y) \equiv \sum_{i=1}^{n} x_i y_i^*$$

Another example is the space $L_2(\mu)$, where $(X,F,\mathfrak{F},\mathfrak{M},\mu)$ is any measure space, with inner product defined by

$$(f,g) \equiv \int fg^* \, d\mu$$

This example specializes to the space $\mathbf{F}^n$ if X is taken to be the set $\{1, \ldots, n\}$ and μ is counting measure on X.

Another interesting instance is obtained by taking μ to be counting measure on the set $X \equiv Z^+$. In this case $L_2(\mu)$, which is denoted by l_2, consists of all sequences $\{x_n\}$ of elements of $\mathbf{F}$ such that $\sum_{n=1}^{\infty} |x_n|^2$ converges. The inner product of $\{x_n\}$ and $\{y_n\}$ is $\sum_{n=1}^{\infty} x_n y_n^*$.

Theorem 5   *An inner product space is a normed linear space, with norm given by*

$$\|x\| \equiv (x,x)^{1/2}$$

in which Hölder's inequality,

$$(4.1) \qquad\qquad |(x,y)| \leq \|x\| \, \|y\|$$

and Minkowski's inequality,

$$(4.2) \qquad \|x + y\| \le \|x\| + \|y\|$$

are valid. Equality holds in (4.2) if and only if $(x,y) = \|x\| \|y\|$, or equivalently, $\|y\|x = \|x\|y$.

Proof   Consider vectors x and y in H. For arbitrary real numbers a and b,

$$(4.3) \quad 0 \le (ax - by, ax - by) = a^2\|x\|^2 + b^2\|y\|^2 - ab((x,y) + (y,x))$$

with equality if and only if $ax - by = 0$. Consider any $\epsilon \ge 0$. Put $a = \max\{\|y\|, \epsilon\}$ and $b = \max\{\|x\|, \epsilon\}$ in (4.3). This gives

$$(4.4) \quad 0 \le 2a^2b^2 - ab((x,y) + (y,x)) = ab(2ab - (x,y) - (y,x))$$

If $\epsilon > 0$, this gives $(x,y) + (y,x) \le 2ab$. Letting $\epsilon \to 0$ gives

$$(x,y) + (y,x) \le 2\|x\| \|y\|$$

In this inequality we replace x with αx, where α is any element of $\mathbf{F}$ with $|\alpha| = 1$, and get

$$\alpha(x,y) + \alpha^*(x,y)^* \le 2\|x\| \|y\|$$

Dividing by 2 and taking the maximum with respect to α gives (4.1).
Using (4.1), we compute

$$(4.5) \quad \|x + y\|^2 = \|x\|^2 + \|y\|^2 + (x,y) + (y,x) \le (\|x\| + \|y\|)^2$$

This is equivalent to (4.2). Equality holds in (4.2), or equivalently, in (4.5), if and only if $(x,y) + (y,x) = 2\|x\| \|y\|$, which is equivalent to $(x,y) = \|x\| \|y\|$.
Taking $\epsilon = 0$ in (4.4) gives

$$\|x\| \|y\| (2\|x\| \|y\| - (x,y) - (y,x)) \ge 0$$

with equality if and only if $\|y\|x - \|x\|y = 0$. Therefore $\|y\|x = \|x\|y$ if $(x,y) = \|x\| \|y\|$. Conversely, if $\|y\|x = \|x\|y$, then

$$\|y\|(x,y) = \|x\|(y,y) = \|x\| \|y\|^2$$

or

$$\|y\|(\|x\|\,\|y\| - (x,y)) = 0$$

Therefore

$$|(\,\|x\|\,\|y\| - (x,y))|^2 \leq 2\|x\|\,\|y\|\,|(\,\|x\|\,\|y\| - (x,y))| = 0$$

and thus $(x,y) = \|x\|\,\|y\|$.

The fact that $\|\quad\|$ is a norm follows from (4.2).

Definition 9 A *Hilbert space* is a complete separable inner product space.

Our next theorem, which is a fundamental geometric property of Hilbert space, makes use of the identity

$$(4.6) \qquad \|x + y\|^2 + \|x - y\|^2 = 2\|x\|^2 + 2\|y\|^2$$

whose proof is left to the reader.

Theorem 6 *If M is a subspace ($=$ closed located linear subset) of a Hilbert space H, to each x in H there exists a unique vector Px in M with*

$$\|x - Px\| = \rho(x,M) \equiv d$$

The vector Px can also be characterized as the unique vector in M such that $(x - Px, z) = 0$ for all z in M. The function P is an operator from H to M, with $P^2 \equiv P$, called the projection onto the subspace M. If M contains a nonzero vector, then P is normable, and $\|P\| = 1$.

Proof Let $\{y_n\}$ be a sequence of elements of M such that

$$(4.7) \qquad \|y_n - x\| \to d \qquad \text{as } n \to \infty$$

For each m and n we compute, using (4.6),

$$\begin{aligned}
\|y_m - y_n\|^2 &= \|y_m - x - (y_n - x)\|^2 \\
&= 2\|y_m - x\|^2 + 2\|y_n - x\|^2 - 4\|\tfrac{1}{2}(y_m + y_n) - x\|^2 \\
&\leq 2(\|y_m - x\|^2 - d^2) + 2(\|y_n - x\|^2 - d^2)
\end{aligned}$$

since $\|z - x\|^2 \geq d^2$ for all z in M. It follows that $\{y_n\}$ is a Cauchy sequence, whose limit we call y. By (4.7),

$$\|y - x\| = d$$

If y' is any element of M with $\|y' - x\| = d$, then by (4.6),

$$\|y - y'\| \leq 2\|y - x\|^2 + 2\|y' - x\|^2 - 4\|\tfrac{1}{2}(y + y') - x\|^2$$
$$= 4(d^2 - \|\tfrac{1}{2}(y + y') - x\|^2) \leq 0$$

Hence $y = y'$. Thus $Px \equiv y$ is the unique closest vector in M to x.
For all z in M and a in $\mathbf{F}$, we have

$$(x - Px + az, x - Px + az) \geq d^2 = (x - Px, x - Px)$$

or $\qquad a^2\|z\|^2 + a(z, x - Px) + a^*(x - Px, z) \geq 0$

This implies that $(x - Px, z) = 0$. Conversely, if y is any vector in M with $(x - y, z) = 0$ for all z in M, then since $y - Px \in M$, we have

$$(y - Px, y - Px) = (x - Px, y - Px) - (x - y, y - Px) = 0$$

and thus $y = Px$. Hence Px is the unique vector in M such that $(x - Px, z) = 0$ for all z in M.
Since

$$(ax + by - (aPx + bPy), z) = a(x - Px, z) + b(y - Py, z) = 0$$

whenever $a, b \in \mathbf{F}$; $x, y \in H$; and $z \in M$, we see that

$$P(ax + by) = aPx + bPy$$

Thus P is linear.
Since $Px \in M$ (for each x in H), it follows that $P^2x = Px$. Also,

$$\|x\|^2 = (x,x) = (x - Px + Px, x - Px + Px) = \|x - Px\|^2 + \|Px\|^2$$

since $(x - Px, Px) = 0$. Therefore $\|Px\| \leq \|x\|$. Hence 1 is a bound for P. On the other hand, $\|Py\| = \|y\| > 0$ if y is any nonzero element of M, and thus P is normable, and $\|P\| = 1$.

Vectors x and y in an inner product space B are *orthogonal*, written $x \perp y$, if $(x,y) = 0$. For orthogonal vectors x and y we have the important equality

$$\|x + y\|^2 = (x,x) + (x,y) + (y,x) + (y,y) = \|x\|^2 + \|y\|^2$$

The *orthogonal complement* $K^\perp$ of a set $K \subset B$ consists of all x in B such that $x \perp y$ for all y in K. It is a closed linear subset of B.

Consider a subspace M of a Hilbert space H. By Theorem 6, every vector x in H has a unique representation $x = Px + (x - Px)$ as the sum of an element Px of M and an element $x - Px$ of $M^\perp$. For each vector y in $M^\perp$ the vector $z \equiv (x - Px) - y \in M^\perp$, and thus $z \perp Px$. Therefore

$$\|x - y\|^2 = \|Px + z\|^2 = \|Px\|^2 + \|z\|^2 \geq \|Px\|^2 = \|x - (x - Px)\|^2$$

and thus $x - Px$ is the closest vector in $M^\perp$ to x. Thus $M^\perp$ is a subspace of H, and the projection of x on $M^\perp$ is $x - Px$.

Definition 10   A sequence $\{x_n\}$ of vectors of a Hilbert space H is *orthonormal* if $x_m \perp x_n$ whenever $m \neq n$ and if for each n either $\|x_n\| = 1$ or $\|x_n\| = 0$. Such a sequence is called an *orthonormal basis* if each vector x can be written uniquely in the form

$$x = \sum_{n=1}^{\infty} a_n x_n$$

where $\{a_n\}$ is a sequence of scalars, called the *coordinates* of x with respect to the orthonormal basis $\{x_n\}$, such that $a_n = 0$ whenever $x_n = 0$.

An orthonormal basis can always be obtained by the so-called Gramm-Schmidt orthogonalization process.

Theorem 7   *Every Hilbert space has an orthonormal basis* $\{x_n\}$. *If* $\{a_n\}$ *are the coordinates of* x, *and* $\{b_n\}$ *are the coordinates of* y, *then*

$$a_n = (x,x_n) \qquad b_n = (y,x_n)$$
$$(x,y) = \sum_{n=1}^{\infty} a_n b_n^* \qquad \|x\|^2 = \sum_{n=1}^{\infty} |a_n|^2$$

Proof   Let $\{y_n\}$ be a countable dense subset of H. We define inductively an orthonormal sequence $\{x_n\}$ such that for each n there exists a linear combination y_n' of $x_1, \ldots , x_n$ with

(4.8) $\|y_n - y_n'\| < n^{-1}$

To start the induction, note that either $\|y_1\| < 1$ or $\|y_1\| > 0$. In the first case write $x_1 \equiv 0$, and in the second case $x_1 \equiv \|y_1\|^{-1} y_1$.

Assume that $x_1, \ldots , x_n$ have been defined. Let M_n be the subset of H consisting of all linear combinations of $x_1, \ldots , x_n$. Then M_n is a finite-dimensional Banach space, and so is locally compact. It is therefore located in H. Hence M_n is a subspace of H. Let P_n be the projection on M_n. Either $\|y_{n+1} - P_n y_{n+1}\| < (n+1)^{-1}$ or $\|y_{n+1} - P_n y_{n+1}\| > 0$. In the first case set $x_{n+1} \equiv 0$. The above condition holds with $y'_{n+1} \equiv P_n y_{n+1}$. In the second case set

$$x_{n+1} \equiv \|y_{n+1} - P_n y_{n+1}\|^{-1}(y_{n+1} - P_n y_{n+1})$$

Condition (4.8) is satisfied with

$$y'_{n+1} \equiv P_n y_{n+1} + \|y_{n+1} - P_n y_{n+1}\|\, x_{n+1} = y_{n+1}$$

The vector x_{n+1} is orthogonal to $x_1, \ldots, x_n$ because $y_{n+1} - P_n y_{n+1} \in M_n^\perp$. Thus the sequence $\{x_n\}$ is orthonormal.

To finish the proof, consider arbitrary vectors x and y in H. For each n we have unique representations

$$P_n x = \sum_{i=1}^{n} a_i x_i \qquad P_n y = \sum_{i=1}^{n} b_i x_i$$

with $a_i = b_i = 0$ whenever $x_i = 0$. Taking inner products with x_k, we get

$$a_k = (P_n x, x_k) = (x, x_k) + (P_n x - x, x_k) = (x, x_k)$$

and similarly $b_k = (y, x_k)$, so that a_k and b_k are independent of n. We have

$$(P_n x, P_n y) = \sum_{i=1}^{n} a_i b_i^*$$

Now the distances of x and y to M_n approach 0 as $n \to \infty$, or $P_n x \to x$ and $P_n y \to y$ as $n \to \infty$. Therefore

$$(x, y) = \lim_{n \to \infty} (P_n x, P_n y) = \sum_{i=1}^{\infty} a_i b_i^*$$

Setting $y = x$ gives

$$\|x\|^2 = (x, x) = \sum_{i=1}^{\infty} |a_i|^2$$

Corollary *An n-dimensional Hilbert space admits a norm-preserving linear map onto* **F**^{n}. *An infinite-dimensional Hilbert space admits a norm-preserving linear map onto* l_2. *(The space H is infinite-dimensional if for every $n > 0$ there exists an n-dimensional subspace of H.)*

Proof Assume that H is n-dimensional, and let $y_1, \ldots, y_n$ be a basis for H. The orthogonalization process of Theorem 7 produces an orthonormal basis $x_1, \ldots, x_n$ for H, with $\|x_i\| = 1$ $(1 \leq i \leq n)$. Let $\sigma: H \to \mathbf{F}^n$ be defined by $\sigma(x) = ((x,x_1), \ldots, (x,x_n))$. Then σ is onto and norm-preserving.

Assume next that H is infinite-dimensional. The orthogonalization process of Theorem 7 produces an orthonormal basis $\{x_n\}$ with $\|x_n\| = 1$ for infinitely many values of n. Leaving out those x_n for which $\|x_n\| = 0$, we may assume that $\|x_n\| = 1$ for *all* n. The map $\sigma: H \to l_2$ defined by $\sigma(x) \equiv \{(x,x_n)\}$ is onto and norm-preserving.

Two operators A and B on a Hilbert space H are called *adjoint* if

$$(Ax,y) = (x,By)$$

for all vectors x and y in H. If A and B are adjoint, then A is uniquely determined by B. If A and A are adjoint, that is, if $(Ax,y) = (x,Ay)$ for all vectors x and y, then A is called *hermitian* or *selfadjoint*. If A is hermitian, then $(Ax,x) = (x,Ax) = (Ax,x)^*$ is a real number for all vectors x.

It is clear that a sum of hermitian operators is hermitian, a real multiple of a hermitian operator is hermitian, and a product of commuting hermitian operators is hermitian.

The identity operator I is hermitian. More generally, let M be a subspace of H, with projection P_M. For all vectors x and y we have

$$(P_M x,y) = (P_M x, P_M y) + (P_M x, y - P_M y)$$
$$= (P_M x, P_M y) = (P_M x, P_M y) + (x - P_M x, P_M y) = (x, P_M y)$$

Thus P_M is hermitian.

Conversely, consider any hermitian operator P on H with $P^2 = P$. Write $M \equiv \{x \in H: Px = x\}$. Clearly M is a closed linear subset of H. For each y in H and z in M,

$$(y - Py, z) = (y,z) - (Py,z) = (y,z) - (y,Pz) = 0$$

since $Pz = z$. Hence $y - Py \perp M$. Since $P(Py) = Py$, we also have $Py \in M$. Thus, for each x in M,

$$\|y - x\|^2 = \|y - Py + Py - x\|^2$$
$$= \|y - Py\|^2 + \|Py - x\|^2 \geq \|y - Py\|^2$$

since $Py - x \in M$ and $y - Py \perp M$. Hence Py is the closest element in M to y. It follows that M is located, and $P = P_M$.

We now prove some lemmas which will help disclose the structure of an arbitrary hermitian operator.

Lemma 8   *Let A be a hermitian operator on a Hilbert space H. For simplicity assume that $\|Ax\| \leq \|x\|$ for all x in H. Let x be any vector of norm 1. Write*

$$y \equiv Ax - (Ax,x)x$$

Then $y \perp x$ (and thus y is the projection of Ax on the orthogonal complement of x). Write

$$x_A \equiv \|x + \tfrac{1}{2}y\|^{-1}(x + \tfrac{1}{2}y)$$

Then

$$(4.9) \qquad (Ax_A,x_A) - (Ax,x) \geq \tfrac{1}{4}\|Ax - (Ax,x)x\|^2$$

and if B is any hermitian operator which commutes with A, then

$$(4.10) \qquad \|Bx_A\| \leq 2\|Bx\|$$

Proof   Set $\lambda \equiv (Ax,x)$ and compute

$$\begin{aligned}
(A(x + \tfrac{1}{2}y),\, x + \tfrac{1}{2}y) &= (\lambda x + y + \tfrac{1}{2}Ay,\, x + \tfrac{1}{2}y) \\
&= \lambda + \tfrac{1}{2}(y,y) + \tfrac{1}{2}(Ay,x) + \tfrac{1}{4}(Ay,y) \\
&= \lambda + \tfrac{1}{2}(y,y) + \tfrac{1}{2}(y,Ax) + \tfrac{1}{4}(Ay,y) \\
&= \lambda + (y,y) + \tfrac{1}{4}(Ay,y) \geq \lambda + \tfrac{3}{4}(y,y)
\end{aligned}$$

Therefore

$$\begin{aligned}
(Ax_A,x_A) - (Ax,x) &\geq (1 + \tfrac{1}{4}(y,y))^{-1}(\lambda + \tfrac{3}{4}(y,y)) - \lambda \\
&= (1 + \tfrac{1}{4}(y,y))^{-1}(\tfrac{3}{4}(y,y) - \tfrac{1}{4}\lambda(y,y)) \geq \tfrac{1}{2}(\tfrac{1}{2}(y,y)) \\
&= \tfrac{1}{4}(y,y) = \tfrac{1}{4}\|Ax - (Ax,x)x\|^2
\end{aligned}$$

Since $x + \tfrac{1}{2}y = x - \tfrac{1}{2}(Ax,x)x + \tfrac{1}{2}Ax$, we also have

$$\begin{aligned}
\|Bx_A\| &\leq \|B(x + \tfrac{1}{2}y)\| = \|Bx - \tfrac{1}{2}(Ax,x)Bx + \tfrac{1}{2}ABx\| \\
&\leq \{1 + \tfrac{1}{2}|(Ax,x)| + \tfrac{1}{2}\}\|Bx\| \leq 2\|Bx\|
\end{aligned}$$

Lemma 9   *Let $A_1, \ldots, A_n$ be commuting hermitian operators on a Hilbert space H, let x be a vector of norm 1, and let ϵ be a positive constant. Then there exists a vector y of norm 1 with*

$$(4.11) \qquad (A_1 y,y) > (A_1 x,x) - \epsilon$$

and

$$\text{(4.12)} \qquad \|A_i y - (A_i y, y)y\| < \epsilon \qquad (1 \leq i \leq n)$$

Proof There is no loss of generality in assuming that 1 is a bound for each of the operators $A_1, \ldots, A_n$. First consider the case $n = 1$. Write $x_1 \equiv x$, $x_2 \equiv x_{A_1}$, $x_3 \equiv (x_2)_{A_1}, \ldots$. Choose N in Z^+ so large that $N\epsilon^2 > 32$. Now either

$$\|A_1 x_k - (A_1 x_k, x_k)x_k\| > \frac{\epsilon}{2} \qquad (1 \leq k \leq N)$$

or

$$\text{(4.13)} \qquad \|A_1 x_k - (A_1 x_k, x_k)x_k\| < \epsilon$$

for some k with $1 \leq k \leq N$. In the former case Lemma 8 gives

$$
\begin{aligned}
(A_1 x_{N+1}, x_{N+1}) &\geq (A_1 x_N, x_N) + \frac{\epsilon^2}{16} \\
&\geq (A_1 x_{N-1}, x_{N-1}) + 2\frac{\epsilon^2}{16} \\
&\geq \cdots \geq (A_1 x, x) + N\frac{\epsilon^2}{16} \\
&> -1 + 2 = 1
\end{aligned}
$$

This contradicts the fact that 1 is a bound for A_1. Thus (4.13) holds for some k. Set $y \equiv x_k$. Then (4.12) is valid, and $(A_1 y, y) \geq (A_1 x, x)$. Thus the lemma is true when $n = 1$.

Assume therefore that $n > 1$, and that the lemma is true for all smaller values of n. Choose the positive integer N so that $N\epsilon^2 > 32$. By the inductive hypothesis, there exists a unit vector u with

$$(A_1 u, u) > (A_1 x, x) - \frac{\epsilon}{2}$$

and $\qquad \|A_i u - (A_i u, u)u\| < \epsilon 2^{-N} \qquad (1 \leq i \leq n-1)$

Write $u_1 \equiv u$, $u_2 \equiv (u_1)_{A_n}$, $u_3 \equiv (u_2)_{A_n}, \ldots$. Either

$$\|A_n u_k - (A_n u_k, u_k)u_k\| > \frac{\epsilon}{2} \qquad (1 \leq k \leq N)$$

or

$$\text{(4.14)} \qquad \|A_n u_k - (A_n u_k, u_k)u_k\| < \epsilon$$

for some k with $1 \leq k \leq N$. In the former case Lemma 8 gives

$$
\begin{aligned}
(A_n u_{N+1}, u_{N+1}) &\geq (A_n u_N, u_N) + \frac{\epsilon^2}{16} \\
&\geq (A_n u_{N-1}, u_{N-1}) + 2\frac{\epsilon^2}{16} \\
&\geq \cdots \geq (A_n u, u) + N\frac{\epsilon^2}{16} \\
&> -1 + 2 = 1
\end{aligned}
$$

This contradiction shows that (4.14) is valid for some value of k. Write $y \equiv u_k$. Then (4.14) gives (4.12) for $i = n$. If $1 \leq i \leq n - 1$, then the operator $A_i - (A_i u, u)I$ commutes with A_n, and thus, by Lemma 8,

$$
\begin{aligned}
(4.15) \quad \|A_i y - (A_i y, y) y\| &\leq \|A_i y - (A_i u, u) y\| \\
&\leq 2\|A_i u_{k-1} - (A_i u, u) u_{k-1}\| \\
&\leq \cdots \leq 2^{k-1}\|A_i u - (A_i u, u) u\| \\
&\leq 2^{N-1} \epsilon 2^{-N} = \frac{\epsilon}{2}
\end{aligned}
$$

and (4.12) holds also for $1 \leq i \leq n - 1$.

Finally, (4.11) follows from (4.15) as follows:

$$
\begin{aligned}
(A_1 y, y) &\geq (A_1 u, u)(y, y) - \|A_1 y - (A_1 u, u) y\| \, \|y\| \\
&\geq (A_1 u, u) - \frac{\epsilon}{2} > (A_1 x, x) - \epsilon
\end{aligned}
$$

A sequence $\{A_n\}$ of operators on a Hilbert space H is said to converge *strongly* to an operator A if

(i) $A_n x \to A x \qquad$ as $n \to \infty \qquad (x \in H)$
and
(ii)   The operators A_n have a common bound c

The following criterion for strong convergence is useful.

Lemma 10   *Let $\{A_n\}$ be a sequence of operators on a Hilbert space H having a common bound c, such that the sequence $\{A_n x\}$ converges for all vectors x in some dense set $\Gamma \subset H$. Then $\{A_n\}$ converges strongly.*

Proof   If x is any vector, and ϵ any positive constant, choose y in Γ with $\|x - y\| < \epsilon$. Then

$$
\|A_m x - A_n x\| \leq 2c\|x - y\| + \|A_m y - A_n y\| < (2c + 1)\epsilon
$$

if m and n are sufficiently large. Thus $\{A_n x\}$ is a Cauchy sequence, whose limit we call Ax. Clearly A is an operator, and $\{A_n\}$ converges to A strongly.

Our next result is the famous spectral theorem, which describes the structure of a sequence of commuting hermitian operators.

Theorem 8 *Let $A \equiv \{A_n\}$ be a sequence of commuting hermitian operators on a Hilbert space H, with common bound 1. Let $\mathfrak{A}$ be the subalgebra over* **R** *of the algebra* Hom (H,H) *generated by the operators A_n and the operator I. Let X be the compact space*

$$X \equiv \prod_{n=1}^{\infty} [-1,1]$$

For each n in Z^+ let π_n be the nth coordinate function on X, defined by

$$\pi_n(\{x_k\}) \equiv x_n$$

Let $\mathfrak{A}_0$ be the subalgebra of $C(X)$ generated by the functions π_n and the function 1. Let $f \to f(A)$ be the unique homomorphism of $\mathfrak{A}_0$ into $\mathfrak{A}$ for which $1(A) \equiv I$ and $\pi_n(A) \equiv A_n$ $(1 \leq n < \infty)$. Then there exists a positive measure μ on X and a homomorphism $\varphi \to \varphi(A)$ of the algebra L_∞ of all bounded measurable real-valued functions φ on X into Hom (H,H), *which extends the homomorphism $f \to f(A)$ from $\mathfrak{A}_0$ onto $\mathfrak{A}$ already defined, so that the following conditions hold.*

(a)   If $|\varphi| \leq c$ a.e., then c is a bound for $\varphi(A)$.

(b)   If a sequence $\{\varphi_n\}$ of elements of L_∞ converges in measure to $\varphi \in L_\infty$, and if $|\varphi_n| \leq c$ a.e. for all n and some $c > 0$, then $\varphi_n(A) \to \varphi(A)$ strongly as $n \to \infty$.

(c)   If $\varphi \geq c > 0$ on a set of positive measure, then $(\varphi(A)x,x) \geq c$ for some unit vector x.

(d)   The operators $\varphi(A)$ are all hermitian, and all commute.

Proof   Let $\{x_n\}$ be an orthonormal basis for H. For each f in $\mathfrak{A}_0$ write

$$\int f \, d\mu \equiv \sum_{k=1}^{\infty} 2^{-k}(f(A)x_k,x_k)$$

Assume that f is a polynomial function of I, π_1, π_2, . . . , π_n. By Lemma 9, for each unit vector x and each $\epsilon > 0$ there exists a unit vector y such that

$$(f(A)x,x) < (f(A)y,y) + \epsilon$$

and
$$\|A_iy - \lambda_iy\|$$

is arbitrarily small, for $1 \leq i \leq n$, where $\lambda_i \equiv (A_iy,y)$. Write $\lambda \equiv \{\lambda_i\}$. Then $\lambda \in X$. Also, $f(\lambda)y$ is arbitrarily near to $f(A)y$, so $f(\lambda)$ is arbitrarily near to $(f(A)y,y)$. Thus we may choose y so that

$$(f(A)x,x) < f(\lambda) + \epsilon$$

Since ϵ is arbitrary,

$$(4.16) \qquad\qquad (f(A)x,x) \leq \|f\| \qquad (\|x\| = 1)$$

Replacing f with f^2, we find that

$$\|f(A)x\|^2 = (f(A)^2x,x) \leq \|f^2\| = \|f\|^2$$

and thus $\|f(A)x\| \leq \|f\|$. Therefore $\|f\|$ is a bound for the operator $f(A)$. Since $\|x_k\| = 0$ or 1 for each k, (4.16) gives

$$\left| \int f \, d\mu \right| \leq \sum_{k=1}^{\infty} 2^{-k}\|f\| \leq \|f\|$$

for all f in $\mathfrak{A}_0$. By the Stone-Weierstrass theorem, $\mathfrak{A}_0$ is dense in $C(X)$. The linear functional $f \to \int f \, d\mu$ can therefore be uniquely extended to $C(X)$, and $|\int f \, d\mu| \leq \|f\|$ for all f in $C(X)$. Thus μ is a measure on X. If $f \in \mathfrak{A}_0$, then

$$\int f^2 \, d\mu = \sum_{k=1}^{\infty} 2^{-k}(f(A)x_k,f(A)x_k) \geq 0$$

Therefore $\int f \, d\mu \geq 0$ whenever $f \geq 0$, and thus μ is positive.

For each f in $\mathfrak{A}_0$ we have

$$(4.17) \quad \|f\|_2^2 = \int f^2 \, d\mu = \sum_{k=1}^{\infty} 2^{-k}(f(A)^2x_k,x_k) = \sum_{k=1}^{\infty} 2^{-k}\|f(A)x_k\|^2$$

Consider any φ in L_∞, and let $\{f_n\}$ be a bounded sequence of elements of $\mathfrak{A}_0$ converging in measure to φ. Then $\{f_n\}$ is a Cauchy sequence in L_2. By (4.17), the sequence $\{f_n(A)x_k\}$ is a Cauchy sequence for each k.

Since $f_n(A)$ is bounded by $\|f_n\|$, the operators $f_n(A)$ have a common bound. Lemma 10 implies that the sequence $\{f_n(A)\}$ converges strongly, to an operator which we denote by $\varphi(A)$. It does not depend on the choice of the sequence f_n, since any other sequence, say $\{f_n'\}$, which is bounded and converges in measure to φ can be alternated with $\{f_n\}$ to form a sequence $f_1, f_1', f_2, f_2', \ldots$, and the corresponding sequence $f_1(A), f_1'(A), \ldots$ converges strongly, to a limit which must equal $\varphi(A)$.

The fact that $\varphi \to \varphi(A)$ is a homomorphism follows from the fact that $f \to f(A)$ is a homomorphism. If $|\varphi| \le c$ a.e., then we can choose the sequence f_n with $\|f_n\| \le c$. Therefore c is a bound for each of the operators $f_n(A)$. Thus (a) holds.

To check (b), note first that c is a bound for $\varphi_n(A)$ $(1 \le n < \infty)$, by (a). Consider any vector x. For each n in Z^+ choose f_n in $\mathfrak{A}_0$ with $\|\varphi_n - f_n\|_1 < n^{-1}$, $\|\varphi_n(A)x - f_n(A)x\| < n^{-1}$, and $\|f_n\| \le c$. Then $\{f_n\}$ converges to φ in measure, and therefore

$$\varphi(A)x = \lim_{n \to \infty} f_n(A)x = \lim_{n \to \infty} \varphi_n(A)x$$

Thus (b) is valid.

To check (c), note first that if $\psi \in L_\infty$ and $\psi \ge 0$ a.e., then $(\psi(A)x,x) = (\psi^{1/2}(A)x, \psi^{1/2}(A)x) \ge 0$ for all x in H. Assume that $\varphi \ge c > 0$ on some set K of positive measure. By the definition of μ,

$$0 < \mu(K) = \int \chi_K \, d\mu = \sum_{k=1}^{\infty} 2^{-k}(\chi_K(A)x_k, x_k)$$

and thus $\chi_K(A)x_k \neq 0$ for some k. Write

$$x \equiv \|\chi_K(A)x_k\|^{-1}\chi_K(A)x_k$$

$$\text{Then} \quad (\varphi(A)x,x) = \|\chi_K(A)x_k\|^{-2}(\varphi(A)\chi_K(A)x_k, x_k)$$
$$\ge \|\chi_K(A)x_k\|^{-2}c(\chi_K(A)x_k, x_k) = c(x,x) = c$$

Since sums and products of commuting hermitian operators are hermitian, the operators in $\mathfrak{A}$ are all hermitian. Therefore the operators $\varphi(A)$ are all hermitian. Each $\varphi(A)$ is a strong limit of operators which commute with the operators in $\mathfrak{A}$. Therefore each $\varphi(A)$ commutes with the operators in $\mathfrak{A}$. Therefore the operators $\varphi(A)$ commute with one another.

Remark By (c), we see that if $|(\varphi(A)x,x)| \le c$ for all unit vectors x, then $|\varphi| \le c$ a.e., which by (b) implies that c is a bound for $\varphi(A)$. Thus, if T is *any* hermitian operator, then c is a bound for T if and only if $|(Tx,x)| \leq c$ for all unit vectors x.

The spectral theorem has many nontrivial consequences. Here are two.

Call a hermitian operator T *positive* if $(Tx,x) \geq 0$ for all x in H. The spectral theorem implies that every positive operator has a positive square root.

For any hermitian operator T and any $\epsilon > 0$ there exist commuting projections $P_1, \ldots, P_n$, with $P_i P_j = 0$ for $1 \leq i < j \leq n$, and real numbers $c_1, \ldots, c_n$ such that the operator $T - \sum_{i=1}^{n} c_i P_i$ is bounded by ϵ. To prove this, we take $T \equiv A_1$ in Theorem 8, so $T = \pi_1(A)$. Choose disjoint integrable sets $S_1, \ldots, S_n$ and constants $c_1, \ldots, c_n$ so that $\left| \pi_1 - \sum_{i=1}^{n} c_i \chi_{S_i} \right| \leq \epsilon$ a.e. (This can be done by the proof of Proposition 4 of Chap. 7.) For $1 \leq i \leq n$ the operator $P_i \equiv \chi_{S_i}(A)$ is a projection (onto the subspace $M_i \equiv \{x \in H : P_i(x) = x\}$), and $T - \sum_{i=1}^{n} c_i P_i$ is bounded by ϵ.

The homomorphism $\varphi \rightarrow \varphi(A)$ of Theorem 8 is often called a *functional calculus* for the hermitian operators $\{A_n\}$. In case the operators in $\mathfrak{A}$ are all normable, the functional calculus can be sharpened considerably. To do this we need a lemma.

Lemma 11 *Let X be a compact space, and σ a map from $C(X, \mathbf{F})$ into $\mathbf{R}^{0+}$ such that $\sigma(1) > 0$, $\sigma(f) \leq \|f\|$, $\sigma(f + g) \leq \sigma(f) + \sigma(g)$,*

$$\sigma(cf) = |c|\sigma(f)$$

and $\sigma(fg) \leq \sigma(f)\sigma(g)$ for all f and g in $C(X, \mathbf{F})$ and all c in $\mathbf{F}$. Then there exists a compact set $K \subset X$ such that $\sigma(f) = \|f\|_K$ for all f in $C(X, \mathbf{F})$.

Proof Let K consist of all x in X such that for each $\epsilon > 0$ there exists h_ϵ in $C(X, \mathbf{F})$, with $\sigma(h_\epsilon) > 0$, for which $h_\epsilon(y) = 0$ whenever $\rho(y,x) \geq \epsilon$. Consider any x in K, any c in $\mathbf{F}$, and any g in $C(X, \mathbf{F})$ such that $g(y) = c$ at all points y belonging to some neighborhood of x. Then $gh_\epsilon = ch_\epsilon$ if ϵ is sufficiently small. Hence

$$|c|\sigma(h_\epsilon) = \sigma(ch_\epsilon) = \sigma(gh_\epsilon) \leq \sigma(g)\sigma(h_\epsilon)$$

or $\sigma(g) \geq |c|$. Now, if f is any element of $C(X, \mathbf{F})$, and k is any positive integer, we can find a function g in $C(X, \mathbf{F})$ such that $\|f - g\| \leq k^{-1}$

and $g(y) = f(x)$ for all y in some neighborhood of x. Hence

$$\sigma(f) \geq \sigma(g) - \sigma(f - g) \geq |f(x)| - k^{-1}$$

Letting $k \to \infty$ gives $\sigma(f) \geq |f(x)|$, for all f in $C(X,\mathbf{F})$ and x in K.

We next show that if f_0 is any element of $C(X,\mathbf{F})$ with $\sigma(f_0) > 0$, and K_0 is any compact support for f_0, then $K_0 \cap K$ is nonvoid. To this end we define inductively a sequence $f_0, f_1, \ldots$ of elements of $C(X,\mathbf{F})$ and a sequence $K_0 \supset K_1 \supset \cdots$ of compact sets, so that $\sigma(f_n) > 0$, K_n supports f_n, and diam $K_n \leq n^{-1}$ for all $n \geq 1$. The function f_0 and the set K_0 have already been defined. Assume therefore that $n \geq 1$ and that $f_0, \ldots, f_{n-1}, K_0, \ldots, K_{n-1}$ have already been defined. Choose elements $g_1, \ldots, g_N$ of $C(X,\mathbf{F})$ whose sum is f_{n-1}, each of which is supported by some compact subset of K_{n-1} of diameter at most n^{-1}. Since

$$\sigma(g_1) + \cdots + \sigma(g_N) \geq \sigma(f_{n-1}) > 0$$

$\sigma(g_i) > 0$ for some i. Set $f_n \equiv g_i$ and take K_n to be a compact support for g_i with $K_n \subset K_{n-1}$ and diam $K_n \leq n^{-1}$. This completes the induction. The sets K_n have a unique point x in common. Clearly $x \in K_0 \cap K$, and thus $K_0 \cap K$ is nonvoid.

To see that K is totally bounded, fix $\epsilon > 0$ and choose functions $f_1, \ldots, f_n$ in $C(X,\mathbf{F})$, with $f_1 + \cdots + f_n = 1$, so that f_i is supported by some compact set K_i of diameter less than ϵ. The set of positive integers i with $1 \leq i \leq n$ is the union of finite subsets S and T, with $\sigma(f_i) < (2n)^{-1}$ for $i \in S$, and $\sigma(f_i) > 0$ for $i \in T$. By the above, for each i in T there exists a point x_i in $K_i \cap K$. We shall show that these points x_i are an ϵ approximation to K. To this end, consider x in K. Then $|f_i(x)| > (2n)^{-1}$ for some i. Therefore $\sigma(f_i) > (2n)^{-1}$, and thus $i \in T$. Since $x \in K_i$ and $x_i \in K_i$ and diam $K_i < \epsilon$, we have $\rho(x,x_i) < \epsilon$. Thus the points x_i are an ϵ approximation to K. Hence K is compact.

Now consider any f in $C(X,\mathbf{F})$. For each $\epsilon > 0$ there exist g and $h = f - g$ in $C(X,\mathbf{F})$ such that $\|g\| \leq \|f\|_K + \epsilon$ and h is supported by a compact set $L \subset -K$. If $\sigma(h) > 0$, then, by the above, $L \cap K$ is nonvoid. This contradiction shows that $\sigma(h) = 0$. Hence

$$\sigma(f) \leq \sigma(g) + \sigma(h) = \sigma(g) \leq \|g\| \leq \|f\|_K + \epsilon$$

Thus $\sigma(f) \leq \|f\|_K$. Since we have already shown that $\sigma(f) \geq |f(x)|$ for all x in K, it follows that $\sigma(f) = \|f\|_K$.

We now state and prove our supplement to the spectral theorem.

Supplement to Theorem 8   *In case H is* nontrivial, *in the sense that it has a nonzero vector, and in case the operators in $\mathfrak{A}$ are all normable, there exists a compact set $K \subset X$ such that $\mu(X - K) = 0$ and*

$$(4.18) \qquad\qquad \|f(A)\| = \|f\|_K$$

for all f in $C(X)$.

Proof   Since $\|f(A)\| \leq \|f\|$ for all f in $\mathfrak{A}$, the operators $f(A)$ are normable for all f in $C(X)$, and $\|f(A)\| \leq \|f\|$. The hypotheses of Lemma 11 are satisfied by the function $\sigma(f) \equiv \|f(A)\|$. Hence there exists a compact set $K \subset X$ satisfying (4.18). If f is any function in $C(X)$ vanishing on K, we have $f(A) = 0$, by (4.18), and thus $\int f \, d\mu = 0$ by the definition of μ. Hence $\mu(X - K) = 0$.

In order to reformulate this supplement, we remark another consequence of Lemma 11, which has considerable interest of its own.

Proposition 10   *Let K be a compact space and Γ the set of all nonzero multiplicative bounded linear functionals u on $C(K,\mathbf{F})$. For each x in K define the element u_x of Γ by $u_x(f) \equiv f(x)$. Then the map $\lambda \colon K \to \Gamma$ defined by $\lambda(x) \equiv u_x$ is a metric equivalence [where Γ has the metric induced by the double norm on $C(K,\mathbf{F})^*$].*

Proof   If $|u(f)| > 1$, and $\|f\| < 1$ for some u in Γ and f in $C(K,\mathbf{F})$, then $\|f^n\| \to 0$ as $n \to \infty$, and $|u(f^n)| = |u(f)|^n \to \infty$ as $n \to \infty$, contradicting the boundedness of u. Therefore $|u(f)| \leq \|f\|$ for all u in Γ and f in $C(K,\mathbf{F})$. By Lemma 11, for each u in Γ there exists a compact set $L \subset K$ such that $|u(f)| = \|f\|_L$ for all f in $C(K,\mathbf{F})$. Since u is multiplicative, L must consist of a single point x, so that $|u(f)| = |f(x)|$. Thus $u(f) = 0$ whenever $f(x) = 0$. Hence, for each f in $C(K,\mathbf{F})$,

$$u(f) = u(f - f(x)) + f(x)u(1) = f(x) = u_x(f)$$

and thus $u = u_x$. Thus the map λ is onto.

Since each f in $C(K,\mathbf{F})$ is uniformly continuous, the map λ is uniformly continuous, by the definition of the double norm on $C(K,\mathbf{F})^*$. To show that λ^{-1} is also uniformly continuous, consider $\epsilon > 0$ and let $x_1, \ldots , x_n$ be an ϵ approximation to K. By Proposition 9, there exists $\delta > 0$ such that $|\rho(x,x_i) - \rho(y,x_i)| < \epsilon$ $(1 \leq i \leq n)$ whenever x and y are points of K with $|||\lambda(x) - \lambda(y)||| \leq \delta$. If we choose i with $\rho(x,x_i) < \epsilon$, it follows that $\rho(y,x_i) < 2\epsilon$, so $\rho(x,y) < 3\epsilon$. Thus λ^{-1} is uniformly continuous. Hence λ is a metric equivalence.

We now give three corollaries, which are actually reformulations of the Supplement to Theorem 8. To do this, we need to define the *uniform neighborhood structure* on Hom (H,H). For each A in Hom (H,H) and each $\epsilon > 0$ the uniform neighborhood $U(A,\epsilon)$ consists of all B in Hom (H,H) such that $A - B$ is bounded by some constant $c < \epsilon$. These neighborhoods define a neighborhood structure. Such concepts as uniformly closed and uniformly separable pertain to this neighborhood structure. The set of normable operators is uniformly closed, and on any linear subset of the set of normable operators the uniform neighborhood structure is induced by the metric $\rho(A,B) \equiv \|A - B\|$.

Corollary 1 *Let $\mathfrak{B}$ be a commutative algebra of normable hermitian operators on a nontrivial Hilbert space H, with $I \in \mathfrak{B}$, which is uniformly separable and uniformly closed. Let the spectrum Σ of $\mathfrak{B}$ consist of all nonzero bounded multiplicative linear functionals on $\mathfrak{B}$. Then each u in Σ is normable, with $\|u\| = 1$, and Σ is compact in the metric induced by the double norm. Moreover, if for each operator U in $\mathfrak{B}$ we let $\gamma(U)$ be the element of $C(\Sigma)$ defined by*

$$(\gamma(U))(u) \equiv u(U)$$

then γ is a norm-preserving isomorphism from the algebra $\mathfrak{B}$ onto the algebra $C(\Sigma)$.

Proof Let $A \equiv \{A_n\}$ be a sequence of elements of $\mathfrak{B}$, with common bound 1, whose linear combinations are uniformly dense in $\mathfrak{B}$. Let X be defined as in Theorem 8, and let K be the compact set defined in the Supplement to Theorem 8. Since $\mu(X - K) = 0$, the element $f(A)$ of $\mathfrak{B}$ is well defined for all f in $C(K)$. By Theorem 10 of Chap. 4, each element of $C(K)$ extends to an element of $C(X)$. By (4.18), it follows that the map $f \to f(A)$ from $C(K)$ to Hom (H,H) is a norm-preserving isomorphism F of $C(K)$ with $\mathfrak{B}$. Since $\Sigma \subset \mathfrak{B}^*$ stands in the same relation to $\mathfrak{B}$ as the space $\Gamma \subset C(K)^*$ of Proposition 10 does to $C(K)$, we see from Proposition 10 that Σ is compact and that each u in Σ is normable with $\|u\| = 1$. Moreover, the map $\gamma \colon \mathfrak{B} \to C(\Sigma)$ stands in the same relation to $\mathfrak{B}$ and Σ as the map $\omega \colon C(K) \to C(\Gamma)$, defined by $\omega(f) \equiv f \circ \lambda^{-1}$, stands to $C(K)$ and Γ. It follows that γ is a norm-preserving isomorphism.

In case H is a complex Hilbert space, this reformulation admits a modification.

Corollary 2 *Let $\mathfrak{B}'$ be a complex commutative algebra of normable linear operators on a nontrivial complex Hilbert space H, with $I \in \mathfrak{B}'$, which is *uniformly separable and uniformly closed, and is* selfadjoint *in the sense that each operator B in $\mathfrak{B}'$ has an adjoint B^* that also lies in $\mathfrak{B}'$. Let the spectrum Σ' of $\mathfrak{B}'$ consist of all nonzero multiplicative bounded linear functionals u' on $\mathfrak{B}'$. Then each u' in Σ' is normable, with $\|u'\| = 1$, Σ' is compact, and the map $\gamma' \colon \mathfrak{B}' \to C(\Sigma', \mathbf{C})$, defined by*

$$\gamma'(U)(u') \equiv u'(U) \qquad (u' \in \Sigma', \ U \in \mathfrak{B}')$$

is a norm-preserving isomorphism of $\mathfrak{B}'$ onto $C(\Sigma', \mathbf{C})$, with

$$(4.19) \qquad \gamma'(U^*) = \gamma'(U)^* \qquad (U \in \mathfrak{B}')$$

Proof Each U in $\mathfrak{B}'$ can be written uniquely in the form $U = U_1 + iU_2$, where $U_1 \equiv \frac{1}{2}(U + U^*)$ and $U_2 \equiv 1/2i(U - U^*)$ are hermitian. Let $\mathfrak{B}$ be the set of all hermitian operators in $\mathfrak{B}'$ and Σ the set of all nonzero multiplicative bounded linear functionals u on $\mathfrak{B}$. By the above, the map $\gamma \colon \mathfrak{B} \to C(\Sigma)$ is a norm-preserving isomorphism.

Let u' be any element of Σ' and U any operator in $\mathfrak{B}$. If $u'(U)$ is not real, then there exists V in $\mathfrak{B}$ with $u'(V) = i$. Thus $u'(V^2 + I) = 0$. Now, $V^2 + I$ has an inverse W in $\mathfrak{B}$, because $v(V^2 + I) = v(V)^2 + 1 > 1$ for all v in Σ. We have $0 = u'(V^2 + I)u'(W) = u'(I) = 1$. This contradiction shows that $u'(U)$ is real for all u' in Σ' and U in $\mathfrak{B}$. Therefore the restriction u of u' to $\mathfrak{B}$ belongs to Σ. Moreover

$$u'(U^*) = u'(U_1 - iU_2) = u'(U_1) - iu'(U_2) = u'(U)^* \qquad (U \in \mathfrak{B}')$$

and thus (4.19) is valid.

Conversely, if u is any element of Σ, define

$$u'(U_1 + iU_2) = u(U_1) + iu(U_2)$$

for all $U = U_1 + iU_2$ in Σ. Since $u \in \Sigma$, to show that $u' \in \Sigma'$ it is enough to note that

$$u'(iU) = u'(iU_1 - U_2) = -u(U_2) + iu(U_1) = iu'(U)$$

for all U in $\mathfrak{B}'$. Thus the map $u' \to u$ is a one-one correspondence of Σ' with Σ. Since both $u' \to u$ and its inverse are clearly continuous, it is a metric equivalence. Therefore Σ' is compact. Now if U is any element of $\mathfrak{B}'$, $(U^*U)^* = U^*U^{**} = U^*U$, and thus $U^*U \in \mathfrak{B}$. Therefore

$$\|U\|^2 = \sup\{\|Ux\|^2 \colon \|x\| \le 1\} = \sup\{(U^*Ux, x) \colon \|x\| \le 1\}$$
$$= \|U^*U\| = \|\gamma(U^*U)\| = \|\gamma'(U^*U)\| = \|\gamma'(U)\|^2$$

or $\|U\| = \|\gamma'(U)\|$, and thus γ' is a norm-preserving isomorphism of $\mathfrak{B}'$ into $C(\Sigma',\mathbf{C})$. It is actually onto: if $f_1 + if_2$ is any element of $C(\Sigma',\mathbf{C})$, then both f_1 and f_2 are in the range of γ, because γ is onto, and therefore $f_1 + if_2$ is in the range of γ'. This isomorphism of $\mathfrak{B}'$ with $C(\Sigma',\mathbf{C})$ implies that each u' in Σ' is normable with $\|u'\| = 1$.

The hypothesis $I \in \mathfrak{B}'$ of Corollary 2 can be relaxed.

Corollary 3 *Let $\tilde{H}$ be a nontrivial complex Hilbert space, and $\widetilde{\mathfrak{B}}$ a nonvoid complex commutative selfadjoint algebra of normable operators on $\tilde{H}$, which is uniformly separable and uniformly closed. Let $\tilde{\Sigma}_0$ consist of all bounded multiplicative homomorphisms of $\widetilde{\mathfrak{B}}$ into $\mathbf{C}$. Then $\tilde{\Sigma}_0$ is compact in the metric induced by the double norm, and thus the spectrum $\tilde{\Sigma} \equiv \tilde{\Sigma}_0 - \{0\}$ of $\widetilde{\mathfrak{B}}$ (which consists of all bounded multiplicative homomorphisms of $\widetilde{\mathfrak{B}}$ onto $\mathbf{C}$) is a locally compact metric space under the metric*

$$(4.20) \qquad \rho(\varphi_1,\varphi_2) \equiv |(\||\varphi_1\||^{-1} - \||\varphi_2\||^{-1})| + \||\varphi_1 - \varphi_2\||$$

Moreover, the map $\tilde{\gamma}\colon \widetilde{\mathfrak{B}} \to C_\infty(\tilde{\Sigma},\mathbf{C})$ defined by

$$\tilde{\gamma}(U)(\varphi) \equiv \varphi(U)$$

is a norm-preserving isomorphism of $\widetilde{\mathfrak{B}}$ onto $C_\infty(\tilde{\Sigma},\mathbf{C})$, with

$$(4.21) \qquad\qquad \tilde{\gamma}(U^*) = \tilde{\gamma}(U)^* \qquad (U \in \widetilde{\mathfrak{B}})$$

Proof Let H be the Hilbert space whose elements are the pairs (x,α), with $x \in \tilde{H}$ and $\alpha \in \mathbf{C}$, with inner product

$$((x_1,\alpha_1), (x_2,\alpha_2)) \equiv (x_1,x_2) + \alpha_1\alpha_2{}^*$$

We identify each x in $\tilde{H}$ with the vector $(x,0)$ of H, and each operator U on $\tilde{H}$ with the operator on H given by

$$U(x,\alpha) \equiv (Ux,0)$$

Let $\mathfrak{B}'$ consist of all operators on H of the form $U + \alpha I$, with $U \in \widetilde{\mathfrak{B}}$ and $\alpha \in \mathbf{C}$. Then $\mathfrak{B}'$ satisfies the hypotheses of Corollary 2. Therefore the conclusions of Corollary 2 are valid. Now each u' in Σ' induces an element $\tilde{u}$ of $\tilde{\Sigma}_0$, obtained by restricting the domain of u' to $\widetilde{\mathfrak{B}}$. Moreover, every $\tilde{u}$ in $\tilde{\Sigma}_0$ arises in this way, and the map $u' \to \tilde{u}$ is one-one. Hence (4.21) is a consequence of (4.19). By the definitions of the double norms on Σ' and $\tilde{\Sigma}_0$, the map $u' \to \tilde{u}$ is a metric equivalence.

Therefore $\tilde{\Sigma}_0$ is compact. Hence $\tilde{\Sigma}$ is locally compact under the metric (4.20). Since γ' is a norm-preserving isomorphism of $\mathfrak{B}'$ onto $C(\Sigma',\mathbf{C})$, it induces a norm-preserving isomorphism of $\widetilde{\mathfrak{B}}$ onto $C_\infty(\Sigma' - \{\omega\}, \mathbf{C})$, where ω is that element of Σ' which vanishes on $\mathfrak{B}$. Since $\tilde{\gamma}$ is the composition of γ' with the norm-preserving isomorphism of $C_\infty(\Sigma' - \{\omega\}, \mathbf{C})$ onto $C_\infty(\tilde{\Sigma},\mathbf{C})$ induced by the metric equivalence $u' \to \tilde{u}$ of Σ' and $\tilde{\Sigma}_0$, it follows that $\tilde{\gamma}$ is a norm-preserving isomorphism.

5. LOCALLY CONVEX SPACES

In many cases a real linear space has a topology which cannot be described by a single seminorm: a family of seminorms is needed. This leads to the concept of a locally convex space.

Definition 11 A *locally convex space* V is a linear space V over $\mathbf{F}$, together with a family $\mathfrak{N}$ of seminorms on V, called *admissible* seminorms, such that if $\| \ \|_1$ and $\| \ \|_2$ are admissible, and if $c > 0$, then every seminorm $\| \ \|$ such that

$$\|x\| \leq c(\|x\|_1 + \|x\|_2) \qquad (x \in V)$$

is admissible. A linear functional $\varphi : V \to \mathbf{F}$ is *bounded* if it is bounded with respect to some $\| \ \|$ in $\mathfrak{N}$, that is, if $|\varphi(x)| \leq c\|x\|$ for all x in V and some $c > 0$ (called a *bound* for φ with respect to $\| \ \|$). A *defining family* of admissible seminorms is a set $\mathfrak{D} \subset \mathfrak{N}$ such that for each $\| \ \|$ in $\mathfrak{N}$ there exist $\| \ \|_1, \ldots, \| \ \|_n$ in $\mathfrak{D}$ and $c > 0$ such that

$$(5.1) \qquad \|x\| \leq c(\|x\|_1 + \cdots + \|x\|_n) \qquad (x \in V)$$

Any family of seminorms on a linear space V is a defining family for a locally convex structure on V: Take $\mathfrak{N}$ to be all seminorms $\| \ \|$ for which (5.1) is satisfied.

The simplest example of a locally convex space is a normed space. In this case the norm on V is the sole element of the defining family.

It is convenient to assume that $\mathfrak{N}$ defines the equality relation on V. This means that

$$x = 0 \qquad \text{if and only if} \qquad \|x\| = 0 \text{ for all } \| \ \| \text{ in } \mathfrak{N}$$

We shall assume that this is always the case.

A simple example of a locally convex space which is not a Banach space is obtained by considering a locally compact space X which is not compact, and taking the family

$$\{\| \quad \|_K : K \text{ is a compact subset of } X\}$$

to be a defining family of seminorms on $C(X)$.

This last example is *σ-normed*, in the sense that there exists a countable family $\{\| \quad \|_n\}$ of defining seminorms. A σ-normed locally convex space has a natural metric ρ, given by

$$\rho(x,y) = \sum_{n=1}^{\infty} 2^{-n} \min \{\|x - y\|_n, 1\}$$

Definition 12   The *dual space* or *adjoint space* of a locally convex space V is the set V^* of all bounded linear functionals on V, with the locally convex structure determined by the defining family $\{\| \quad \|_x\}_{x \in V}$ of seminorms given by

$$\|u\|_x \equiv |u(x)| \qquad (u \in V^*)$$

Definition 13   A subset A of the dual V^* of a locally convex space V is *bounded* if it is bounded with respect to some $\| \quad \|$ in $\mathfrak{N}$. This means that there exists $c > 0$ such that

$$|u(x)| \leq c\|x\| \qquad (x \in V, u \in A)$$

In studying a bounded subset A of the dual V^* of a locally convex space V, there is no loss of generality in taking V to be a Banach space, and in taking A to be a subset of the *unit sphere*

$$S^* \equiv \{u \in V^* : |u(x)| \leq \|x\| \text{ for all } x \text{ in } V\}$$

of V^*. We now consider the case $A = S^*$.

Theorem 9   *The unit sphere S^* of the dual V^* of a separable Banach space V is compact in the double norm, and the normable elements are dense. Each x in V defines a uniformly continuous function $u \to u(x)$ on S^*. For any two choices of the countable dense subset of V the corresponding double norms give rise to equivalent metrics on S^*.*

Proof   The proof that any choice $\{x_n'\}$ of the countable dense subset of V gives rise to a double norm $\||| \quad \|||'$ which induces a metric on S^* equivalent to the metric induced by $\||| \quad \|||$ is a special case of Proposition 9.

If x is any element of V, we can choose the sequence $\{x_n'\}$ with $x_1' = x$. Since $|||\ \ |||$ and $|||\ \ |||'$ induce equivalent metrics on S^*, the function $u \to u(x)$, which is uniformly continuous on S^* in the metric induced by $|||\ \ |||'$, is uniformly continuous in the metric induced by $|||\ \ |||$.

To see that S^* is complete, consider a Cauchy sequence $\{u_n\}$. Since each of the functions $u \to u(x)$ is uniformly continuous, $\{u_n(x)\}$ is a Cauchy sequence for each x in V, whose limit we call $u(x)$. Clearly $u \in S^*$, and $|||u_n - u||| \to 0$ as $n \to \infty$. Therefore S^* is complete.

To see that S^* is totally bounded, consider $\epsilon > 0$. Choose m so large that

$$|||u - v||| \le \sum_{k=1}^{m} 2^{-k}(1 + \|x_k\|)^{-1}|u(x_k) - v(x_k)| + \tfrac{1}{3}\epsilon$$

for all u and v in S^*. Since B is separable, there exists a finite-dimensional subspace B_0 of B such that the distances of $x_1, \ldots, x_m$ to B_0 are less than $\tfrac{1}{6}\epsilon$. Choose $x_1', \ldots, x_m'$ in B_0 with

$$\|x_k' - x_k\| < \tfrac{1}{6}\epsilon \qquad (1 \le k \le m)$$

Since B_0 is a finite-dimensional Banach space, its unit sphere is compact. Therefore every element of B_0^* is normable. It follows that B_0^* is a finite-dimensional Banach space, whose unit sphere S_0^* is therefore compact. Since S_0^* is compact, there exist vectors $v_1{}^0, \ldots, v_N{}^0$ in S_0^* such that, for each v^0 in S_0^*,

$$\sum_{k=1}^{m} 2^{-k}(1 + \|x_k\|)^{-1}|v^0(x_k') - v_i{}^0(x_k')| < \tfrac{1}{3}\epsilon$$

for some value of i $(1 \le i \le N)$. We may assume that $\|v_i{}^0\| < 1$ $(1 \le i \le N)$. By the Hahn-Banach theorem, there exist normable linear functionals $v_i \in S^*$ $(1 \le i \le N)$ such that $v_i(x) = v_i{}^0(x)$ for all x in B_0, in particular for $x = x_k'$ $(1 \le i \le m)$. For each v in S^* let v^0 be the restriction of v to B_0, so that $v^0 \in S_0^*$. Choosing i as above, we get

$$|||v - v_i||| \le \sum_{k=1}^{m} 2^{-k}(1 + \|x\|_k)^{-1}|v(x_k) - v_i(x_k)| + \tfrac{1}{3}\epsilon$$

$$\le \sum_{k=1}^{m} 2^{-k}(1 + \|x_k\|)^{-1}\{|v^0(x_k') - v_i{}^0(x_k')| + 2\|x_k' - x_k\|\} + \tfrac{1}{3}\epsilon$$

$$< \tfrac{1}{3}\epsilon + \tfrac{1}{3}\epsilon + \tfrac{1}{3}\epsilon = \epsilon$$

Thus $v_1, \ldots, v_N$ is an ϵ approximation to S^* in the double norm. It follows that S^* is totally bounded. Hence S^* is compact.

As discussed in Chap. 3, Brouwer contends that every real-valued function on a compact space is continuous. Although Brouwer's proof is unsatisfactory, his contention has not been disproved. Therefore, if we wish to examine linear functionals on B^*, it is reasonable to require that they be continuous on S^*. The following result describes the form of all such linear functionals.

Theorem 10 *Each linear functional φ on the dual B^* of a separable Banach space B which is continuous on S^* with respect to the metric ρ induced by the double norm comes from a vector x_0 in B:*

$$(5.2) \qquad \varphi(u) = u(x_0) \qquad (u \in B^*)$$

Proof It is sufficient to show that for each positive constant c there exists x_1 in B with $|\varphi(u) - u(x_1)| \le c$ for all u in S^*, for if this is true, we can successively choose vectors $x_1, x_2, \ldots$ in B with

$$|\varphi(u) - u(x_1 + \cdots + x_n)| < 2^{-n}$$

for all u in S^*. Then

$$|u(x_n)| \le |\varphi(u) - u(x_1 + \cdots + x_{n-1})| + |\varphi(u) - u(x_1 + \cdots + x_n)|$$
$$\le 2^{-n+2}$$

for all u in S^*, and thus $\|x_n\| \le 2^{-n+2}$, by the Corollary to Theorem 3. The series $\sum_{n=1}^{\infty} x_n$ therefore converges in norm to a vector x_0 in B which satisfies (5.2).

Assume therefore that c is given. In case $|\varphi(u)| \le c$ for all u in S^*, we may take $x_1 = 0$. We may assume therefore that $\varphi(u_0) \ne 0$ for some normable vector u_0 in S^*, with $\|u_0\| = 1$. Normalize φ so that $\varphi(u_0) = 1$. Since φ is continuous on S^*, for each $\epsilon > 0$ there exists a finite-dimensional subspace B_0 of B and $\delta > 0$ such that whenever $u \in S^*$ and $|u(x)| < \delta$ for all x in $B_0 \cap S$, then $|\varphi(u)| < \epsilon/2$. Write

$$N^* \equiv \{u \in S^*\colon \varphi(u) = 0\}$$
and
$$N_t^* \equiv \{u \in S^*\colon |\varphi(u)| \le t\}$$

for each $t > 0$. There exist arbitrarily small $t > 0$ such that N_t^* is compact. For each such t let $u_1, \ldots, u_m$ be a t approximation to

N_i^*. For $1 \le i \le m$ write $u_i' \equiv (1 + t)^{-1}(u_i - \varphi(u_i)u_0)$. Then $u_i' \in N^*$, and $|||u_i - u_i'||| \le 2t$. Therefore $u_1', \ldots, u_m'$ is a $3t$ approximation to N^*. Since t can be arbitrarily small, N^* is totally bounded and hence compact. Thus for each x in B_0 the quantity

$$\|x\|_0 \equiv \sup \{|u(x)| : u \in N^*\}$$

exists, and defines a seminorm $\| \quad \|_0$ on B_0, with $\| \quad \|_0 \le \| \quad \|$. Write

$$\beta \equiv \inf \{\|x\|_0 : x \in B_0, \|x\| = 1\}$$

Either $\beta < \epsilon$ or $\beta > 0$. First consider the case $\beta > 0$. Then $\|x\|_0 \ge \beta\|x\|$ for all x in B_0. Thus norms $\| \quad \|$ and $\| \quad \|_0$ on B_0 are equivalent, and B_0 is a finite-dimensional Banach space relative to $\| \quad \|_0$. Each u in N^*, when restricted to B_0, is in the unit sphere S_0^* of B_0^*, by the definition of $\| \quad \|_0$. Since N^* is compact and the map from N^* to S_0^* is continuous, the image K of N^* in S_0^* is totally bounded.

Let u be any element of S_0^*. If $\rho(u,K) > 0$, by the separation theorem there exists a bounded linear functional ψ on B_0^* such that

$$\psi(u) > \sup \{|\psi(v)| : v \in K\}$$

Since B_0 is finite-dimensional, ψ is induced by a vector x in B_0, and thus

$$u(x) > \sup \{|v(x)| : v \in K\} \equiv \|x\|_0$$

This contradicts the fact that $u \in S_0^*$. Therefore $\rho(u,K) = 0$. In other words, $\bar{K} = S_0^*$.

Since $\|u_0\| = 1$,

$$\sup \{|u_0(x)| : x \in B_0, \|x\|_0 \le 1\} \le \sup \{|u_0(x)| : x \in B, \|x\| \le \beta^{-1}\}$$
$$\le \beta^{-1}\|u_0\| = \beta^{-1}$$

Thus βu_0 is in S_0^*; there therefore exists u in N^* such that

$$|(\beta u_0 - u)(x)| < 2\delta \qquad (x \in B_0 \cap S)$$

Since $\frac{1}{2}(\beta u_0 - u) \in S^*$, it follows that

$$|\varphi(\tfrac{1}{2}(\beta u_0 - u))| < \frac{\epsilon}{2}$$

Since $\varphi(u) = 0$, this gives

$$\beta = \beta\varphi(u_0) = 2\varphi(\tfrac{1}{2}\beta u_0) < \epsilon$$

Thus, in any case, $\beta < \epsilon$. We may therefore choose x_0 in B_0 such that $\|x_0\| = 1$ and

$$|u(x_0)| < \epsilon \qquad (u \in N^*)$$

Let a be a positive constant such that $|\varphi(u)| \leq a$ for all u in S^*. Choose u in S^* with $u(x_0) = \frac{1}{2}$. Then $(1 + a)^{-1}(u - \varphi(u)u_0) \in N^*$, and thus

$$\begin{aligned}
\tfrac{1}{2} = u(x_0) &\leq |\varphi(u)u_0(x_0)| + (1 + a)|(1 + a)^{-1}(u - \varphi(u)u_0)(x_0)| \\
&\leq a|u_0(x_0)| + (1 + a)\epsilon
\end{aligned}$$

Therefore $|u_0(x_0)| > \frac{1}{3}a^{-1}$ if ϵ is sufficiently small. Write $x_1 \equiv u_0(x_0)^{-1}x_0$. Then $u_0(x_1) = 1$. Since also $(1 + a)^{-1}|(u - \varphi(u)u_0)(x_0)| < \epsilon$, for all u in S^*, we have

$$\begin{aligned}
|u(x_1) - \varphi(u)| &= |u(x_1) - \varphi(u)u_0(x_1)| \\
&\leq 3a|(u - \varphi(u)u_0)(x_0)| < 3a(1 + a)\epsilon < c
\end{aligned}$$

if ϵ is sufficiently small, as was to be proved.

In addition to the examples already given, we give one more important instance of a locally convex space. Let $\mathfrak{D}$ consist of all infinitely differentiable functions $f: \mathbf{R} \to \mathbf{F}$ having compact support. Clearly $\mathfrak{D}$ is a linear space over $\mathbf{F}$. To each sequence $N \equiv \{n_k\}$ of positive integers corresponds a seminorm $\| \quad \|_N$ on $\mathfrak{D}$, given by

$$\|f\|_N = \sum_{k=1}^{\infty} n_k \sup \{|f(x)| + |f'(x)| + \cdots + |f^{(n_k)}(x)| : |x| \geq k - 1\}$$

where, of course, the sum is actually finite. The space $\mathfrak{D}$ is given the locally convex structure defined by these seminorms. The dual space $\mathfrak{D}^*$ of $\mathfrak{D}$ is called the space of distributions on $\mathbf{R}$.

6. EXTREME POINTS

The vertices of a convex polygon are extreme, in the sense that each line segment lying in the polygon which contains a vertex has that vertex as an end point. The notion of an extreme point can be defined in a very general situation.

Definition 14 Let B be a real Banach space and K a compact convex subset of B. A point x_0 of K is an *extreme point* if for each $\epsilon > 0$ there exists $\delta > 0$ such that $\|x - y\| \le \epsilon$ whenever x, $y \in K$ and $\|\frac{1}{2}(x + y) - x_0\| \le \delta$.

Our geometric intuition leads us to hope that every point is an average of extreme points. This is not quite true. The key to the situation is the following lemma.

Lemma 12   *Let K be a compact convex subset of a real normed linear space B. For each w in B^* write*

$$M(w,K) \equiv \sup \{w(x) : x \in K\}$$

and for each α in $\mathbf{R}$ write

$$S(w,\alpha,K) \equiv \{x \in K : w(x) \ge \alpha\}$$

Then for arbitrary u and v in B^* there exists w in B^*, which is arbitrarily close to u, and $c < M(w,K)$ such that the variations

$$\operatorname{diam}\{u(x) : x \in S(w,c,K)\} \qquad \operatorname{diam}\{v(x) : x \in S(w,c,K)\}$$

of u and v on the set $S(w,c,K)$ are arbitrarily small.

Proof   For a given $\epsilon > 0$ choose constants r and s, with

$$M(u,K) - \epsilon^2 < r < M(u,K) \qquad \text{and} \qquad s = \tfrac{1}{2}(r + M(u,K))$$

so that the sets $S_r \equiv S(u,r,K)$ and $S_s \equiv S(u,s,K)$ are compact, and

$$M(v,S_r) - M(v,S_s) < \epsilon^2$$

Let x_0 be any point in S_s with $v(x_0) > M(v,S_r) - \epsilon^2$. For each x in K with $u(x) < r$ the point

$$x' \equiv (u(x_0) - r)(u(x_0) - u(x))^{-1}x + (r - u(x))(u(x_0) - u(x))^{-1}x_0$$

lies in K, and $u(x') = r$. Therefore

$$M(v,S_r) \ge v(x') \ge (u(x_0) - r)(u(x_0) - u(x))^{-1}v(x)$$
$$+ (r - u(x))(u(x_0) - u(x))^{-1}(M(v,S_r) - \epsilon^2)$$

or

$$\tag{6.1} M(v,S_r) \ge v(x) - (r - u(x))(u(x_0) - r)^{-1}\epsilon^2$$
$$\ge v(x) - (r - u(x))(s - r)^{-1}\epsilon^2$$

Write
$$w \equiv u + \epsilon^{-1}(s - r)v$$

Since $s - r < \epsilon^2$, by taking ϵ arbitrarily small we can make w arbitrarily close to u. Since $M(w,K) \geq w(x_0) > s + \epsilon^{-1}(s - r)(M(v,S_r) - \epsilon^2)$, there exists $c > s + \epsilon^{-1}(s - r)(M(v,S_r) - \epsilon^2)$ such that $S(w,c,K)$ is compact. For each x in K with $u(x) < r$, inequality (6.1) then gives

$$
\begin{aligned}
w(x) &= u(x) + \epsilon^{-1}(s - r)v(x) \\
&\leq u(x) + (r - u(x))\epsilon + \epsilon^{-1}(s - r)M(v,S_r) < c
\end{aligned}
$$

if $\epsilon < 1$, which we assume. It follows that $u(x) \geq r$ whenever $x \in S(w,c,K)$. Hence $v(x) \leq M(v,S_r)$ whenever $x \in S(w,c,K)$. On the other hand, for x in $S(w,c,K)$ we also have

$$
u(x) \leq M(u,K) \leq r + \epsilon^2
$$

and
$$
\begin{aligned}
v(x) &= \epsilon(s - r)^{-1}(w(x) - u(x)) \\
&\geq \epsilon(s - r)^{-1}(c - M(u,K)) \\
&= \epsilon(s - r)^{-1}(c - s) + \epsilon(s - r)^{-1}(s - M(u,K)) \\
&> M(v,S_r) - \epsilon^2 - \epsilon
\end{aligned}
$$

Thus the variations of u and v on $S(w,c,K)$ can be made arbitrarily small.

We now prove the existence of extreme points.

Lemma 13 *Let K be a compact convex subset of a real normed linear space B. Let u be any element of B^* and δ any positive constant. Then there exists an extreme point x_0 of K and an element v of B^* such that*

    (i)   *$v - u$ is bounded by δ*
    (ii) *$v(x_0) = M(v,K)$*
    (iii) *For each $\epsilon > 0$ there exists $\alpha < M(v,K)$ such that $S(v,\alpha,K)$ is compact and has diameter at most ϵ.*

Proof Let $\{u_n\}$ be a sequence of elements of the unit sphere S^* of B^* which is dense in S^* (relative to the double norm). Using Lemma 12, we define by induction a sequence $\{v_n\}$ of elements of B^*, with $v_1 \equiv u$, and a sequence $\{\alpha_n\}$ of real numbers, with $M(v_n,K) - n^{-1} < \alpha_n < M(v_n,K)$ for all n, having the following properties:

    (a)   The sets $S(v_n,\alpha_n,K)$ are all compact
    (b) $S(v_{n+1},\alpha_{n+1},K) \subset S(v_n, \frac{1}{3}\alpha_n + \frac{2}{3}M(v_n,K),K)$
    (c)   The variation of u_n on $S(v_n,\alpha_n,K)$ is at most n^{-1}, for all $n \geq 2$
    (d) $|v_m(x) - v_n(x)| < \frac{1}{4}(M(v_n,K) - \alpha_n)$ for all points x in K and all positive integers $m \geq n$
    (e) $v_{n+1} - v_n$ is bounded by $2^{-n-1}\delta$

From (b) it follows that $S(v_n,\alpha_n,K) \supset S(v_{n+1},\alpha_{n+1},K)$ for all n. By (e), $\{v_n\}$ converges to an element v of B^* such that $v - u = v - v_1$ is bounded by δ. Thus (i) holds.

We now show that

$$(6.2) \qquad \operatorname{diam} S(v_n,\alpha_n,K) \to 0 \text{ as } n \to \infty$$

To this end, let ϵ be any positive constant. Let $x_1, \ldots, x_N$ be an ϵ approximation to K. Choose n so large that $n^{-1} < \epsilon$. By the Hahn-Banach theorem, for $1 \le i,\ j \le N$ there exists w in S^* such that $w(x_i) - w(x_j) > \|x_i - x_j\| - \epsilon$. Since the sequence $\{u_k\}$ is dense in S^*, and the function $w \to w(x_i) - w(x_j)$ is continuous on S^*, we may assume that $w = u_k$ for some value of k with $k > n$. Thus there exists $m > n$ such that for each i and j with $1 \le i, j \le N$ we have $u_k(x_i) - u_k(x_j) > \|x_i - x_j\| - \epsilon$ for some value of k with $n \le k \le m$. Let x and y be any points of $S(v_m,\alpha_m,K)$. Choose i and j with $\|x_i - x\| < \epsilon$, $\|x_j - y\| < \epsilon$. Choose k as above. Then

$$\begin{aligned}
\|x - y\| &\le \|x_i - x_j\| + 2\epsilon \le u_k(x_i) - u_k(x_j) + 3\epsilon \\
&\le u_k(x) - u_k(y) + 5\epsilon \le k^{-1} + 5\epsilon \le 6\epsilon
\end{aligned}$$

by (c). Thus the diameter of $S(v_m,\alpha_m,K)$ is at most 6ϵ, and thus (6.2) holds.

Since K is compact, (6.2) implies that $\bigcap_{n=1}^{\infty} S(v_n,\alpha_n,K)$ consists of a single point x_0. By (e), we see that $v - v_n$ is bounded by $2^{-n}\delta$, for all n. Letting r be any positive constant such that $\|x\| \le r$ for all x in K, we therefore have

$$\begin{aligned}
v(x_0) &\ge v_n(x_0) - 2^{-n}\delta\|x_0\| \ge \alpha_n - 2^{-n}\delta r \\
&\ge M(v_n,K) - n^{-1} - 2^{-n}\delta r \\
&\ge M(v,K) - 2^{-n}\delta r - n^{-1} - 2^{-n}\delta r
\end{aligned}$$

for all positive integers n. Letting $n \to \infty$ gives (ii).

To prove (iii), note first that $v_n(x_0) \ge \frac{1}{3}\alpha_n + \frac{2}{3}M(v_n,K)$, by (b), since $x_0 \in S(v_{n+1},\alpha_{n+1},K)$. Using (d), we get

$$\begin{aligned}
M(v,K) = v(x_0) &> v_n(x_0) - \tfrac{1}{3}(M(v_n,K) - \alpha_n) \\
&\ge \tfrac{1}{3}\alpha_n + \tfrac{2}{3}M(v_n,K) - \tfrac{1}{3}(M(v_n,K) - \alpha_n) \\
&= \tfrac{2}{3}\alpha_n + \tfrac{1}{3}M(v_n,K)
\end{aligned}$$

Also, by (d), we see that $S(v, \frac{2}{3}\alpha_n + \frac{1}{3}M(v_n,K)) \subset S(v_n,\alpha_n,K)$. Since $\operatorname{diam} S(v_n,\alpha_n,K) \to 0$ as $n \to \infty$, it follows that (iii) is valid.

It remains to show that x_0 is an extreme point. Let ϵ be any positive constant. Choose $\alpha < v(x_0)$ so that diam $S(v,\alpha,K) < \epsilon$. Choose the constant $c > 0$ so small that $|v(x_0) - v(x)| < \frac{1}{2}(v(x_0) - \alpha)$ for all x in K with $\|x_0 - x\| < c$. Let x and y be any elements of K with $\|x_0 - \frac{1}{2}(x + y)\| < c$. Then

$$\tfrac{1}{2}(v(x_0) - v(x)) + \tfrac{1}{2}(v(x_0) - v(y)) = v(x_0 - \tfrac{1}{2}(x + y))$$
$$< \tfrac{1}{2}(v(x_0) - \alpha)$$

Since both the terms on the left are nonnegative, this gives $v(x) > \alpha$ and $v(y) > \alpha$. Therefore x and y lie in the set $S(v,\alpha,K)$, and thus $\|x - y\| < \epsilon$. Thus x_0 is extreme.

The next result is called the Krein-Milman theorem.

Theorem 11 *Let x be a point in a compact convex subset K of a real Banach space B. Then there exist extreme points $x_1, \ldots, x_n$ of K and positive constants $\alpha_1, \ldots, \alpha_n$ whose sum is 1, such that $\alpha_1 x_1 + \cdots + \alpha_n x_n$ is arbitrarily near to x.*

Proof Let ϵ be any positive constant. Let x_1 be any extreme point of K. We define inductively a finite sequence $x_1, \ldots, x_n$ of extreme points as follows. Once x_k has been constructed, let K_k be the convex set spanned by $x_1, \ldots, x_k$; that is

$$K_k \equiv \left\{ \alpha_1 x_1 + \cdots + \alpha_k x_k : \alpha_i \geq 0, \sum_{i=1}^{k} \alpha_i = 1 \right\}$$

Since K_k is totally bounded, the distance d_k of x to K_k exists. Either $d_k < \epsilon$ or $d_k > \frac{1}{2}\epsilon$. In the first case, stop the construction and take $n = k$. In the second case, by the separation theorem there exists u in S^* with

$$u(x) - \sup \{u(y) : y \in K_k\} > \tfrac{1}{2}\epsilon$$

By Lemma 13, there exists an extreme point x_{k+1} with

$$u(x_{k+1}) - \sup \{u(y) : y \in K_k\} > \tfrac{1}{2}\epsilon$$

Thus the distance of x_{k+1} to K_k is greater than $\frac{1}{2}\epsilon$. Our construction thus guarantees that $\|x_i - x_j\| > \frac{1}{2}\epsilon$ for all i and j $(i \neq j)$, as long as the construction proceeds. Since K is totally bounded, the construction must stop at some stage. Therefore $d_k < \epsilon$ for some k. Take $n = k$. Then

$$\|\alpha_1 x_1 + \cdots + \alpha_n x_n - x\| < \epsilon$$

for some choice of $\alpha_1, \ldots, \alpha_n$. This completes the proof.

Theorem 11 actually holds for a compact convex subset K of a real separable σ-normed linear space V. To see this, let $\{\| \quad \|_n\}$ be any defining sequence of seminorms on V, and for each n let c_n be a positive constant such that $\|x\|_n \leq c_n$ for all x in K. Let V_0 consist of all x in V such that the sequence $\{c_n^{-1}\|x\|_n\}$ is bounded. For each x in V_0 write

$$\|x\| \equiv \sum_{n=1}^{\infty} 2^{-n} c_n^{-1} \|x\|_n$$

The completion of V_0 with respect to the norm $\| \quad \|$ is a Banach space B, and K is a compact convex subset of B, for which Theorem 11 is valid.

PROBLEMS

1.   Give an example of a closed linear subset V of a finite-dimensional Banach space B that is *not* finite-dimensional. Construct a bounded linear functional on V that is *not* normable.

2.   With V and B as in Prob. 1, construct a bounded linear functional on V that can *not* be extended to B.

3.   The sequential L_p space, $L_p(Z)$, consists of all sequences $\{a_n\}$ of elements of $\mathbf{F}$ such that $\|\{a_n\}\| \equiv (\Sigma|a_n|^p)^{1/p}$ exists. Show that for $p = 1$ the dual of $L_p(Z)$ can be identified with all bounded sequences of elements of $\mathbf{F}$. Show that for $p > 1$ the dual of $L_p(Z)$ can be identified with all sequences $\{b_n\}$ of elements of $\mathbf{F}$ such that $\sum_{k=1}^{n} |a_k|^q$ is a bounded function of n. What about the general case $p \geq 1$?

4.   Show that the complex L_p spaces are uniformly convex, for $p > 1$, and exhibit the form of the normable linear functionals.

5.   Let K be a closed located convex subset of a uniformly convex Banach space B. For each x in B show that there exists a unique point y in K with $\|x - y\| = \rho(x,K)$.

6.   Prove the uniform boundedness theorem: If $\{A_n\}$ is a sequence of bounded linear operators from a Banach space B_1 to a normed linear space B_2, and $\{x_n\}$ is a sequence of unit vectors in B_1 such that $\|A_n x_n\| \rightarrow \infty$ as $n \rightarrow \infty$, then there exists a vector x in B_1 such that the sequence $\{\|A_n x\|\}$ is unbounded. (*Hint:* Write

$$U_k \equiv \{x \in B : \|A_n x\| > k \text{ for some } n\}$$

Show that the sets U_k are open and dense in B, and apply the Baire category theorem.)

7. Prove that there exists a continuous function f on $X \equiv [-\pi,\pi]$ whose Fourier series at the point 0, $\sum_{k=-\infty}^{\infty} \int_{-\pi}^{\pi} f(x)e^{ikx}\,dx$, diverges. [*Hint:* Define the linear functional u_n on $C(X)$ by

$$u_n(f) \equiv \sum_{k=-n}^{n} \int_{-\pi}^{\pi} f(x)e^{ikx}\,dx$$

Show that there exists a sequence $\{f_n\}$ of unit vectors in $C(X)$ with $\|u_n(f_n)\| \to \infty$ as $n \to \infty$.]

8. Call a Banach space B *reflexive* if the set $V \subset B^*$ of normable linear functionals on B is closed with respect to addition, and if every normable linear functional on V (relative to the norm $\|\ \ \|$) is induced by an element of B. Is every uniformly convex Banach space reflexive? (This is an unsolved problem.)

9. Let K be a closed convex subset of a Banach space B, such that $ax \in K$ whenever $a \in \mathbf{F}$, $|a| \leq 1$, and $x \in K$. Show that K is located if and only if $K' \equiv \{u \in B^*: |u(x)| \leq 1 \text{ for all } x \text{ in } K\}$ is a located subset of B^* (relative to the double norm).

10. Construct a bounded linear operator on a Hilbert space H whose adjoint does *not* exist.

11. Let A be a hermitian operator on a Hilbert space H. Let c be a bound for A. Show that there exists a dense set $S \subset [-c,c]$ and a map $\lambda \to P_\lambda$ from S to the set of projection operators such that

(a) $P_\lambda P_\mu = P_\lambda \qquad (\lambda,\ \mu \in S,\ \lambda \leq \mu)$
(b) $-c \in S,\ c \in S,\ P_{-c} = 0,\ P_c = I$
(c) $A = \int_{-c}^{c} \lambda\,dP_\lambda$, in the sense that for all x and y in H the function

$(P_\lambda x,y)$ is of bounded variation, and $(Ax,y) = \int_{-c}^{c} \lambda\,d(P_\lambda x,y)$.

(*Hint:* Use the spectral theorem.)

12. A bounded linear operator U on a Hilbert space H is *unitary* if U^* exists and $UU^* = U^*U = I$. Show that a unitary operator U preserves norms.

13. Let U be a unitary operator on a Hilbert space H, and x any vector in H. For each positive integer n write

$$x_n \equiv 1/n(x + Ux + \cdots + U^{n-1}x)$$

Show that $\{\|x_n\|\}$ is an essentially decreasing sequence, in the sense that for each n and each $\epsilon > 0$ there exists N such that $\|x_m\| \le \|x_n\| + \epsilon$ for all $m \ge N$.

14. Continuing Prob. 13, prove the *mean ergodic theorem* of von Neumann: $\{x_n\}$ converges if and only if $\{\|x_n\|\}$ converges.

15. Let B be a Banach space. Discuss the relation of the following statements: (a) The Banach space B is finite-dimensional. (b) All elements of B^* are normable.

16. A *Banach dual* is a normed linear space V and a compact convex set $K \subset V$ such that (a) $au \in K$ for all a in $\mathbf{F}$ with $|a| \le 1$ and all u in K, and (b) for each nonzero u in V there exist positive constants c_1 and c_2 with $c_1 u \in K$ and $\rho(c_2 u, K) > 0$. Let B be the set of all linear maps f from V to $\mathbf{F}$ that are continuous on K, and for each f in B write $\|f\| \equiv \sup \{|f(u)| : u \in K\}$. Show that B is a Banach space. Identify V with B^* and K with S^*.

17. Give an example of a compact convex subset of $\mathbf{R}^2$ with at most three extreme points such that (a) the set of extreme points is *not* located, and (b) *not* every point of K is a linear combination of extreme points.

18. Find the extreme points of the unit sphere of $C(X)^*$.

NOTES

Some of the Banach spaces of classical analysis (for example, L_∞) are *not* really Banach spaces, because the norm is *not* constructively defined. L_∞ can be realized as a dense subset of the dual of L_1.

Note the care with which a finite basis is defined (Definition 4).

One might wish to complete Theorem 2 by determining the form of all bounded linear functionals on L_p—not just the normable ones. Perhaps there is no better answer than that given by Theorem 9.

It is *not* always possible to choose the normable linear functional ν of Theorem 4 so that $\|\nu\| = \|\lambda\|$.

It would be interesting to find a good constructive substitute for the closed-graph theorem. This theorem says that a linear transformation T from a Banach space B_1 into a Banach space B_2 is continuous if and only if its graph is closed (as a subset of $B_1 \times B_2$).

Since a Hilbert space may be neither finite-dimensional nor infinite-dimensional, an orthonormal basis (Definition 10) must be allowed to have some of its vectors equal to 0.

The map $f \rightarrow \lambda_f$, where λ_f is the linear functional given by $\lambda_f(g) \equiv (g,f)$, from a Hilbert space H to its dual H^*, need not be onto, and in general is not, as Prob. 3 shows. This raises interesting questions about the constructive interpretation of various classical results, which we have managed to avoid.

The definition of strong convergence (preceding Lemma 10) could have been phrased in terms of the double norm on Hom (H,H), and Lemma 10 could have been derived from Proposition 9. It is simpler to work directly.

A Banach space B is classically called reflexive if $B = B^{**}$. Since B^* is not in general a Banach space, this definition does not make constructive sense. Problem 8 is an attempt to find a constructive substitute for reflexivity.

Notice that Definition 14 (the constructive definition of an extreme point) constitutes a strengthening of the classical version.

CHAPTER 10

LOCALLY COMPACT ABELIAN GROUPS

Section 1 constructs Haar measure on a locally compact group G, by a method of H. Cartan. Certain least upper bounds must be proved to exist in order to make the classical proof constructive. This adds length to the classical treatment. In Sec. 2 convolution is defined and the group algebra is studied. Specializing to a commutative group G, we prove the fundamental fact that the convolution operator of an integrable function is normable. The chapter closes with a study of the dual group G^*. The spectral theorem is used to establish basic properties of the Fourier transform, in particular the inversion theorem. By standard methods, we use the inversion theorem to get the Pontryagin duality theorem.

Many groups of special mathematical interest, for instance, various important groups of geometrical transformations, have a locally compact metric with respect to which the group operations are continuous. Such groups are called *locally compact groups*. Every locally compact group has a left-invariant measure. We construct this measure, and use it for a detailed study of a locally compact commutative group. In particular, we construct the dual group, prove the duality theorem, and study the Fourier transform.

1. HAAR MEASURE

In this section we define the concept of a locally compact group, prove the remarkable fact that it admits a left-invariant measure, and study convolution on the group.

Definition 1 *A locally compact group* is a set G which is both a group and a locally compact metric space, such that the operation

$$(x,y) \rightarrow x^{-1}y$$

from $G \times G$ to G is continuous.

Setting $y = e$, we see that the operation $x \rightarrow x^{-1}$ is continuous. Since the operation $(x,y) \rightarrow xy$ is the product of the operations $(x,y) \rightarrow (x^{-1},y)$ and $(x,y) \rightarrow x^{-1}y$, this operation is also continuous. In particular, for each x in G the maps $y \rightarrow xy$ and $y \rightarrow yx$ are continuous. Since the inverse maps are continuous too, each of these maps is a homeomorphism of G with itself.

Since a continuous function from one locally compact space into another takes bounded sets into bounded sets, the sets

$$KL \equiv \{xy: x \in K, y \in L\} \qquad K^{-1}L \equiv \{x^{-1}y: x \in K, y \in L\}$$

and
$$KL^{-1} \equiv \{xy^{-1}: x \in K, y \in L\}$$

are bounded subsets of G whenever K and L are bounded subsets of G. By the same token, KL, $K^{-1}L$, and KL^{-1} are totally bounded whenever K and L are totally bounded.

For brevity we let $\rho(x)$ be the distance $\rho(x,e)$ from an arbitrary element x of G to the identity element e.

Proposition 1 *Let K be a bounded subset of a locally compact group G. For each $\epsilon > 0$ there exists $\delta > 0$ such that $\rho(x,y) \leq \epsilon$ whenever x and y*

are elements of K with $\rho(x^{-1}y) \leq \delta$, and for each $\delta > 0$ there exists $\epsilon > 0$ such that $\rho(x^{-1}y) \leq \delta$ whenever x and y are elements of K with $\rho(x,y) \leq \epsilon$.

Proof Consider $\delta > 0$. Since $K \times K$ is bounded, the map $(x,y) \to x^{-1}y$ is uniformly continuous on $K \times K$. It follows that

$$\rho(x^{-1}y) = \rho(x^{-1}y) - \rho(x^{-1}x)$$

will be less than δ whenever $x \in K$, $y \in K$, and $\rho(x,y)$ is sufficiently small, say at most ϵ.

Consider conversely any $\epsilon > 0$. Since multiplication $(x,y) \to xy$ is uniformly continuous on $K \times K$, we can choose δ so small that

$$\rho(x,y) = \rho(x,x(x^{-1}y)) \leq \epsilon$$

whenever $x, y \in K$ and $\rho(x^{-1}y) \leq \delta$.

Corollary *A subset K of a locally compact group G is totally bounded if and only if for each $\delta > 0$ there exist $x_1, \ldots, x_n$ in K such that for each x in K we have $\rho(x_i^{-1}x) \leq \delta$ for some i $(1 \leq i \leq n)$.*

Proof If K is totally bounded, then K is bounded. By Proposition 1, there exists $\epsilon > 0$ such that $\rho(x^{-1}y) \leq \delta$ whenever $x \in K$, $y \in K$, and $\rho(x,y) \leq \epsilon$. Therefore any ϵ approximation $x_1, \ldots, x_n$ to K has the given property.

Assume, conversely, that the points $x_1, \ldots, x_n$ exist for each $\delta > 0$. Each of the sets $\{x \in G : \rho(x) < \delta\}$ is bounded. Since the map $x \to x_i x$ takes bounded sets into bounded sets, each of the sets

$$\{x \in G : \rho(x_i^{-1}x) < \delta\}$$

is bounded. Therefore K is bounded. Hence if ϵ is any positive constant, there exists $\delta > 0$ such that $\rho(x,y) < \epsilon$ whenever $x, y \in K$ and $\rho(x^{-1}y) \leq \delta$. The points $x_1, \ldots, x_n$ are an ϵ approximation to K. Thus K is totally bounded.

Lemma 1 *Every test function f on G is uniformly continuous.*

Proof Let K be any compact support for f, and r any positive constant such that

$$K_r \equiv \{x : \rho(x,K) \leq r\}$$

is compact. Since f is continuous on K_r, for each $\epsilon > 0$ there exists δ, with $0 < \delta < \frac{1}{2}r$, such that $|f(x) - f(y)| \leq \epsilon$ whenever x, $y \in K_r$ and $\rho(x,y) \leq \delta$. Consider any points x and y of G with $\rho(x,y) \leq \delta$. Either $\rho(x,K) \leq r - \delta$ or $\rho(x,K) > \delta$. In the first case $\rho(y,K) \leq r$. Therefore x, $y \in K_r$, and thus $|f(x) - f(y)| \leq \epsilon$. In the second case $\rho(y,K) > 0$, and thus $|f(x) - f(y)| = |0 - 0| = 0$. Thus f is uniformly continuous.

For each $f\colon G \to \mathbf{R}$ and each s in G we define the *left translate* $T(s)f$ of f by s as follows:

$$(T(s)f)(x) \equiv f(sx)$$

The right translate is defined similarly. A measure μ on G is called *left-invariant*, or invariant under left translations, if

$$\int T(s)f \, d\mu = \int\!\!\int f \, d\mu$$

for all test functions f and all s in G. Right-invariant measures are defined similarly.

Our immediate goal is to construct a positive left-invariant measure. We begin by defining certain quantities which approximate to a left-invariant measure.

Definition 2   Let $C^+ \equiv C(G)^+$ consist of all nonnegative elements of $C \equiv C(G)$, and $\mathcal{P}$ consist of all nonzero elements of C^+. For each f in C^+ and each φ in $\mathcal{P}$ we consider the set A of all finite linear combinations

$$(1.1) \qquad\qquad S \equiv \Sigma c_i T(s_i)\varphi$$

where the c_i are nonnegative constants and the s_i are points of G, for which $f \leq S$. We write

$$(f\colon \varphi) \equiv \inf \{\Sigma c_i \colon S \in A\}$$

whenever the infimum exists.

Lemma 2   *To prove the existence of $(f\colon \varphi)$ only elements s_i in a certain compact set $Y \subset G$ need be considered.*

Proof   Let K and L be compact supports for f and φ, respectively. Choose $\epsilon > 0$ so that the set

$$Y \equiv \{x\colon \rho(x,LK^{-1}) \leq \epsilon\}$$

is compact. For each element $S = \Sigma c_i T(s_i)\varphi$ of A, partition the set of indices i into free subsets M and N, so that $s_i \in Y$ for all i in M and $\rho(s_i, LK^{-1}) > 0$ for all i in N. For each x in K and each i in N, we have $\rho(s_i x, L) > 0$; therefore $\varphi(s_i x) = 0$. Hence

$$f \leq \sum_{i \in M} c_i T(s_i)\varphi$$

as was to be proved.

Lemma 3 *For each f in C^+ and each φ in $\mathcal{P}$ the quantity $(f: \varphi)$ exists.*

Proof Let K be any compact support for f. Choose Y as above. Since $\varphi \in \mathcal{P}$, it follows that $\varphi(t) > 0$ for some t in G. Choose $\gamma > 0$ so that $\varphi(tx) > \gamma$ whenever $\rho(x) < \gamma$. Choose $t_1, \ldots, t_m$ in K^{-1} so that for each x in K we have $\rho(t_j x) < \gamma$ for some j, with $1 \leq j \leq m$. Then $\varphi(tt_j x) > \gamma$. Hence

$$(1.2) \qquad\qquad 1 \leq \gamma^{-1} \sum_{j=1}^{m} \varphi(tt_j x) \qquad (x \in K)$$

Therefore
$$f \leq \gamma^{-1}\|f\| \sum_{j=1}^{m} T(tt_j)\varphi$$

Thus to show that $(f: \varphi)$ exists, we consider the subset A_0 of A consisting of all S for which (i) $\Sigma c_i \leq \gamma^{-1}\|f\|m$ and (ii) each s_i belongs to Y. It will be enough to show that for each $\epsilon > 0$ there exists a subfinite set $B \subset A$ such that for each $S \equiv \Sigma c_i T(s_i)\varphi$ in A_0 there exists $S' \equiv \Sigma c_i' T(s_i')\varphi$ in B with

$$(1.3) \qquad\qquad \Sigma c_i' \leq c_i + \epsilon$$

To this end, write $\delta \equiv (2 + \gamma^{-2}\|f\|m^2 + \gamma^{-1}m)^{-1}\epsilon$. By Lemma 1, we see that φ is uniformly continuous. By Proposition 1, there exist $x_1, \ldots, x_N$ in Y such that for each s in Y there exists k $(1 \leq k \leq N)$ with $\varphi(sy) \leq \varphi(x_k y) + \delta$ for all y in K. Choose J in Z^+, with $\delta^{-1}N\gamma^{-1}\|f\|m \leq J$, and let Ω consist of all N-tuples $M \equiv (M_1, \ldots, M_N)$ of non-negative integers such that $0 \leq M_k \leq J$ for all k. For each M in Ω we write

$$(1.4) \quad S' \equiv \delta N^{-1} \sum_{k=1}^{N} (M_k + 2) T(x_k)\varphi + \delta\gamma^{-1}(\gamma^{-1}\|f\|m + 1) \sum_{j=1}^{m} T(tt_j)\varphi$$

Partition Ω into finite subsets Ω_1 and Ω_2 so that (i) $f \leq S'$ for all M in Ω_1, and (ii) for each M in Ω_2 there exists y in K with $f(y) > S'(y) - \delta$.

The set

$$B \equiv \{S' : M \in \Omega_1\}$$

is a subfinite subset of A.

Let $S \equiv \Sigma c_i T(s_i)\varphi$ be any element of A_0. By the choice of $x_1, \ldots, x_N$ there exist nonnegative constants $a_1, \ldots, a_N$ with

$$(1.5) \qquad\qquad \Sigma a_k = \Sigma c_i$$

and

$$(1.6) \qquad S(y) \le \Sigma a_k \varphi(x_k y) + \delta \Sigma c_i \qquad (y \in K)$$

By (1.2) and the inequality $\Sigma c_i \le \gamma^{-1}\|f\|m$, inequality (1.6) gives

$$(1.7) \quad f(y) + \delta \le S(y) + \delta$$
$$\le \Sigma a_k \varphi(x_k y) + \delta\gamma^{-1}(\gamma^{-1}\|f\|m + 1)\sum_{j=1}^{m} \varphi(t t_j y)$$

for all y in K. Since $a_k \le \Sigma c_i \le \gamma^{-1}\|f\|m \le \delta N^{-1}J$ for all k, there exists M in Ω such that $\delta N^{-1}M_k \le a_k \le \delta N^{-1}(M_k + 2)$ for all k. Let $S' = \Sigma c_i' T(s_i')\varphi$ be given by (1.4). By (1.7), we see that $f(y) + \delta \le S'(y)$ for all y in K. Therefore $M \in \Omega_1$, so that $S' \in B$. Also,

$$\Sigma c_i' = \delta N^{-1}\Sigma(M_k + 2) + \delta\gamma^{-1}(\gamma^{-1}\|f\|m + 1)m$$
$$\le \Sigma a_k + \delta(2 + \gamma^{-2}\|f\|m^2 + \gamma^{-1}m)$$

By (1.5) and the definition of δ, inequality (1.3) is valid.

For later use we note some simple properties of the function $(f : \varphi)$. If $f \in \mathcal{P}$, then $(f : \varphi) > 0$. If $f \in C^+$, $g \in \mathcal{P}$, and $\varphi \in \mathcal{P}$, then

$$(f : \varphi) \le (f : g)(g : \varphi)$$

Hence, if $f \in \mathcal{P}$, then

$$(1.8) \qquad (g : f)^{-1} \le (f : \varphi)(g : \varphi)^{-1} \le (f : g)$$

As a function of f, $(f : \varphi)$ is *homogeneous*,

$$(\lambda f : \varphi) = \lambda(f : \varphi) \qquad (\lambda \in \mathbf{R}^{0+})$$

and *subadditive*,

$$(f_1 + f_2 : \varphi) \le (f_1 : \varphi) + (f_2 : \varphi)$$

It is also *left-invariant*, in the sense that

$$(T(s)f:\varphi) = (f:\varphi) \qquad (s \in G)$$

Now fix a particular function f_0 in $\mathcal{P}$, for the rest of this section, for purposes of normalization. For each f in C^+ and each φ in $\mathcal{P}$ we write

$$I_\varphi(f) \equiv (f:\varphi)(f_0:\varphi)^{-1}$$

In case $f \in \mathcal{P}$,

$$(1.9) \qquad (f_0:f)^{-1} \leq I_\varphi(f) \leq (f:f_0)$$

by (1.8).

A function $\varphi \in \mathcal{P}$ is *small of order* c (where c is a positive constant) if $\varphi(x) = 0$ whenever $\rho(x) \geq c$.

The following lemma shows that I_φ is approximately linear if φ is sufficiently small.

Lemma 4   *Let ϵ and M be positive constants and $f_1, \ldots, f_n$ elements of C^+. Then there exists $c > 0$ such that, whenever $\lambda_1, \ldots, \lambda_n$ are real numbers with $0 \leq \lambda_j \leq M$ $(1 \leq j \leq n)$ and φ is small of order c, we have*

$$\Sigma\lambda_j I_\varphi(f_j) \leq I_\varphi(\Sigma\lambda_j f_j) + \epsilon$$

Proof   Since

$$I_\varphi(\Sigma\lambda_j f_j) \leq I_\varphi(\Sigma M f_j)$$

the quantities $I_\varphi(\Sigma\lambda_j f_j)$ are bounded independently of the values of the λ_j. To prove the lemma it is therefore enough to show that for each $\epsilon > 0$ we can choose $c > 0$ so that

$$\Sigma\lambda_j I_\varphi(f_j) \leq (I_\varphi(\Sigma\lambda_j f_j) + \epsilon)(1 + \epsilon)$$

whenever φ is small of order c.

Let g be any element of $\mathcal{P}$ such that $g(x) = 1$ for all x in some compact set K which supports each of the functions f_j. Write

$$\delta \equiv \epsilon(g:f_0)^{-1}$$

By (1.9),

$$\delta I_\varphi(g) \leq \delta(g:f_0) = \epsilon$$

The functions

$$h_j \equiv f_j(\Sigma\lambda_j f_j + \delta g)^{-1} \qquad (1 \leq j \leq n)$$

have a common modulus of continuity, independent of the values of $\lambda_1, \ldots, \lambda_n$. We may therefore choose c so small that

$$(1.10) \qquad |h_j(s^{-1}) - h_j(x)| < n^{-1}\epsilon \qquad (1 \le j \le n, \ \rho(sx) < 2c)$$

Consider now a function φ in $\mathcal{P}$ which is small of order c, and an inequality

$$\Sigma \lambda_j f_j + \delta g \le \Sigma c_i T(s_i)\varphi$$

For each x in K and each i either $\rho(s_i x) > c$ or $\rho(s_i x) < 2c$. In the first case $\varphi(s_i x) = 0$. Therefore

$$(1.11) \qquad\qquad \Sigma \lambda_j f_j(x) + \delta g(x) \le \Sigma' c_i \varphi(s_i x)$$

where Σ' denotes summation over certain terms i for which $\rho(s_i x) < 2c$. For each j $(1 \le j \le n)$ and each such i we have $h_j(x) \le h_j(s_i^{-1}) + n^{-1}\epsilon$, by (1.10). This gives

$$\begin{aligned}
f_j(x) &= h_j(x) \left(\sum \lambda_i f_i(x) + \delta g(x) \right) \\
&\le h_j(x) \sum{}' c_i \varphi(s_i x) \\
&\le \sum_i (h_j(s_i^{-1}) + n^{-1}\epsilon) c_i \varphi(s_i x)
\end{aligned}$$

Therefore $$I_\varphi(f_j) \le \sum_i c_i (h_j(s_i^{-1}) + n^{-1}\epsilon)(f_0 : \varphi)^{-1}$$

Multiplying by λ_j and summing with respect to j, we get

$$\begin{aligned}
\sum \lambda_j I_\varphi(f_j) &\le \sum_i c_i \left(\sum_{j=1}^{n} \lambda_j h_j(s_i^{-1}) + \epsilon \right) (f_0 : \varphi)^{-1} \\
&\le \sum_i c_i (1 + \epsilon)(f_0 : \varphi)^{-1}
\end{aligned}$$

Since $\sum_{i=1}^{m} c_i (f_0 : \varphi)^{-1}$ can be made arbitrarily close to $I_\varphi(\Sigma \lambda_j f_j + \delta g)$, it follows that

$$\begin{aligned}
\Sigma \lambda_j I_\varphi(f_j) &\le I_\varphi(\Sigma \lambda_j f_j + \delta g)(1 + \epsilon) \\
&\le (I_\varphi(\Sigma \lambda_j f_j) + \delta I_\varphi(g))(1 + \epsilon) \\
&\le (I_\varphi(\Sigma \lambda_j f_j) + \epsilon)(1 + \epsilon)
\end{aligned}$$

as desired.

Lemma 5 *Let f be a function in C^+ and ϵ a positive constant. Choose $\delta > 0$ so that $|f(x) - f(y)| \leq \epsilon$ whenever $\rho(x^{-1}y) \leq \delta$. Then if g is any element of $\mathcal{P}$ which is small of order δ, and if $\alpha > \epsilon$, there exist $s_1, \ldots, s_n$ in G and $c_1, \ldots, c_n$ in $\mathbf{R}^{0+}$ with*

$$(1.12) \qquad |f(x) - \Sigma c_i g(s_i x)| \leq \alpha \qquad (x \in G)$$

Proof For all x and s in G,

$$(1.13) \qquad (f(x) - \epsilon)g(s^{-1}x) \leq f(s)g(s^{-1}x) \leq (f(x) + \epsilon)g(s^{-1}x)$$

Let $\tilde{g}$ be the element of $\mathcal{P}$ defined by $\tilde{g}(x) \equiv g(x^{-1})$, for all x in G. Write

$$\eta \equiv \tfrac{1}{2}(\alpha - \epsilon)(f : \tilde{g})^{-1}$$

Choose $c > 0$ so small that $|g(x) - g(y)| \leq \eta$ whenever $\rho(xy^{-1}) \leq c$. Let the compact set K be a support for f. Let $s_1, \ldots, s_n$ be elements of K^{-1} such that for each x in K we have $\rho(s_i x) < \tfrac{1}{2}c$ for some i. Let $h_1, \ldots, h_n$ be elements of $\mathcal{P}$ such that (a) $\Sigma h_i(x) = 1$ for all x in K, and (b) $h_i(x) = 0$ whenever $\rho(s_i x) \geq c$. Consider a particular function h_i $(1 \leq i \leq n)$ and a particular s in G. If $h_i(s) = 0$, then

$$(1.14) \qquad h_i(s)f(s)\{g(s^{-1}x) - \eta\} \leq h_i(s)f(s)g(s_i x)$$
$$\leq h_i(s)f(s)\{g(s^{-1}x) + \eta\}$$

while if $h_i(s) > 0$, then $\rho(s_i s) < c$, and thus $|g(s_i x) - g(s^{-1}x)| \leq \eta$, and (1.14) is valid in this case also. By continuity, (1.14) is valid for all s in G.

Summation of the inequalities (1.14) with respect to i and use of (1.13) gives

$$(1.15) \qquad (f(x) - \epsilon)g(s^{-1}x) - \eta f(s) \leq f(s)(g(s^{-1}x) - \eta)$$
$$= (\Sigma h_i(s)f(s))(g(s^{-1}x) - \eta)$$
$$\leq \Sigma h_i(s)f(s)g(s_i x)$$
$$\leq \Sigma h_i(s)f(s)(g(s^{-1}x) + \eta)$$
$$= f(s)(g(s^{-1}x) + \eta)$$
$$\leq (f(x) + \epsilon)g(s^{-1}x) + \eta f(s)$$

Let φ be any element of $\mathcal{P}$. Consider the quantities in (1.15) as functions of s, and apply I_φ. In view of the equality

$$I_\varphi(g(s^{-1}x)) = I_\varphi(\tilde{g}(x^{-1}s)) = I_\varphi(\tilde{g}(s))$$

we obtain

$$(1.16) \quad (f(x) - \epsilon)I_\varphi(\tilde{g}) - \eta I_\varphi(f) \leq I_\varphi(\Sigma h_i f g(s_i x))$$
$$\leq (f(x) + \epsilon)I_\varphi(\tilde{g}) + \eta I_\varphi(f)$$

Since
$$I_\varphi(f)I_\varphi(\tilde{g})^{-1} \leq (f : \tilde{g}) = \frac{\alpha - \epsilon}{2\eta}$$

dividing (1.16) by $I_\varphi(\tilde{g})$ gives

$$(1.17) \quad f(x) - \beta \leq I_\varphi(I_\varphi(\tilde{g})^{-1}\Sigma g(s_i x)h_i f) \leq f(x) + \beta$$

where $\beta \equiv \epsilon + \frac{1}{2}(\alpha - \epsilon) < \alpha$.

Write $f_i \equiv h_i f$ $(1 \leq i \leq n)$. Choose k in Z^+ so large that $k^{-1} < \alpha - \beta$ and $g(s_i x)(f_0 : \tilde{g}) \leq k$ for all i and all x. Then

$$\lambda_i(x) \equiv g(s_i x)I_\varphi(\tilde{g})^{-1} \leq g(s_i x)(f_0 : \tilde{g}) \leq k \qquad (1 \leq i \leq n, x \in G)$$

By Lemma 4, we can specialize φ so that

$$\Sigma \lambda_i(x)I_\varphi(f_i) \leq I_\varphi(\Sigma \lambda_i(x)f_i) + k^{-1}$$

for all x in G. Write

$$c_i \equiv I_\varphi(\tilde{g})^{-1}I_\varphi(f_i) \qquad (1 \leq i \leq n)$$

To show that (1.12) is valid, consider any x in G. Then

$$I_\varphi(\Sigma g(s_i x)I_\varphi(\tilde{g})^{-1}h_i f) = I_\varphi(\Sigma \lambda_i(x)f_i) \leq \Sigma \lambda_i(x)I_\varphi(f_i)$$
$$= \Sigma g(s_i x)I_\varphi(\tilde{g})^{-1}I_\varphi(f_i) = \Sigma c_i g(s_i x)$$
$$\leq I_\varphi(\Sigma \lambda_i(x)f_i) + k^{-1}$$
$$< I_\varphi(\Sigma g(s_i x)I_\varphi(\tilde{g})^{-1}h_i f) + \alpha - \beta$$

Together with (1.17) this gives (1.12).

Definition 3   The fact that f and g are elements of C^+ such that there exist $s_1, \ldots, s_n, t_1, \ldots, t_m$ in G and $c_1, \ldots, c_n, d_1, \ldots, d_m$ in $\mathbf{R}^{0+}$ with $\Sigma c_i = \Sigma d_j > 0$ and

$$(1.18) \qquad \Sigma c_i T(s_i)f \leq \Sigma d_j T(t_j)g$$

is expressed by

$$f \prec g$$

The fact that f and g are elements of C^+ such that there exist $f_1, \ldots, f_N$ in C^+ with $f < f_1 < \cdots < f_N < g$ is expressed by

$$f < < g$$

Lemma 6 *If $f < < g$, then for each $\epsilon > 0$ there exists $c > 0$ such that $I_\varphi(f) \le I_\varphi(g) + \epsilon$ whenever φ is small of order c.*

Proof There is no loss of generality in assuming that $f < g$, so that (1.18) holds. Clearly,

$$I_\varphi(\Sigma c_i T(s_i)f) \le I_\varphi(\Sigma d_j T(t_j)g)$$

By Lemma 4, there exists $c > 0$ such that

$$\begin{aligned}
\Sigma c_i I_\varphi(f) &= \Sigma c_i I_\varphi(T(s_i)f) \\
&\le \Sigma d_j I_\varphi(T(t_j)g) + \epsilon\Sigma c_i \\
&= \Sigma d_j I_\varphi(g) + \epsilon\Sigma c_i
\end{aligned}$$

Division by Σc_i gives the desired inequality.

As a consequence of this lemma we see that if $t_1 f < < t_2 f$ (with $f \in \mathcal{P}$), then $t_1 \le t_2$.

Lemma 7 *Consider $\epsilon > 0$, f and g in $\mathcal{P}$, and h in $\mathcal{P}$ such that*

$$h(f + g) = f + g$$

Then there exists $t > 0$ such that $f < < tg + \epsilon h$ and $tg < < f + \epsilon h$.

Proof We write $\delta \equiv \min \{1, \epsilon, (f + h : f_0)^{-1} (f_0 : g)^{-1}\epsilon\}$. By Lemma 5, there exists φ in $\mathcal{P}$, nonnegative real numbers $c_1, \ldots, c_n, d_1, \ldots, d_m$, and elements $s_1, \ldots, s_n, t_1, \ldots, t_m$ of G with

$$f \le \Sigma c_i T(s_i)\varphi \le f + \delta h \quad \text{and} \quad g \le \Sigma d_j T(t_j)\varphi \le g + \delta h$$

Thus whenever $f + h \le \Sigma c_k' T(s_k')\varphi$, we have

$$\Sigma c_i \varphi < \Sigma c_i T(s_i)\varphi < \Sigma c_k' T(s_k')\varphi < \Sigma c_k'\varphi$$

and thus $\Sigma c_i \le \Sigma c_k'$, by Lemma 6. Hence

$$\Sigma c_i \le (f + h : \varphi) \le (f + h : f_0)(f_0 : \varphi)$$

Also,
$$\Sigma d_j \ge (g : \varphi) \ge (f_0 : \varphi) : (f_0 : g)^{-1}$$

Putting
$$t \equiv \Sigma c_i (\Sigma d_j)^{-1}$$

gives
$$t \le (f + h : f_0)(f_0 : g)$$

so $t\delta \leq \epsilon$. Therefore

$$f < \Sigma c_i \varphi = t\Sigma d_j \varphi < t(g + \delta h) < tg + \epsilon h$$

and
$$tg < t\Sigma d_j\varphi = \Sigma c_i \varphi < f + \epsilon h$$

as was to be proved.

Theorem 1   *There exists a positive left-invariant measure μ on G, such that $\int f\, d\mu > 0$ for all f in $\mathcal{P}$. If ν is any positive left-invariant measure on G, then $\nu = c\mu$ for some c in $\mathbf{R}^{0+}$.*

Proof   Fix an element g of $\mathcal{P}$. Consider f in C^+, and choose h in $\mathcal{P}$ so that $h(f + g) = f + g$. For each k in Z^+ choose $t_k > 0$ with

$$(1.19) \qquad f << t_k g + k^{-1}h \qquad \text{and} \qquad t_k g << f + k^{-1}h$$

Let t_k' be a second solution to these equations. By Lemma 6, there exists φ in $\mathcal{P}$ with

$$I_\varphi(f) \leq t_k I_\varphi(g) + k^{-1}(1 + I_\varphi(h))$$

and
$$t_k I_\varphi(g) \leq I_\varphi(f) + k^{-1}(1 + I_\varphi(h))$$

or
$$|t_k - I_\varphi(f)I_\varphi(g)^{-1}| \leq k^{-1}(1 + I_\varphi(h))I_\varphi(g)^{-1} \leq \lambda k^{-1}$$

where $\lambda \equiv (1 + (h:f_0))(f_0:g)$; and the same for t_k'. Hence $|t_k - t_k'| \leq 2\lambda k^{-1}$. In particular, $|t_k - t_j| \leq 2\lambda k^{-1}$ whenever $j \geq k$. Thus $\{t_k\}$ is a Cauchy sequence, whose limit we denote by $\int f\, d\mu$. Clearly $\int f\, d\mu$ does not depend on the choice of h. It is also clear that $\int \alpha f\, d\mu = \alpha \int f\, d\mu$ for all $\alpha \geq 0$.

Consider elements f_1 and f_2 in C^+. By the above, φ can be chosen to make the quantities $\int f_1\, d\mu$, $\int f_2\, d\mu$, and $\int (f_1 + f_2)\, d\mu$, respectively, arbitrarily near to $I_\varphi(f_1)I_\varphi(g)^{-1}$, $I_\varphi(f_2)I_\varphi(g)^{-1}$, and $I_\varphi(f_1 + f_2)I_\varphi(g)^{-1}$, and to make $I_\varphi(f_1 + f_2)$ arbitrarily near to $I_\varphi(f_1) + I_\varphi(f_2)$. Therefore

$$\int (f_1 + f_2)\, d\mu = \int f_1\, d\mu + \int f_2\, d\mu$$

For each test function f choose functions f_1 and f_2 in C^+ with $f = f_1 - f_2$, and define

$$\int f\, d\mu \equiv \int f_1\, d\mu - \int f_2\, d\mu$$

If also $f = f_1' - f_2'$, for functions f_1' and f_2' in C^+, then $f_1 + f_2' = f_1' + f_2$, so that $\int f_1\, d\mu + \int f_2'\, d\mu = \int f_1'\, d\mu + \int f_2\, d\mu$. Therefore $\int f\, d\mu$ does not depend on the choice of f_1 and f_2. It follows that $f \rightarrow \int f\, d\mu$ is a linear functional from $C(G)$ to $\mathbf{R}$. Clearly $\int f\, d\mu \leq \int h\, d\mu$ whenever $f \leq h$. Therefore μ is a positive measure on G.

If $f \in C^+$ and $s \in G$ then any solution t_k of equations (1.19) also works for $T(s)f$. Therefore $\int f \, d\mu = \int T(s) f \, d\mu$, so that μ is left-invariant.

By the definitions, $\int g \, d\mu = 1$. For each f in $\mathcal{P}$ we have $g \leq \Sigma c_i T(s_i) f$, for certain $c_1, \ldots, c_n$ in $\mathbf{R}^{0+}$ and certain $s_1, \ldots, s_n$ in G. Since μ is positive and left-invariant, it follows that

$$1 = \int g \, d\mu \leq \Sigma c_i \int f \, d\mu$$

so $\int f \, d\mu > 0$.

It remains to show that any positive left-invariant measure ν is a multiple of μ. For all f_1 and f_2 in C^+ with $f_1 < < f_2$ we have $\int f_1 \, d\nu \leq \int f_2 \, d\nu$, by the invariance and positivity of ν. In particular,

$$\int f \, d\nu \leq t_k \int g \, d\nu + k^{-1} \int h \, d\nu$$

and

$$t_k \int g \, d\nu \leq \int f \, d\nu + k^{-1} \int h \, d\nu$$

for all k. Letting $k \to \infty$ gives $\int f \, d\nu = \int f \, d\mu \int g \, d\nu$, or $\nu = (\int g \, d\nu)\mu$.

The measure μ of Theorem 1 is called *Haar measure*.

Group multiplication induces an important operation on test functions, called *convolution*. In defining this operation we work with complex-valued test functions. Unless otherwise stated, all functions on G from this point on will be permitted to have complex values. For each test function f define the test function $\tilde{f}$ by $\tilde{f}(x) \equiv f(x^{-1})^*$. (This agrees with the definition given previously for real f.) In case G is commutative, the functional $f \to \int \tilde{f} \, d\mu$ (when restricted to real test functions) is a positive left-invariant measure on G. By Theorem 1, there exists $c > 0$ such that $\int \tilde{f} \, d\mu = c \int f \, d\mu$ for all real test functions f. Since $\tilde{\tilde{f}} = f$, we have

$$\int (f + \tilde{f}) \, d\mu = c \int (f + \tilde{f}) \, d\mu$$

Therefore $c = 1$, and thus

$$(1.20) \qquad\qquad \int \tilde{f} \, d\mu = \int f \, d\mu$$

for all real test functions f. For complex test functions f we have $\int \tilde{f} \, d\mu = (\int f \, d\mu)^*$.

In general, a group G satisfying (1.20) is called *unimodular*. Thus every commutative locally compact group is unimodular.

For each unimodular group G,

$$\int f(xs) \, d\mu(x) = \int f(x^{-1}s) \, d\mu(x) = \int f((s^{-1}x)^{-1}) \, d\mu(x)$$
$$= \int f(x^{-1}) \, d\mu(x) = \int f(x) \, d\mu(x)$$

Thus Haar measure on a unimodular group is right-invariant.

By the Stone-Weierstrass theorem, any test function $f(x,y)$ on the product $G \times G$ of a locally compact group G with itself can be uniformly approximated by finite sums $\Sigma f_i(x)g_i(y)$, where the f_i and g_i are test functions on G whose support lies in some compact set $K \subset G$ depending only on f. It follows that

$$(1.21) \qquad \int \left(\int f(x,y) \, d\mu(x) \right) d\mu(y) = \int \left(\int f(x,y) \, d\mu(y) \right) d\mu(x)$$

The *convolution* $f * g$ of test functions f and g is defined by

$$(1.22) \qquad (f * g)(x) \equiv \int f(y)g(y^{-1}x) \, d\mu(y) \qquad (x \in G)$$

It is clear that $f * g$ is a test function.

If f, g, and h are test functions, then

$$\begin{aligned}
((f * g) * h)(x) &= \int (f * g)(y)h(y^{-1}x) \, d\mu(y) \\
&= \int\int f(z)g(z^{-1}y)h(y^{-1}x) \, d\mu(z) \, d\mu(y) \\
&= \int\int f(z)g(z^{-1}y)h((z^{-1}y)^{-1}z^{-1}x) \, d\mu(y) \, d\mu(z) \\
&= \int\int f(z)g(y)h(y^{-1}z^{-1}x) \, d\mu(y) \, d\mu(z) \\
&\qquad\qquad\qquad\qquad\qquad \text{(by left invariance of } \mu) \\
&= \int f(z)(g * h)(z^{-1}x) \, d\mu(z) = (f * (g * h))(x)
\end{aligned}$$

Thus convolution is associative. In case G is commutative,

$$\begin{aligned}
(f * g)(x) &= \int f(y)g(y^{-1}x) \, d\mu(y) = \int f(y^{-1})g(yx) \, d\mu(y) \\
&= \int f(y^{-1}x)g(y) \, d\mu(y) = (g * f)(x)
\end{aligned}$$

and thus convolution is commutative.

Proposition 2 *Let f and g be test functions on a unimodular group G. Let p, q, and r be constants with $p \geq 1$, $q \geq 1$, $r \geq 1$, and*

$$p^{-1} + q^{-1} = 1 + r^{-1}$$

Then

$$(1.23) \qquad\qquad\qquad \|f * g\|_r \leq \|f\|_p \|g\|_q$$

Proof It is sufficient to consider the case $f \geq 0$ and $g \geq 0$. The numbers $\alpha \equiv r^{-1}$, $\beta \equiv p^{-1} - r^{-1}$, and $\gamma \equiv q^{-1} - r^{-1}$ are positive and sum to 1. Hölder's inequality gives

$$\begin{aligned}
(f * g)(x) &= \int \{f(y)^p g(y^{-1}x)^q\}^\alpha f(y)^{p\beta} g(y^{-1}x)^{q\gamma} \, d\mu(y) \\
&\leq \left(\int f(y)^p g(y^{-1}x)^q \, d\mu(y) \right)^\alpha \|f\|_p^{p\beta} \|g\|_q^{q\gamma}
\end{aligned}$$

Therefore

$$\|f * g\|_r \leq \|f\|_p^{p\beta}\|g\|_q^{q\gamma}\left(\int\int f(y)^p g(y^{-1}x)^q\, d\mu(y)\, d\mu(x)\right)^{1/r}$$
$$= \|f\|_p^{p\beta}\|g\|_q^{q\gamma}(\|f\|_p^p\|g\|_q^q)^{1/r} = \|f\|_p\|g\|_q$$

Proposition 2 extends to more general functions.

Proposition 3 *Let p, q, and r be constants with $p \geq 1$, $q \geq 1$, $r \geq 1$,
$p^{-1} + q^{-1} = 1 + r^{-1}$. Let f and g be functions in L_p and L_q, respectively,
on the unimodular group G. Then the integral (1.22) exists for almost all x
in G, the function $f * g$ so defined belongs to L_r, and (1.23) is valid.*

Proof It is enough to consider the case $f \geq 0$ a.e. and $g \geq 0$ a.e.
In case both f and g are test functions, the result is already established.
Next consider the case in which f is a test function and g is arbitrary.
Let $\{g_n\}$ be a sequence of test functions converging to g in L_q. By (1.23),
we see that $\{f * g_n\}$ is a Cauchy sequence in L_r, whose limit we call h.
By Theorem 4 of Chap. 7, we may assume that $\{f * g_n\}$ converges to h
almost everywhere. Since for each x in G the linear functional

$$F \to \int f(y)F(y^{-1}x)\, d\mu(y)$$

on L_q is bounded, we have

$$(f * g)(x) = \lim_{n \to \infty} (f * g_n)(x) = h(x) \text{ a.e.}$$

Also,
$$\|f * g\|_r = \|h\|_r = \lim_{n \to \infty} \|f * g_n\|_r$$
$$\leq \|f\|_p \lim_{n \to \infty} \|g_n\|_q = \|f\|_p\|g\|_q$$

Next consider the case in which f is a simple function that vanishes on
the complement of a compact set and g is arbitrary. Let $\{f_n\}$ be a
uniformly bounded sequence of test functions having a common support
and converging to f almost everywhere. Then $\{(f_n * g)(x)\}$ converges to
$(f * g)(x)$ for all x, and $\{f_n * g\}$ is a Cauchy sequence in L_r. Hence
$f * g \in L_r$, and

$$\|f * g\|_r \leq \lim_{n \to \infty} \|f_n * g\|_r \leq \|f\|_p\|g\|_q$$

Next consider the general case. Let $\{f_n\}$ be a nondecreasing sequence of
nonnegative simple functions converging to f in L_p. By the case just
considered, $\{f_n * g\}$ is a Cauchy sequence in L_r, whose limit we denote

by h. We may assume that $\{(f_n * g)(x)\}$ converges to $h(x)$ for almost all x. For such an x we see by Theorem 5 of Chap. 7 that $(f * g)(x) = h(x)$. Thus $f * g$ is defined almost everywhere, $f * g \in L_r$, and

$$\|f * g\|_r = \|h\|_r = \lim_{n \to \infty} \|f_n * g\|_r \leq \|f\|_p \|g\|_q$$

The reader may check that in case $f \in L_1$ and $g \in L_p$ the proof of Proposition 3 holds for an arbitrary locally compact group (not necessarily unimodular).

2. CONVOLUTION OPERATORS

In the sequel G will be a locally compact abelian group. We have seen that if $f \in L_1(G)$ and $g \in L_2(G)$, then $f * g \in L_2(G)$ and $\|f * g\| \leq \|f\|_1 \|g\|_2$. Thus $g \to f * g$ is a bounded linear operator $T(f)$ on $L_2(G)$, and

$$(2.1) \qquad \|T(f)g\|_2 \leq \|f\|_1 \|g\|_2$$

Since convolution is commutative, the operators $T(f)$ commute with one another. Our first task is to show that these operators are normable. We need a lemma.

Lemma 8 *For each compact integrable neighborhood U of e in G there exist compact integrable neighborhoods V and W of e such that $UV \subset W$ and $\mu(W)\mu(V)^{-1}$ is arbitrarily near to 1. (A neighborhood of e is any set containing e as an interior point.)*

Proof Since U^2 is totally bounded, there exists a finite subset F of G with $FU \supset U^2$. By induction we see that $F^iU \supset U^{i+1}$ for all i in Z^+. Fix the positive integer n and let $V_1, \ldots, V_n$ be compact integrable sets with $U^{2i+1} \supset V_i \supset U^{2i}$ $(1 \leq i \leq n)$. Let F have k elements. Then F^{2n} has at most $(2n)^k$ elements (since G is commutative). Since $F^{2n}U \supset U^{2n+1} \supset V_n$, it follows that

$$(\mu(V_n)\mu(V_{n-1})^{-1})(\mu(V_{n-1})\mu(V_{n-2})^{-1}) \cdots (\mu(V_2)\mu(V_1)^{-1})$$
$$= \mu(V_n)\mu(V_1)^{-1} < (2n)^k \mu(U)\mu(U)^{-1} = (2n)^k$$

Therefore $\qquad \mu(V_{i+1})\mu(V_i)^{-1} < (2n)^{k/(n-1)}$

for some i $(1 \leq i \leq n - 1)$. Since $UV_i \subset UU^{2i+1} = U^{2i+2} \subset V_{i+1}$, taking $V = V_i$ and $W = V_{i+1}$ and taking n to be sufficiently large satisfies the requirements of the lemma.

Theorem 2 *For each f in $L_1(G)$ the operator $T(f)$ is normable.*

Proof Since the test functions are dense in $L_1(G)$, because of (2.1) we may assume that f is a test function.

Since the test functions are dense in $L_2(G)$, to show $\|T(f)\|$ exists it is enough to consider expressions $\|f * h\|_2 \|h\|_2^{-1}$, where h is a test function with $\|h\|_2 > 0$. Let U be a compact integrable neighborhood of e which supports f. By Lemma 8, for each $\delta > 0$ there exist compact integrable neighborhoods V and W of e with $U^{-1}V \subset W$ and $\mu(V)\mu(W)^{-1} > 1 - \delta$. We compute

$$\mu(W)^{-1}\int \|\chi_W T(y)h\|_2^2 \, d\mu(y) = \mu(W)^{-1}\iint \chi_W(x)|h(yx)|^2 \, d\mu(x) \, d\mu(y)$$
$$= \mu(W)^{-1}\iint \chi_W(x)|h(yx)|^2 \, d\mu(y) \, d\mu(x)$$
$$= \mu(W)^{-1}\int \chi_W(x)\|h\|_2^2 \, d\mu(x) = \|h\|_2^2$$

Similarly,

$$\mu(W)^{-1}\int \|\chi_V T(y)(f * h)\|_2^2 \, d\mu(y) = \mu(W)^{-1}\mu(V)\|f * h\|_2^2$$
$$> (1 - \delta)\|f * h\|_2^2$$

It follows that there exists y in G such that

$$\|\chi_V T(y)(f * h)\|_2^2 > (1 - \delta)\|f * h\|_2^2 \|h\|_2^{-2} \|\chi_W T(y)h\|_2^2$$

Since $U^{-1}V \subset W$, we have

$$\chi_V T(y)(f * h) = \chi_V(f * T(y)h) = \chi_V(f * \chi_W T(y)h)$$

Hence $\|f * \chi_W T(y)h\|_2 \|\chi_W T(y)h\|_2^{-1} > (1 - \delta)^{1/2}\|f * h\|_2 \|h\|_2^{-1}$

Since δ is an arbitrary positive constant, to show that $\|T(f)\|$ exists it is thus enough to show that for each compact integrable neighborhood W of e the supremum

$$c \equiv \sup \{\|f * h\|_2 : h \in S\}$$

exists, where S is the unit sphere of $L_2(W)$. By Theorem 9 of Chap. 9, each h in S gives rise to an element λ_h of the unit sphere S^* of the dual of $L_2(W)$, defined by

$$\lambda_h(g) \equiv \int gh \, d\mu \qquad (g \in L_2(W))$$

and in fact every normable element of S^* arises in this way. In the metric induced by the double norm S^* is compact and $\Gamma \equiv \{\lambda_h : h \in S\}$ is dense. Therefore Γ is totally bounded. Hence to show that c exists it is enough to show that the function $\lambda_h \to \|f * h\|_2$ is uniformly continuous on Γ. Let ω be any modulus of continuity for f, in the sense that $|f(x) - f(y)| \leq \epsilon$ whenever $\epsilon > 0$, $x \in G$, $y \in G$, and $\rho(y^{-1}x) \leq \omega(\epsilon)$. Then

$$|(f * h)(x) - (f * h)(y)| = |\int h(z)(f(z^{-1}x) - f(z^{-1}y))\, d\mu(z)|$$
$$\leq \epsilon \int |h(z)|\, d\mu(z) \leq \epsilon \|h\|_2 \mu(W)^{\frac{1}{2}} \leq \epsilon \mu(W)^{\frac{1}{2}}$$

whenever $h \in S$ and $\rho(y^{-1}x) \leq \omega(\epsilon)$. The functions $f * h$ are thus test functions with a common support and a common modulus of continuity. Hence to make $\|f * h_1 - f * h_2\|_2$ small it is enough to make $|f * h_1 - f * h_2|$ small on an appropriate finite set. Since for each x the map $\lambda_h \to (f * h)(x)$ is uniformly continuous on Γ, it follows that $\lambda_h \to \|f * h\|_2$ is uniformly continuous on Γ.

3. THE CHARACTER GROUP

If a function f of $L_1(G)$ approximates a point x of G, in the sense that f vanishes outside some small neighborhood of x, and if g approximates y, then $f * g$ approximates xy. In other words, the operation of convolution in $L_1(G)$ can be regarded as a smoothing out of the operation of multiplication in G. In some cases the smoothed operation is easier to handle. Because of this and because the set $L_1(G)$ has a linear structure, it is often more convenient to work with $L_1(G)$ than with G itself.

The set $L_1(G)$, with multiplication defined by convolution, and with the usual addition, is an algebra over $\mathbf{C}$, called the *group algebra* of G. The *norm* $\| \quad \|_1$ on $L_1(G)$ has the properties

$$\|f + g\|_1 \leq \|f\|_1 + \|g\|_1 \qquad \|f * g\|_1 \leq \|f\|_1\|g\|_1 \qquad \|cf\|_1 = |c|\,\|f\|_1$$

for all f and g in $L_1(G)$ and all c in $\mathbf{C}$. (An algebra with such a norm is called a *normed algebra*.)

Recall that for each f in $L_1(G)$ the element $\tilde{f}$ of $L_1(G)$ is defined by $\tilde{f}(x) = f(x^{-1})^*$. The map $f \to \tilde{f}$ is a norm-preserving map of $L_1(G)$ onto itself, which preserves sums and convolutions and is antilinear in the sense that $\widetilde{cf} = c^*\tilde{f}$ [for all c in $\mathbf{C}$ and f in $L_1(G)$].

The group algebra $L_1(G)$ admits the norm-preserving translation operators $T(x)$, already introduced. We compute

$$(T(x)(f * g))(z) = \int f(y)g(y^{-1}xz)\, d\mu(y) = (f * T(x)g)(z)$$

Therefore

$$(3.1) \quad T(x)(f * g) = f * T(x)g = T(x)f * g \qquad (x \in G, f, g \in L_1(G))$$

Lemma 9 *For each f in $L_1(G)$ the function $x \to T(x)f$ from G to $L_1(G)$ is continuous.*

Proof In case $f \in C(G)$, we see that f is uniformly continuous by Lemma 1, and therefore $x \to T(x)f$ is continuous. For a general element f of $L_1(G)$, and an arbitrary $\epsilon > 0$, choose g in $C(G)$ with $\|f - g\|_1 \le \epsilon$. For each bounded subset of K of G we have $\|T(x)g - T(y)g\|_1 \le \epsilon$ whenever $x, y \in K$ and $\rho(x,y)$ is sufficiently small. Hence

$$\|T(x)f - T(y)f\|_1 \le \|T(x)f - T(x)g\|_1 + \|T(x)g - T(y)g\|_1$$
$$+ \|T(y)g - T(y)f\|_1$$
$$\le \epsilon + \epsilon + \epsilon = 3\epsilon$$

Therefore $x \to T(x)f$ is continuous.

A continuous function $\alpha: G \to \mathbf{C}$ is called a *character* if $|\alpha(x)| = 1$ for all x in G and if

$$\alpha(xy) = \alpha(x)\alpha(y)$$

for all x and y in G.

For each character α we have $\alpha(e) = \alpha(e^2) = \alpha(e)^2$, and thus $\alpha(e) = 1$. Since $|\alpha(x)| = 1$, we see that $\alpha(x^{-1}) = \alpha(x)^{-1} = \alpha(x)^*$, for all x in G.

The *product* $\alpha\beta$ of characters α and β is defined by $(\alpha\beta)(x) = \alpha(x)\beta(x)$. This product makes the set G^* of all characters into a group, called the *character group* or *dual group* of G. The identity element of G^* is the function 1, and the inverse of a character α is α^*.

For each α in G^* and each f in $L_1(G)$ we write

$$\alpha(f) \equiv \int f(x)\alpha(x)\,d\mu(x)$$

Proposition 4 *For each α in G^* the map $f \to \alpha(f)$ is a nonzero homomorphism of $L_1(G)$ into $\mathbf{C}$, with $\alpha(\tilde{f}) = \alpha(f)^*$ and $|\alpha(f)| \le \|f\|_1$ for all f in $L_1(G)$.*

Proof Clearly $\alpha(f + g) = \alpha(f) + \alpha(g)$ and $\alpha(cf) = c\alpha(f)$. We compute

$$\alpha(f * g) = \int (f * g)(x)\alpha(x)\,d\mu(x)$$
$$= \int\int f(y)g(y^{-1}x)\,d\mu(y)\alpha(x)\,d\mu(x)$$
$$= \int\int g(y^{-1}x)\alpha(y^{-1}x)\,d\mu(x)f(y)\alpha(y)\,d\mu(y)$$
$$= \alpha(f)\alpha(g)$$

Thus α is a homomorphism, and

$$\alpha(\tilde{f}) = \int f(x^{-1})^*\alpha(x)\,d\mu(x) = \int f(x^{-1})^*\alpha(x^{-1})^*\,d\mu(x)$$
$$= \left(\int f(x^{-1})\alpha(x^{-1})\,d\mu(x)\right)^* = \alpha(f)^*$$

Since α is continuous, and $\alpha(e) = 1$, there exists a neighborhood U of e such that the real part of $\alpha(x)$ is greater than $\frac{1}{2}$ for all x in U. Thus if f is a nonnegative test function which vanishes off u, the real part of $\alpha(f)$ is positive, and thus α is nonzero. Finally $|\alpha(f)| \leq \|f\|_1$ because $|\alpha(x)| = 1$ for all x.

To show that, conversely, every bounded nonzero homomorphism of $L_1(G)$ into $\mathbf{C}$ is induced by an element of G^*, we need a lemma.

Lemma 10 *If λ is any bounded homomorphism of $L_1(G)$ into $\mathbf{C}$, then*

$$(3.2) \quad \lambda(f * g) = \int f(x)\lambda(T(x^{-1})g)\, d\mu(x) \qquad (f \in C(G),\, g \in L_1(G))$$

Proof Since both sides of (3.2) depend continuously on the element g of $L_1(G)$, it is enough to prove (3.2) for test functions g. By the definitions of the integrals involved, we can find constants $c_1, \ldots, c_n$ and elements $x_1, \ldots, x_n$ of G such that $f * g$ is arbitrarily closely approximated in the norm $\|\ \|_1$ by the function $F \equiv \sum_{i=1}^{n} c_i f(x_i) T(x_i^{-1})g$, and $\int f(x)\lambda(T(x^{-1})g)\, d\mu(x)$ is arbitrarily closely approximated by the sum $S \equiv \sum_{i=1}^{n} c_i f(x_i)\lambda(T(x_i^{-1})g)$. Since λ is bounded, $\lambda(f * g)$ is arbitrarily closely approximated by $\lambda(F) = S$. Thus both sides of (3.2) can be arbitrarily closely approximated by the same sum S.

We now show that G^* corresponds exactly to the bounded nonzero homomorphisms λ of $L_1(G)$ into $\mathbf{C}$.

Theorem 3 *If $\lambda: L_1(G) \to \mathbf{C}$ is any bounded nonzero homomorphism, then there exists a unique α in G^* with $\alpha(f) = \lambda(f)$ for all f in $L_1(G)$.*

Proof For each x in G write

$$\alpha(x) \equiv \lambda(T(x^{-1})f_0)\lambda(f_0)^{-1}$$

where f_0 is any element of $L_1(G)$ with $\lambda(f_0) \neq 0$. This does not depend on the choice of f_0, for if f_1 is another element of $L_1(G)$ with $\lambda(f_1) \neq 0$, then, by (3.1),

$$\lambda(f_0)\lambda(T(x^{-1})f_1) = \lambda(f_0 * T(x^{-1})f_1) =$$
$$\lambda(T(x^{-1})f_0 * f_1) = \lambda(T(x^{-1})f_0)\lambda(f_1)$$

By Lemma 9, we see that α is a continuous function on G. For each x in G with $\alpha(x) \neq 0$ and each positive integer n, we have

$$
\begin{aligned}
|\alpha(x)|^n &= \left| \frac{\lambda(T(x^{-1})f_0)}{\lambda(f_0)} \frac{\lambda(T(x^{-2})f_0)}{\lambda(T(x^{-1})f_0)} \cdots \frac{\lambda(T(x^{-n})f_0)}{\lambda(T(x^{-n+1})f_0)} \right| \\
&= \left| \frac{\lambda(T(x^{-n})f_0)}{\lambda(f_0)} \right| \leq \|T(x^{-n})f_0\|_1 |\lambda(f_0)|^{-1} \\
&= \|f_0\|_1 |\lambda(f_0)|^{-1}
\end{aligned}
$$

It follows that

$$(3.3) \qquad\qquad |\alpha(x)| \leq 1 \qquad (x \in G)$$

We now show that α is a homomorphism. Since λ is nonzero, $\lambda(f) \neq 0$ for all f belonging to some dense subset of $L_1(G)$. Thus for each x in G there exists f_1 in $L_1(G)$, with $\lambda(f_1) \neq 0$, which is arbitrarily near to $T(x)f_0$. Hence $T(x^{-1})f_1$ is arbitrarily close to f_0. We may thus take f_1 so close to $T(x)f_0$ that $\lambda(T(x^{-1})f_1) \neq 0$. Then for each y in G we have

$$
\begin{aligned}
\alpha(xy) &= \frac{\lambda(T(x^{-1}y^{-1})f_1)}{\lambda(f_1)} = \frac{\lambda(T(x^{-1}y^{-1})f_1)}{\lambda(T(x^{-1})f_1)} \frac{\lambda(T(x^{-1})f_1)}{\lambda(f_1)} \\
&= \alpha(x)\alpha(y)
\end{aligned}
$$

Thus α is a homomorphism. It is nonzero, since for all x we have $\alpha(x)\alpha(x^{-1}) = \alpha(e) = 1$. In fact we have $|\alpha(x)| = |\alpha(x^{-1})^{-1}| \geq 1$, by (3.3). Together with (3.3), this gives $|\alpha(x)| = 1$. Therefore $\alpha \in G^*$.

For each f in $C(G)$ we have (by Lemma 10)

$$
\begin{aligned}
\alpha(f) &\equiv \int f(x)\alpha(x)\, d\mu(x) \\
&= \int f(x)\lambda(T(x^{-1})f_0)\lambda(f_0)^{-1}\, d\mu(x) \\
&= \lambda(f_0)^{-1} \int f(x)\lambda(T(x^{-1})f_0)\, d\mu(x) \\
&= \lambda(f_0)^{-1}\lambda(f * f_0) = \lambda(f)
\end{aligned}
$$

By continuity, $\alpha(f) = \lambda(f)$ for all f in $L_1(G)$.

If x is any element of G and ϵ any positive constant, by the continuity of α there exists a neighborhood U of x such that (a) $\alpha(y)$ is arbitrarily near to $\alpha(x)$, for all y in U, and (b) $|\alpha(x) - \lambda(f)| < \epsilon$ for all nonnegative elements f of $L_1(G)$ that vanish almost everywhere on $G - U$ and satisfy the normalization condition $\int f\, d\mu = 1$. It follows that α is uniquely determined by λ.

Corollary *If $\lambda: L_1(G) \to \mathbf{C}$ is a bounded nonzero homomorphism, then $\lambda(\check{f}) = \lambda(f)^*$ for all f in $L_1(G)$, and λ is normable with $\|\lambda\| = 1$.*

Proof If α is chosen as above, then $\lambda(\check{f}) = \alpha(\check{f}) = \alpha(f)^* = \lambda(f)^*$. Also, $|\lambda(f)| = |\alpha(f)| \le \|f\|$. Moreover, if f is any nonnegative test function with $\int f\, d\mu = 1$ whose support is contained in an arbitrarily small neighborhood of e, then $\lambda(f) = \alpha(f)$ is arbitrarily near to $\alpha(e) = 1$. Therefore λ is normable, and $\|\lambda\| = 1$.

A further hold on the structure of G^* is obtained by establishing a connection with homomorphisms of operator algebras, and applying the spectral theorem. The key step is the following.

Proposition 5 *For each α in G^* and each f in $L_1(G)$ we have*

$$(3.4) \qquad\qquad |\alpha(f)| \le \|T(f)\|$$

Proof Since both sides of (3.4) are continuous functions of f, there is no loss of generality in taking f to be a test function. Let U be a compact neighborhood of e which supports f.

By Lemma 8, for each $\epsilon > 0$ there exist compact neighborhoods V and W of e with $\mu(V)\mu(W)^{-1} \ge 1 - \epsilon$ and $U^{-1}V \subset W$. The function $g \equiv \chi_W \alpha^{-1}$ is in $L_2(G)$, and $\|g\|_2 = \mu(W)^{1/2}$. If $x \in V$, then $y^{-1}x \in W$ for all y in U, and thus

$$\begin{aligned}
(f * g)(x) &= \int_U f(y)g(y^{-1}x)\, d\mu(y) \\
&= \int_U f(y)\alpha(x^{-1}y)\, d\mu(y) \\
&= \alpha(x^{-1})\alpha(f)
\end{aligned}$$

Thus $|f * g| = |\alpha(f)|$ on V, and thus

$$\begin{aligned}
\|T(f)g\|_2 &= \|f * g\|_2 \ge |\alpha(f)|\mu(V)^{1/2} \\
\|T(f)\| &\ge \|T(f)g\|_2 \|g\|_2^{-1} \\
&\ge |\alpha(f)|(\mu(V)\mu(W)^{-1})^{1/2} \\
&\ge |\alpha(f)|(1 - \epsilon)^{1/2}
\end{aligned}$$

Hence

Since ϵ is arbitrary, (3.4) is valid.

Let $\mathfrak{A}$ be the algebra of all operators of the form $T(f)$, with $f \in L_1(G)$. Then $f \to T(f)$ is a bounded homomorphism of $L_1(G)$ onto $\mathfrak{A}$. Since $L_1(G)$ is separable, $\mathfrak{A}$ is separable. For all f in $L_1(G)$ and all g_1 and g_2 in

$L_2(G)$ we have

$$\begin{aligned}
(T(f)g_1,g_2) &= \iint f(y)g_1(y^{-1}x)g_2(x)^* \, d\mu(y) \, d\mu(x) \\
&= \iint f(y^{-1}x)g_1(y)g_2(x)^* \, d\mu(y) \, d\mu(x) \\
&= \int (\int \tilde{f}(yx^{-1})g_2(x) \, d\mu(x))^* g_1(y) \, d\mu(y) \\
&= (g_1, T(\tilde{f})g_2)
\end{aligned}$$

Therefore $\qquad\qquad\qquad T(f)^* = T(\tilde{f})$

and thus $\mathfrak{A}$ is selfadjoint. The spectral theorem does not apply directly to $\mathfrak{A}$, because $\mathfrak{A}$ may not be uniformly closed. Because of this we first pass to the *uniform closure* $\bar{\mathfrak{A}}$ of $\mathfrak{A}$, consisting of all operators U on $L_2(G)$ such that for each $\epsilon > 0$ there exists U_ϵ in $\mathfrak{A}$ such that $U - U_\epsilon$ is bounded by ϵ. The operators in $\bar{\mathfrak{A}}$ are normable, and $\bar{\mathfrak{A}}$ is commutative and selfadjoint, as well as uniformly closed. Let Σ be the spectrum of $\bar{\mathfrak{A}}$. By Corollary 3 of Theorem 8 of Chap. 9 (page 283), Σ is locally compact, and the map γ_0 from $\bar{\mathfrak{A}}$ to $C_\infty(\Sigma,\mathbf{C})$ defined by $\gamma_0(U)(u) \equiv u(U)$ is a normpreserving isomorphism.

For each f in $L_1(G)$ and each α in G^* we write

$$\alpha(T(f)) \equiv \alpha(f)$$

By (3.4), we see that $|\alpha(T(f))| \le \|T(f)\|$ for all $T(f)$ in $\mathfrak{A}$. Thus $\alpha(T(f))$ defines a bounded homomorphism of $\mathfrak{A}$ onto $\mathbf{C}$, and so extends to a bounded homomorphism $U \to \alpha(U)$ of $\bar{\mathfrak{A}}$ onto $\mathbf{C}$, that is, to an element of Σ. Conversely, if $u \in \Sigma$, then $f \to u(T(f))$ is a bounded homomorphism of L_1 onto $\mathbf{C}$, and so comes from an element α of G^*. Thus we may identify G^* with Σ, via the map which takes each α in G^* into the element $U \to \alpha(U)$ of Σ. This permits us to give G^* the metric of Σ, so that G^* is a locally compact metric space. To write the metric explicitly, consider a dense sequence $\{f_n\}$ in $L_1(G)$, so that $\{T(f_n)\}$ is dense in $\bar{\mathfrak{A}}$. By the definition of the double norm on Σ, we have

$$\rho^*(\alpha,\beta) = |||\alpha - \beta||| + |(|||\alpha|||^{-1} - |||\beta|||^{-1})|$$

for all α and β in G^*, where

$$|||\alpha||| \equiv \sum_{n=1}^{\infty} 2^{-n}|\alpha(f_n)|(1 + \|T(f_n)\|)^{-1}$$

for all α in G^*, and $|||\alpha - \beta|||$ is defined similarly.

Combining the norm-preserving isomorphism γ_0 from $\bar{\mathfrak{A}}$ to $C_\infty(\Sigma,\mathbf{C})$ with the map which identifies G^* with Σ, we obtain a norm-preserving isomorphism γ of $\bar{\mathfrak{A}}$ with $C_\infty(G^*,\mathbf{C})$ given by

$$(3.5) \qquad\qquad \gamma(U)(\alpha) \equiv \alpha(U) \qquad (U \in \bar{\mathfrak{A}},\ \alpha \in G^*)$$

In particular, for each f in $L_1(G)$,

$$(3.6) \qquad\qquad \gamma(T(f))(\alpha) \equiv \alpha(T(f)) \equiv \alpha(f)$$

The function $\hat{f} \equiv \gamma(T(f))$ is called the *Fourier transform* of f. For notational convenience, we also write $\hat{f} \equiv \mathfrak{F}(f)$. The Fourier transform $\mathfrak{F}$ is an isomorphism of the algebra $L_1(G)$ into $C_\infty(G^*,\mathbf{C})$, which is norm-decreasing:

$$\|\mathfrak{F}(f)\|_\infty \equiv \|\gamma(T(f))\|_\infty = \|T(f)\| \leq \|f\|_1$$

It is given by the formula

$$(3.7) \quad \hat{f}(\alpha) = \alpha(f) = \int f(x)\alpha(x)\,d\mu(x) \qquad (f \in L_1(G),\ \alpha \in G^*)$$

For later use, we shall wish to express the Fourier transform of a translate $T(x^{-1})f$ in terms of the Fourier transform of f. We have

$$\begin{aligned}
\mathfrak{F}(T(x^{-1})f)(\alpha) &= \int f(x^{-1}y)\alpha(y)\,d\mu(y) \\
&= \int f(x^{-1}y)\alpha(x^{-1}y)\alpha(x)\,d\mu(y) \\
&= \alpha(x)(\mathfrak{F}(f))(\alpha)
\end{aligned}$$

or

$$(3.8) \qquad\qquad \mathfrak{F}(T(x^{-1})f) = \hat{x}\,\mathfrak{F}(f)$$

where $\hat{x}$ is the function $\alpha \to \alpha(x)$ on G^*.

The metric ρ^* on G^* is not easy to work with. It is helpful to relate the metric ρ^* to the behavior of the characters α on certain subsets of G. First we give a criterion for boundedness.

Lemma 11 *For each $\epsilon > 0$ and each subset K of G we define*

$$N^*(K,\epsilon) \equiv \{\alpha \in G^*\colon |\alpha(x) - 1| \leq \epsilon \text{ for all } x \text{ in } K\}$$

Then $N^*(K,\epsilon)$ is a bounded subset of G^*, for each real number ϵ, with $0 < \epsilon < 1$, and each neighborhood K of e in G. Conversely, if M^* is any bounded subset of G^* and ϵ any positive constant, there exists a neighborhood K of e in G such that $M^* \subset N^*(K,\epsilon)$.

Proof　First, consider any real number ϵ, with $0 < \epsilon < 1$, and any neighborhood K of e in G. Then $\operatorname{Re} \alpha(x) \geq \delta > 0$, for all x in K and α in $N^*(K,\epsilon)$, where δ is some absolute constant. Let f be any nonnegative test function supported by K, with $\int f \, d\mu = 1$. Then

$$\operatorname{Re} \alpha(f) = \int f(x) \operatorname{Re} \alpha(x) \, d\mu(x) \geq \delta \qquad (\alpha \in N^*(K,\epsilon))$$

Hence $|||\alpha|||$ is bounded away from 0 on $N^*(K,\epsilon)$. Therefore $N^*(K,\epsilon)$ is bounded.

Conversely, consider any bounded set $M^* \subset G^*$ and any $\epsilon > 0$. Then $|||\alpha|||$ is bounded away from 0 on M^*. Let $\{f_n\}$ be the sequence from $L_1(G)$ we used in defining the double norm $||| \quad |||$ on Σ. Then $|\alpha(f_1)| + \cdots + |\alpha(f_n)|$ is bounded away from 0 on M^*, for some n in Z^+. Thus to show the neighborhood K of 0 exists, it is sufficient to take M^* to be the set

$$M^* \equiv \{\alpha \in G^* : |\alpha(f)| \geq r\}$$

where r is a positive constant and f is an element of $L_1(G)$. By Lemma 9, for each $\delta > 0$ there exists a neighborhood K of e in G such that $\|T(x)f - f\|_1 \leq \delta$ for all x in K. For each α in M^* and x in K we have

$$\begin{aligned}
|\alpha(x) - 1| = |\alpha(x)^* - 1| &\leq r^{-1}|\alpha(x)^* - 1|\,|\alpha(f)| \\
&= r^{-1}|\alpha(T(x)f - f)| \leq r^{-1}\delta
\end{aligned}$$

Thus, if δ is sufficiently small, M^* will be a subset of $N^*(K,\epsilon)$.

Our next lemma characterizes the metric ρ^* on bounded subsets M^* of G^*.

Lemma 12　*Let M^* be any bounded subset of G^*. For each $\epsilon > 0$ there exists $\delta > 0$ and a compact set $K \subset G$ such that $\rho^*(\alpha,\beta) \leq \epsilon$ whenever $\alpha, \beta \in M^*$ and $\|\alpha - \beta\|_K \leq \delta$. Conversely, for each $\delta > 0$ and each compact set $K \subset G$ there exists $\epsilon > 0$ such that $\|\alpha - \beta\|_K \leq \delta$ whenever $\alpha, \beta \in M^*$ and $\rho^*(\alpha,\beta) \leq \epsilon$.*

Proof　Consider any $\epsilon > 0$. By the definition of the metric ρ^*, to make $\rho^*(\alpha,\beta) \leq \epsilon$ (for arbitrary elements α and β of M^*) it is enough to make the quantities $|(\alpha - \beta)(f_1)|, \ldots, |(\alpha - \beta)(f_n)|$ each less than some sufficiently small $r > 0$, where $f_1, \ldots, f_n$ are certain test functions on G. To do this it is enough to make $\|\alpha - \beta\|_K \leq \delta$, where K is a common support for $f_1, \ldots, f_n$ and δ is a small enough positive constant.

Assume, conversely, that K and δ are given. Since M^* is a bounded subset of G^*, there exists a positive constant r and a finite subset F of the unit sphere of $L_1(G)$ such that for each α in M^* there exists f in F with $|\alpha(f)| \geq 2r$. By Lemma 11, there exists a neighborhood V of e in G with $|\alpha(y) - \beta(y)| \leq \delta/2$ for all α and β in M^* and all y in V. Let A be a finite subset of G such that $e \in A$ and $K \subset AV$. Choose $\epsilon > 0$ so small that $|\alpha(T(x)f) - \beta(T(x)f)| \leq \min\{r, r^2\delta/2\}$ for all x in A, f in F, and α and β in M^* with $\rho^*(\alpha,\beta) \leq \epsilon$. Consider any α and β in M^* with $\rho^*(\alpha,\beta) \leq \epsilon$. Choose f in F with $|\alpha(f)| \geq 2r$. Then $|\beta(f)| \geq 2r - |\alpha(f) - \beta(f)| \geq r$. Hence for all x in A and y in V we have

$$
\begin{aligned}
|\alpha(xy) - \beta(xy)| &\leq |\alpha(x)|\,|\alpha(y) - \beta(y)| + |\beta(y)|\,|\alpha(x) - \beta(x)| \\
&\leq \frac{\delta}{2} + |\alpha(T(x)f)\alpha(f)^{-1} - \beta(T(x)f)\beta(f)^{-1}| \\
&\leq \frac{\delta}{2} + |\alpha(f)|^{-1}|\alpha(T(x)f) - \beta(T(x)f)| \\
&\qquad\qquad\qquad + |\beta(T(x)f)|\,|\alpha(f)^{-1} - \beta(f)^{-1}| \\
&\leq \frac{\delta}{2} + (2r)^{-1}r^2\frac{\delta}{2} + \|f\|_1|\alpha(f)|^{-1}|\beta(f)|^{-1}|\alpha(f) - \beta(f)| \\
&\leq \frac{\delta}{2} + \frac{\delta}{4} + (2r)^{-1}r^{-1}r^2\frac{\delta}{2} = \delta
\end{aligned}
$$

Therefore $\|\alpha - \beta\|_K \leq \delta$.

The set G^* is an abelian group under pointwise multiplication: The *product $\alpha\beta$* of points α and β in G^* is defined by

$$
(\alpha\beta)(x) \equiv \alpha(x)\beta(x) \qquad (x \in G)
$$

The last two lemmas allow us to show that it is a locally compact abelian group.

Theorem 4 *The dual G^* of a locally compact abelian group G is a locally compact abelian group.*

Proof Let M^* be any bounded subset of G^*. We must show that for each $\epsilon > 0$ there exists $\epsilon' > 0$ such that

$$
\rho^*(\alpha\beta, \alpha_0\beta_0) \leq \epsilon
$$

whenever $\alpha, \beta, \alpha_0, \beta_0 \in M^*$; $\rho^*(\alpha,\alpha_0) \leq \epsilon'$; and $\rho^*(\beta,\beta_0) \leq \epsilon'$. Note first that $(M^*)^2$ is a bounded subset of G^*, by Lemma 11. Therefore to prove the existence of ϵ', by Lemma 12 it is enough to show that for each compact set $K \subset G$ and each $\delta > 0$ there exists a compact set $K' \subset G$ and a constant $\delta' > 0$ such that $\|\alpha\beta - \alpha_0\beta_0\|_K \leq \delta$ whenever $\|\alpha - \alpha_0\|_{K'} \leq \delta'$ and $\|\beta - \beta_0\|_{K'} \leq \delta'$. This is straightforward: Take $K' = K$ and take δ' to be very small.

4. DUALITY AND THE FOURIER TRANSFORM

The full duality between G and its character group G^* is established by inverting the Fourier transform $\mathfrak{F}$. The possibility of doing this is already contained in the fact, derived from the spectral theorem, that the map γ of $\bar{\mathfrak{A}}$ onto $C_\infty(G^*,\mathbf{C})$ given by (3.5) is a norm-preserving isomorphism. The inverse γ^{-1} of γ is also a norm-preserving isomorphism. For each f in $L_1(G)$ we have $\gamma^{-1}(\hat{f}) = T(f)$, by the definition of $\hat{f}$. In other words,

$$\gamma^{-1}(\hat{f})g = f * g \qquad (g \in L_2(G))$$

This identity forms the point of departure for inverting the Fourier transform $\mathfrak{F}$. To each φ in $C(G^*,\mathbf{C})$ we shall associate a function $\bar{\varphi}$ on G, such that

$$(4.1) \qquad\qquad \gamma^{-1}(\varphi)g = \bar{\varphi} * g \qquad (g \in L_2(G))$$

If φ is of the form $\hat{f}$, for some f in $L_1(G)$, then $\bar{\varphi}$ will turn out to be equal to f. The map $\varphi \to \bar{\varphi}$ is called the *inverse Fourier transform* $\mathfrak{F}^*$.

Proposition 6 *For each φ in $C(G^*,\mathbf{C})$ there exists a unique function $\bar{\varphi} \equiv \mathfrak{F}^*(\varphi)$ in $C_\infty(G,\mathbf{C}) \cap L_2(G)$ satisfying (4.1).*

Proof Notice first that if g and h are in $L_2(G)$ the convolution

$$(g * h)(x) \equiv \int g(y)h(y^{-1}x) \, d\mu(y)$$

is well defined, and belongs to $C_\infty(G,\mathbf{C})$. The integral defining $(g * h)(x)$ exists for all x, because the product of two functions in L_2 is integrable. If f and g are test functions, then $f * g$ is a test function, and thus $f * g \in C_\infty(G,\mathbf{C})$. By an approximation argument, $f * g \in C_\infty(G,\mathbf{C})$ for all f and g in $L_2(G)$.

Let $K^* \subset G^*$ be a compact support for φ. Choose h in $C(G,\mathbf{C})$ with

$$|\hat{h}(\alpha)| \equiv |\int h(x)\alpha(x) \, d\mu(x)| > 1 \qquad (\alpha \in K^*)$$

This can be done by taking a compact neighborhood V of e in G such that $\alpha(x)$ is uniformly very near to 1 for all x in V and α in K^*, and taking h to be a nonnegative function supported by V, with $\int h \, d\mu = 2$. There exists a unique $\psi \in C(G^*,\mathbf{C})$ supported by K^* with $\varphi = \hat{h}^2\psi$. Clearly $\|\psi\|_\infty \leq \|\varphi\|_\infty$. [We write $\|\psi\|_\infty$ and $\|\varphi\|_\infty$ rather than $\|\psi\|$ and $\|\varphi\|$ to emphasize that norms in the space $C_\infty(G^*,\mathbf{C})$ are what is meant.] For each g in $C(G,\mathbf{C})$ we have

$$\begin{aligned}
\gamma^{-1}(\varphi)g &= \gamma^{-1}(\hat{h})\gamma^{-1}(\psi)\gamma^{-1}(\hat{h})g \\
&= T(h)\gamma^{-1}(\psi)T(h)g \\
&= h * (\gamma^{-1}(\psi)T(g)h) = h * (T(g)\gamma^{-1}(\psi))h \\
&= h * g * \gamma^{-1}(\psi)h = (h * \gamma^{-1}(\psi)h) * g
\end{aligned}$$

Write
$$\bar{\varphi} \equiv h * \gamma^{-1}(\psi)h$$

Then (4.1) holds for all g in $C(G,\mathbf{C})$. Since $h \in C(G,\mathbf{C})$ and $\gamma^{-1}(\psi)h \in L_2(G)$, it follows that $\bar{\varphi} \in C_\infty(G,\mathbf{C}) \cap L_2(G)$. Finally, $\bar{\varphi}$ is unique because 0 is the only element of $L_2(G)$ whose convolution with an arbitrary element of $L_2(G)$ vanishes.

For later use we estimate $\|\bar{\varphi}\|_\infty$ and $\|\bar{\varphi}\|_2$:

$$\|\bar{\varphi}\|_\infty \leq \|h\|_2\|\gamma^{-1}(\psi)h\|_2 \leq \|\gamma^{-1}(\psi)\| \, \|h\|_2^2 = \|\psi\|_\infty\|h\|_2^2 \leq C\|\varphi\|_\infty$$

and
$$\begin{aligned}
\|\bar{\varphi}\|_2 &\leq \|h\|_1\|\gamma^{-1}(\psi)h\|_2 \leq \|h\|_1\|h\|_2\|\gamma^{-1}(\psi)\| \\
&= \|h\|_1\|h\|_2\|\psi\|_\infty \leq C'\|\varphi\|_\infty
\end{aligned}$$

where C and C' are constants depending only on K^*.

If φ and ψ are in $C(G^*,\mathbf{C})$, then for all g in $L_2(G)$ we have

$$\overline{\varphi\psi} * g = \gamma^{-1}(\varphi\psi)g = \gamma^{-1}(\varphi)\gamma^{-1}(\psi)g = \bar{\varphi} * (\bar{\psi} * g) = (\bar{\varphi} * \bar{\psi}) * g$$

and thus

$$(4.2) \qquad\qquad \overline{\varphi\psi} = \bar{\varphi} * \bar{\psi}$$

Similarly, $\overline{\varphi^*} * g = \gamma^{-1}(\varphi^*)g = \gamma^{-1}(\varphi)^*g = \tilde{\bar{\varphi}} * g$, and thus

$$(4.3) \qquad\qquad \overline{\varphi^*} = \tilde{\bar{\varphi}}$$

The next two lemmas contain formulas which will be needed later.

Lemma 13 *For each φ in $C(G^*,\mathbf{C})$ and each x in G we have*

$$(4.4) \qquad\qquad \mathfrak{F}^*(\hat{x}\varphi) = T(x^{-1})\mathfrak{F}^*(\varphi)$$

Proof Choose a sequence $\{f_n\}$ of functions in $C(G,\mathbf{C})$ so that $\|\varphi - \hat{f}_n\|_\infty \to 0$ as $n \to \infty$. To do this it is enough to have $\|\gamma^{-1}(\varphi) - T(f_n)\| \to 0$ as $n \to \infty$, since γ is a norm-preserving isomorphism and

$$\gamma(\gamma^{-1}(\varphi) - T(f_n)) = \varphi - \hat{f}_n$$

This can be done because $\mathfrak{A}$ is dense in $\bar{\mathfrak{A}}$. Since $\|\varphi \hat{x}_n - f_n \hat{x}_n\|_\infty \to 0$ as $n \to \infty$, for each g in $L_2(G)$ we have

$$
\begin{aligned}
\mathfrak{F}^*(\hat{x}\varphi) * g &= \gamma^{-1}(\hat{x}\varphi)g = \lim_{n\to\infty} \gamma^{-1}(\hat{x}\hat{f}_n)g \\
&= \lim_{n\to\infty} \gamma^{-1}(\mathfrak{F}(T(x^{-1})f_n))g = \lim_{n\to\infty} T(T(x^{-1})f_n)g \\
&= \lim_{n\to\infty} (T(x^{-1})f_n) * g \\
&= T(x^{-1}) \lim_{n\to\infty} f_n * g = T(x^{-1}) \lim_{n\to\infty} \gamma^{-1}(\hat{f}_n)g \\
&= T(x^{-1})\gamma^{-1}(\varphi)g = T(x^{-1})(\bar{\varphi} * g) = (T(x^{-1})\bar{\varphi}) * g
\end{aligned}
$$

Therefore $\mathfrak{F}^*(\hat{x}\varphi) = T(x^{-1})\bar{\varphi}$.

Lemma 14 *For each α in G^* and each φ in $C(G^*,\mathbf{C})$ we have*

$$(4.5) \qquad\qquad \mathfrak{F}^*(T(\alpha)\varphi) = \alpha\mathfrak{F}^*(\varphi)$$

Proof For each f in $L_1(G)$ and each β in G^* we have

$$\mathfrak{F}(\alpha f)(\beta) = \int f(x)\alpha(x)\beta(x)\,d\mu(x) = \hat{f}(\alpha\beta)$$

or

$$(4.6) \qquad\qquad \mathfrak{F}(\alpha f) = T(\alpha)\mathfrak{F}(f)$$

For each positive integer n choose f_n in $L_1(G)$ with $\|\varphi - \hat{f}_n\|_\infty < n^{-1}$. Then for each g in $C(G,\mathbf{C})$ we have, by (4.6),

$$
\begin{aligned}
(4.7) \quad \mathfrak{F}^*(T(\alpha)\varphi) * g &= \gamma^{-1}(T(\alpha)\varphi)g = \lim_{n\to\infty} \gamma^{-1}(T(\alpha)\mathfrak{F}(f_n))g \\
&= \lim_{n\to\infty} \gamma^{-1}(\mathfrak{F}(\alpha f_n))g = \lim_{n\to\infty} (\alpha f_n) * g
\end{aligned}
$$

Now for each x in G we have

$$
\begin{aligned}
((\alpha f_n) * g)(x) &= \int \alpha(y)f_n(y)g(y^{-1}x)\,d\mu(y) \\
&= \alpha(x)\int f_n(y)g(y^{-1}x)\alpha(y^{-1}x)^*\,d\mu(y) \\
&= \alpha(x)(f_n * \alpha^*g)(x)
\end{aligned}
$$

or $(\alpha f_n) * g = \alpha(f_n * \alpha^*g)$. Similarly, $(\alpha\bar{\varphi}) * g = \alpha(\bar{\varphi} * \alpha^*g)$. Combined with (4.7), this gives

$$\mathfrak{F}^*(T(\alpha)\varphi) * g = \lim_{n \to \infty} \alpha(f_n * \alpha^* g) = \alpha(\bar{\varphi} * \alpha^* g) = (\alpha\bar{\varphi}) * g$$

This is equivalent to (4.5).

The inverse Fourier transform $\mathfrak{F}^*$ gives an explicit formula for Haar measure on G^*.

Proposition 7   *For each φ in $C(G^*,\mathbf{C})$ write*

$$(4.8) \qquad \int \varphi \, d\mu^* \equiv \bar{\varphi}(e)$$

Then μ^* is a Haar measure on G^*.

Proof   Clearly μ^* is a linear functional on $C(G^*,\mathbf{C})$. For each φ we have

$$(4.9) \qquad \int |\varphi|^2 \, d\mu^* = \int \varphi\varphi^* \, d\mu^* = (\bar{\varphi} * \tilde{\bar{\varphi}})(e) = \int |\bar{\varphi}|^2 \, d\mu \geq 0$$

Therefore μ^* is a positive measure on G^*. For each α in G^* we have, by (4.5),

$$\int T(\alpha)\varphi \, d\mu^* = \mathfrak{F}^*(T(\alpha)\varphi)(e) = (\alpha\bar{\varphi})(e) = \bar{\varphi}(e) = \int \varphi \, d\mu^*$$

Thus μ^* is a Haar measure on G^*.

The formula (4.8) for μ^* leads to a formula for the inverse Fourier transform, analogous to the formula (3.7) for the Fourier transform:

$$\bar{\varphi}(x) = (T(x)\bar{\varphi})(e) = \mathfrak{F}^*(\hat{x}^*\varphi)(e) = \int \hat{x}^*\varphi \, d\mu^* \qquad (\varphi \in C(G^*,\mathbf{C}))$$

We extend the map $\varphi \to \bar{\varphi}$ to all of the set $L_1(G^*)$ by defining

$$\bar{\varphi}(x) \equiv \int \hat{x}^*\varphi \, d\mu^* \qquad (\varphi \in L_1(G^*),\ x \in G)$$

Then $\bar{\varphi} \in C_\infty(G,\mathbf{C})$ for all φ in $L_1(G^*)$, by continuity (since $\bar{\varphi} \in C_\infty(G,\mathbf{C})$ whenever $\varphi \in C(G^*,\mathbf{C})$). The proof, already given, that

$$\mathfrak{F}(f * g) = \mathfrak{F}(f)\mathfrak{F}(g)$$

shows that $\mathfrak{F}^*(\varphi * \psi) = \mathfrak{F}^*(\varphi)\mathfrak{F}^*(\psi)$ for all φ and ψ in $L_1(G^*)$.
By (4.9),

$$\|\varphi\|_2 = \|\bar{\varphi}\|_2$$

for all φ in $C(G^*,\mathbf{C})$. By approximation, this equality extends to all φ in $L_1(G^*) \cap L_2(G^*)$. Thus $\mathfrak{F}^*$ is linear on $L_1(G^*) \cap L_2(G^*)$, and preserves the norm $\| \ \|_2$. It therefore extends to a norm-preserving linear map, which we also call $\mathfrak{F}^*$, from $L_2(G^*)$ into $L_2(G)$. Thus $\mathfrak{F}^*$ is defined on $L_1(G^*)$ *and* on $L_2(G^*)$, and on the intersection the definitions agree.

Theorem 5 $\mathfrak{F}^*$ *maps* $L_2(G^*)$ *onto* $L_2(G)$.

Proof   It is enough to show that the image of $L_2(G^*)$ is dense in $L_2(G)$. To prepare the way, consider φ in $C(G^*,\mathbf{C})$ and f in $C(G,\mathbf{C})$. Then for each g in $L_2(G)$ we get

$$\overline{\varphi \hat{f}} * g = \gamma^{-1}(\varphi \hat{f})g = \gamma^{-1}(\varphi)f * g = (\bar{\varphi} * f) * g$$

or $\overline{\varphi \hat{f}} = \bar{\varphi} * f$.

Now consider any h in $C(G,\mathbf{C})$, and any $\epsilon > 0$. Choose f in $C(G,\mathbf{C})$ with $\|f * h - h\|_2 \le \epsilon$. Since $\hat{h} \in C_\infty(G^*,\mathbf{C})$, there exists φ in $C(G^*,\mathbf{C})$ with $\|\hat{h} - \varphi\|_\infty \le \epsilon \|f\|_2^{-1}$. Then

$$\begin{aligned}
\|h - \overline{\varphi \hat{f}}\|_2 &= \|h - f * \bar{\varphi}\|_2 \le \|h - f * h\|_2 + \|f * (h - \bar{\varphi})\|_2 \\
&\le \epsilon + \|\gamma^{-1}(\hat{h} - \varphi)f\|_2 \\
&\le \epsilon + \|\hat{h} - \varphi\|_\infty \|f\|_2 \le 2\epsilon
\end{aligned}$$

Since ϵ is arbitrary and $\varphi \hat{f} \in C(G^*,\mathbf{C})$, we see that h is in the closure of $\mathfrak{F}^*(L_2(G^*))$. Since $C(G,\mathbf{C})$ is dense in $L_2(G)$, it follows that $\mathfrak{F}^*(L_2(G^*))$ is dense in $L_2(G)$.

Our next goal is to prove the Fourier inversion theorem, which states that if $f \in L_1(G) \cap L_2(G)$, then $\hat{f} \in L_2(G^*)$ and $\bar{\hat{f}} = f$. To do this we need some means of constructing functions f in $L_1(G) \cap L_2(G)$ with $\hat{f} \in L_2(G^*)$ and $\bar{\hat{f}} = f$. This is given by the following lemmas.

Lemma 15   *If* $\varphi \in C(G^*,\mathbf{C})$ *and* $\bar{\varphi} \in L_1(G)$, *then* $\hat{\bar{\varphi}} = \varphi$.

Proof   We have

$$\gamma^{-1}(\hat{\bar{\varphi}}) = T(\bar{\varphi}) = \gamma^{-1}(\varphi)$$

so $\hat{\bar{\varphi}} = \varphi$.

Lemma 16   *If* $f \in L_1(G)$ *and* φ *is an element of* $C(G^*,\mathbf{C})$ *with* $\bar{\varphi} \in L_1(G)$ *then* $h \equiv f * \bar{\varphi} \in L_1(G) \cap L_2(G)$, $\hat{h} = \hat{f}\varphi \in L_2(G^*)$, *and* $\bar{\hat{h}} = h$.

Proof   Since $\bar{\varphi} \in L_1(G) \cap L_2(G)$, it follows that $h \in L_1(G) \cap L_2(G)$. Also,

$$\gamma^{-1}(\hat{h}) = T(f * \bar{\varphi}) = T(f)T(\bar{\varphi}) = \gamma^{-1}(\hat{f})\gamma^{-1}(\varphi) = \gamma^{-1}(\hat{f}\varphi)$$

so $\hat{h} = \hat{f}\varphi$, and $\hat{h} \in C(G^*,\mathbf{C}) \subset L_2(G^*)$. Moreover,

$$T(\bar{\hat{h}}) = \gamma^{-1}(\hat{h}) = T(h)$$

or $\bar{\hat{h}} = h$.

Lemma 17   *If S is dense in $L_2(G)$, then*

$$S^2 \equiv \{f^2 : f \in S\}$$

is dense in $L_1(G)$.

Proof   For each g in $L_1(G)$ and each $\epsilon > 0$ there exists h in $L_2(G)$ with $\|g - h^2\|_1 \leq \epsilon$. Choose f in S with $\|f - h\|_2 \leq \epsilon$. Then

$$\begin{aligned}
\|g - f^2\|_1 &\leq \|g - h^2\|_1 + \|f^2 - h^2\|_1 \\
&\leq \epsilon + (\|f\|_2 + \|h\|_2)(\|f - h\|_2) \\
&\leq \epsilon + (2\|h\|_2 + \epsilon)\epsilon
\end{aligned}$$

Since ϵ is arbitrary, g is in the closure of S^2.

Lemma 18   *The set*

$$S = \{f * \bar{\varphi} : f \in L_1(G),\ \varphi \in C(G^*,\mathbf{C}),\ \bar{\varphi} \in L_1(G)\}$$

is dense in $L_1(G) \cap L_2(G)$ [in the sense that for each g in $L_1(G) \cap L_2(G)$ and each $\epsilon > 0$ there exists h in S with $\|g - h\|_1 \leq \epsilon$ and $\|g - h\|_2 \leq \epsilon$].

Proof   Clearly $\{\bar{\psi} : \psi \in C(G^*,\mathbf{C})\}$ is dense in $L_2(G)$. By Lemma 17, we see that $\{\overline{\psi * \psi} : \psi \in C(G^*,\mathbf{C})\}$ is dense in $L_1(G)$. Moreover, for each f in $L_1(G) \cap L_2(G)$ there exists g in $L_1(G)$ such that $\|f - f * g\|_1$ and $\|f - f * g\|_2$ are arbitrarily small. There also exists $\varphi \equiv \psi * \psi$ in $C(G^*,\mathbf{C})$ such that $\bar{\varphi}$ is arbitrarily near to g in $L_1(G)$. Hence $\|f - f * \bar{\varphi}\|_1$ and $\|f - f * \bar{\varphi}\|_2$ are arbitrarily small.

We now prove the inversion theorem.

Theorem 6   *For each h in $L_1(G) \cap L_2(G)$ we have $\hat{h} \in L_2(G^*)$ and $\bar{\hat{h}} = h$.*

Proof   First consider the case where $h = f * \bar{\varphi}$ belongs to the dense set S of Lemma 18. By Lemma 16, we see that $\hat{h} \in L_2(G^*)$ and $\bar{\hat{h}} = h$.

Next consider an arbitrary h in $L_1(G) \cap L_2(G)$. For each positive integer n there exists h_n in S with $\|h - h_n\|_1 < n^{-1}$ and $\|h - h_n\|_2 < n^{-1}$, by Lemma 18. Thus $\{\hat{h}_n\}$ converges pointwise to $\hat{h}$. Since

$$\|\hat{h}_n - \hat{h}_m\|_2 = \|\bar{\hat{h}}_n - \bar{\hat{h}}_m\|_2 = \|h_n - h_m\|_2$$

for all m and n, we see that $\{\hat{h}_n\}$ is a Cauchy sequence in $L_2(G^*)$, which therefore converges to $\hat{h}$ in $L_2(G^*)$. Moreover,

$$\bar{\hat{h}} = \lim_{n \to \infty} \bar{\hat{h}}_n = \lim_{n \to \infty} h_n = h$$

in the metric of $L_2(G)$.

Our final task is to prove the famous Pontryagin duality theorem, which states that the natural map $x \to \hat{x}$ of G onto G^{**} is both a homeomorphism and an isomorphism. To do this we must first prove two lemmas, analogs of Lemmas 11 and 12, that characterize the metric ρ on G in terms of the behavior of the elements x of G on certain subsets of G^*. The first is a criterion for boundedness.

Lemma 19 *For each $\epsilon > 0$ and each $K^* \subset G^*$ we define*

$$N(K^*,\epsilon) \equiv \{x \in G: |\alpha(x) - 1| \le \epsilon \text{ for all } \alpha \text{ in } K^*\}$$

Then $N(K^*,\epsilon)$ is bounded, for each real number ϵ, with $0 < \epsilon < 1$, and each neighborhood K^* of the identity element 1 of G^*. Conversely, if M is any bounded subset of G and ϵ any positive constant, there exists a neighborhood K^* of 1 in G^* such that $M \subset N(K^*,\epsilon)$.

Proof Consider a real number ϵ, with $0 < \epsilon < 1$, and a neighborhood K^* of 1 in G^*. Let K_1 be any compact neighborhood of e in G. By Lemma 11, $N^*(K_1,\frac{1}{2})$ is a bounded subset of G^*. By Lemma 12, there exists a compact set $K \subset G$ and $\delta > 0$ such that $\alpha \in K^*$ whenever $\alpha \in N^*(K_1,\frac{1}{2})$ and $\alpha \in N^*(K,\delta)$. We may assume that $\delta \le \frac{1}{2}$ and $K_1 \subset K$. Then $N^*(K,\delta) \subset K^*$.

By Lemma 8, there exist compact integrable neighborhoods V and W of e in G such that $KV \subset W$ and $\mu(W - V) \le (\delta/4)\mu(W)$. Write $f \equiv \chi_W$. Consider y in K. Then $T(y)f = \chi_{y^{-1}W}$. Since $V \subset y^{-1}W$, this gives

$$\|f - T(y)f\|_1 = \mu(W - y^{-1}W) + \mu(y^{-1}W - W)$$
$$= 2\mu(W - y^{-1}W) \le 2\mu(W - V) \le \frac{\delta}{2}\mu(W) = \frac{\delta}{2}\|f\|_1$$

for all y in K. Since $\|\hat{f}\|_\infty \leq \|f\|_1$ and $\hat{f}(1) = \int f\, d\mu = \|f\|_1$, we have $\|T(f)\| = \|\hat{f}\|_\infty = \|f\|_1$.

Take g in $C(G)$ with $\|g\|_2 = 1$ and $\|f * g\|_2 > 2^{-1/2}\|T(f)\|$. Let L_1 be any compact support for g. Let L be any compact set containing $WW^{-1}L_1L_1^{-1}$. We shall show that $N(K^*,\epsilon) \subset L$. To this end, consider any x in $N(K^*,\epsilon)$. Assume that $x \in -L$. If y is any point of G with $(f * g)(y) \neq 0$, then $y \in WL_1$. If also $(f * g)(xy) \neq 0$, then $xy \in WL_1$, and thus $x \in WW^{-1}L_1L_1^{-1} \subset L$. This contradiction shows that if $(f * g)(y) \neq 0$, then $(f * g)(xy) = 0$. Therefore $f * g$ and $T(x)f * g$ are orthogonal elements of $L_2(G)$. Hence

$$\|T(f - T(x)f)\|^2 \geq \|f * g - T(x)f * g\|_2^2$$
$$= 2\|f * g\|_2^2 > 2(2^{-1/2}\|T(f)\|)^2 = \|T(f)\|^2$$

There therefore exists α in G^* with

$$(4.10) \qquad |\alpha(f) - \alpha(x)^*\alpha(f)| = |\alpha(f - T(x)f)| > \|T(f)\|$$

and thus

$$(4.11) \qquad |1 - \alpha(x)| = |1 - \alpha(x)^*| > |\alpha(f)^{-1}|\,\|T(f)\| \geq 1$$

By (4.10),

$$|\alpha(f)| \geq \tfrac{1}{2}|\alpha(f) - \alpha(x)^*\alpha(f)| > \tfrac{1}{2}\|T(f)\|$$

Hence, for all y in K,

$$\frac{\delta}{2}\,\|T(f)\| \geq \|f - T(y)f\|_1 \geq |\alpha(f - T(y)f)| = |\alpha(f)|\,|1 - \alpha(y)|$$
$$\geq \tfrac{1}{2}\|T(f)\|\,|1 - \alpha(y)|$$

and thus $|1 - \alpha(y)| \leq \delta$. It follows that $\alpha \in N^*(K,\delta) \subset K^*$. Therefore (4.11) contradicts the fact that $x \in N(K^*,\epsilon)$. This contradiction arises from the assumption $x \in -L$. Therefore $x \in L$. Hence $N(K^*,\epsilon) \subset L$, and thus $N(K^*,\epsilon)$ is bounded.

Consider, conversely, any bounded subset M of G and any $\epsilon > 0$. Let K be any compact subset of G, with $M \subset K$, and M^* any compact neighborhood of 1 in G^*. By Lemma 12, there exists $\delta > 0$ such that if $\alpha \in M^*$ and $\rho^*(\alpha,1) \leq \delta$, then $\|\alpha - 1\|_K \leq \epsilon$. Thus $K^* \equiv \{\alpha \in M^*: \rho^*(\alpha,1) \leq \delta\}$ is a neighborhood of 1 in G^*, and $M \subset N(K^*,\epsilon)$.

Our next lemma characterizes the metric ρ on bounded subsets of G.

Lemma 20 *Let M be any bounded subset of G. For each $\epsilon > 0$ there exists $\delta > 0$ and a compact set $K^* \subset G^*$ such that $\rho(x,y) \leq \epsilon$ whenever x, $y \in M$ and $\|\hat{x} - \hat{y}\|_{K*} \leq \delta$. Conversely, for each $\delta > 0$ and each compact set $K^* \subset G^*$ there exists $\epsilon > 0$ such that $\|\hat{x} - \hat{y}\|_{K*} \leq \delta$ whenever x, $y \in M$ and $\rho(x,y) \leq \epsilon$.*

Proof Consider $\epsilon > 0$. By Proposition 1, there exists a compact neighborhood L of e in G such that $\rho(x,y) \leq \epsilon$ whenever x, $y \in M$ and $xy^{-1} \in L$. Let L_0 be a compact neighborhood of e in G with $L_0{}^2 L_0{}^{-2} \subset L$. By the proof of Theorem 2, there exist test functions f and g on G, supported by L_0, such that $\|g\|_2 = 1$ and $\|f * g\|_2 > 2^{-1/2}\|T(f)\|$. Choose a neighborhood K of e in G so that $\|f - T(y)f\|_1 \leq \frac{1}{4}\|T(f)\|$ for all y in K. Write $\delta \equiv \frac{1}{2}$. By the proof of Lemma 19, for each x in $-L$ there exists α in G^* with $|1 - \alpha(x)| > 1$ and $|1 - \alpha(y)| \leq \frac{1}{2}$ for all y in K. Hence $\alpha \in K^*$, where K^* is any compact set containing the bounded subset $N^*(K,\frac{1}{2})$ of G^*. It follows that $N(K^*,\delta) \subset L$.

Consider elements x and y of M, with $\|\hat{x} - \hat{y}\|_{K*} \leq \delta$. Then $\|\hat{x}\hat{y}^{-1} - 1\|_{K*} \leq \delta$, or $xy^{-1} \in N(K^*,\delta)$. Hence $xy^{-1} \in L$. By the choice of L, we have $\rho(x,y) \leq \epsilon$.

Conversely, consider $\delta > 0$ and a compact set $K^* \subset G^*$. By Lemma 11, there exists a neighborhood K of e in G such that $K^* \subset N^*(K,\delta)$. Choose $\epsilon > 0$ so small that $xy^{-1} \in K$ whenever x, $y \in M$ and $\rho(x,y) \leq \epsilon$. Then for all x and y in M with $\rho(x,y) \leq \epsilon$ and all α in K^* we have

$$|\alpha(x) - \alpha(y)| = |\alpha(xy^{-1}) - 1| \leq \delta$$

Hence $\|\hat{x} - \hat{y}\|_{K*} \leq \delta$.

We can now prove the Pontryagin duality theorem.

Theorem 7 *The mapping $x \to \hat{x}$, which we call i, from G to G^{**} is both a homeomorphism and an isomorphism of G onto G^{**}.*

Proof For each x in G, we see that $\hat{x}$ is a uniformly continuous function on bounded subsets M^* of G^*, by the second part of Lemma 12. Therefore $\hat{x} \in G^{**}$.

Comparison of Lemmas 11 and 19 shows that a set $M \subset G$ is bounded if and only if its image $i(M) \subset G^{**}$ is bounded. Comparison of Lemmas 12 and 20 shows that on each bounded set $M \subset G$ the metric ρ of G is equivalent to the metric $(x,y) \to \rho^{**}(\hat{x},\hat{y})$ induced on M by the metric ρ^{**} of G^{**}. Hence the map i is a homeomorphism of G onto the locally compact subset $i(G)$ of G^{**}. To show that i is a homeomorphism of G with G^{**}, we must show that $i(G) = G^{**}$, or that $\rho^{**}(u,i(G)) = 0$ for all u in G^{**}. Assume that $\rho^{**}(u, i(G)) > 0$. Then there exists $h \geq 0$ in $C(G^{**})$, with $h(u) > 0$, and a compact neighborhood K^{**} of $\hat{e}$ in G^{**} such that $xy \in -i(G)$ whenever $h(x) > 0$ and $y \in K^{**}$. Let g be any nonnegative test function on G^{**}, with $g(\hat{e}) > 0$, that is supported by K^{**}. Then $(g * h)(u) > 0$ and $g * h$ vanishes on $i(G)$. Let φ be the inverse Fourier transform of $g * h$, so that $\varphi = \bar{g}\bar{h} \in L_1(G^*) \cap L_2(G^*)$. Also, $\|\varphi\|_2 = \|g * h\|_2 > 0$. For all x in G we have

$$\bar{\varphi}(x) = \int \hat{x}^* \varphi \, d\mu^* = \hat{\varphi}(\hat{x}^*) = (g * h)(\hat{x}^*) = 0$$

since $\hat{x}^* \in i(G)$. Hence $0 = \|\bar{\varphi}\|_2 = \|\varphi\|_2 > 0$. This contradiction proves that $\rho^{**}(u, i(G)) = 0$.

It remains to show that i is an isomorphism. Since i is bijective, it is enough to show that it is a homomorphism. This follows from the identity $\widehat{xy} = \hat{x}\hat{y}$.

PROBLEMS

1. Let G and H be locally compact groups. Show that $G \times H$ is a locally compact group, and construct Haar measure on $G \times H$ in terms of the Haar measures on G and H.

2. Show that any left-invariant measure on a locally compact group G is a multiple of Haar measure (not necessarily positive).

3. Let A be an integrable set of positive measure in a locally compact group G. Show that AA^{-1} contains a neighborhood of e.

4. Exhibit Haar measure on the locally compact group of all 2×2 real matrices $\begin{pmatrix} a & 0 \\ b & c \end{pmatrix}$, with $ac \neq 0$. Show that this group is not unimodular.

5. Prove that Proposition 3, for the case $p = 1$, holds for an arbitrary G (not necessarily unimodular).

6. Does Lemma 8 hold for an arbitrary (not necessarily commutative) locally compact group?

7. Show that $\mathbf{R}$ can be identified with its own character group and $\{z : |z| = 1\}$ with the character group of Z.

8. Find the character group of the additive group D of all rational numbers x such that $2^n x$ is an integer if n is sufficiently large (D has the discrete metric ρ defined by $\rho(x, y) \equiv 1$ if $x \neq y$).

9. Let μ and ν be finite measures on G (by which we mean that $|\int f \, d\mu| + |\int f \, d\nu|$ is bounded as f runs over the set of test functions with $\|f\| \leq 1$). Define the *convolution* $\mu * \nu$ of μ and ν by

$$\int f \, d(\mu * \nu) = \int\int f(xy) \, d\mu(x) \, d\nu(y)$$

Show that $\mu * \nu$ is a finite measure. Show that convolution of finite measures is associative, and commutative if G is commutative.

10. Prove that there exists an identity element for convolution in $L_1(G)$ if and only if G is discrete (which means that for each x in G the set $\{x\}$ is open).

11. Prove that G^* is compact if and only if G is discrete.

12. A continuous function $\varphi\colon G \to \mathbf{C}$ is *positive definite* if

$$\int\int \varphi(y^{-1}x) f(x) f(y)^* \, d\mu(x) \, d\mu(y) \geq 0$$

for all f in $C(G,\mathbf{C})$. If h is any element of $C(G,\mathbf{C})$, show that $h * \bar{h}$ is positive definite.

13. Let φ be any positive definite function on G. Prove *Bochner's theorem*, that there exists a positive finite measure ν on G^* such that $\varphi(x) = \int \alpha(x) \, d\nu(\alpha)$ for all x in G.

14. A continuous real-valued function f on G is *almost periodic* if for each $\epsilon > 0$ there exists a compact subset K of G such that for each x in G there exists y in K with $|f(xz) - f(yz)| \leq \epsilon$ for all z in G. Show that an almost-periodic function f is uniformly continuous on all of G, in the sense that for each $\epsilon > 0$ there exists $\delta > 0$ such that $|f(x) - f(y)| < \epsilon$ whenever $\rho(y^{-1}x) < \delta$.

15. Show that f is almost periodic if and only if for each $\epsilon > 0$ there exist points $x_1, \ldots, x_n$ in G such that for each x in G there exists j $(1 \leq j \leq n)$ with $|f(xz) - f(x_j z)| \leq \epsilon$ for all z in G.

NOTES

The existence of $(f\colon \varphi)$ is trivial classically.

The proof of Lemma 8 is due to Bernard Kripke.

Theorem 2 is a key fact, but it has no classical content.

It would be possible to use Lemmas 11 and 12 to construct directly a metric on G^* homeomorphic to the metric ρ^* induced by the identification G^* with Σ.

CHAPTER 11

COMMUTATIVE BANACH ALGEBRAS

The theory of commutative Banach algebras is the only instance in this book of a classical theory whose constructive version seems forced and unnatural. For this reason, we do not carry the development far. A substitute is found for the classical result that every ideal is contained in a maximal ideal. As corollaries, we get substitutes for other classical results, in particular the compactness of the spectrum and the standard expression for the spectral norm. It is indicated in the exercises that, as far as the theory is carried, it has the same applications to analysis as the corresponding classical results.

Many function spaces form an algebra: the elements can be multiplied, as well as added. To regard such a function space as a Banach space is to ignore some of its structure. To recover this structure in an abstract setting, we introduce the notion of a Banach algebra.

Definition 1 A (complex) Banach algebra $\mathfrak{A}$ is an algebra over the complex numbers, with a multiplicative unit e, which is endowed with a norm $\|\ \ \|$ satisfying the following conditions:

  (a) $\|e\| = 1$
  (b) $\mathfrak{A}$ is a Banach space relative to the norm $\|\ \ \|$
  (c) $\|xy\| \leq \|x\|\ \|y\|$ for all x and y in $\mathfrak{A}$

Examples come readily to mind. The simplest examples, of course, are the function algebras $C(X,\mathbf{C})$, for an arbitrary compact space X, endowed with the usual norm. More generally, any closed subalgebra of $C(X,\mathbf{C})$ that contains the constant functions is a Banach algebra. Various important algebras of analytic functions arise in this way.

Another important example of a Banach algebra comes from the group algebra $L_1(G)$ of a locally compact abelian group G, endowed with the L_1 norm $\|\ \ \|_1$. Now, $L_1(G)$ itself may not be a Banach algebra, because $L_1(G)$ may not have an identity element for convolution. However (by a simple construction which we shall not present), an identity element can be ajoined to $L_1(G)$, and the resulting algebra is a Banach algebra.

Another example is given by an arbitrary algebra of normable linear operators from a complex Banach space B into itself that contains the identity operator and is complete and separable with respect to the norm.

For group algebras of locally compact abelian groups and for self-adjoint commutative operator algebras, the spectrum Σ has already played a vital role. Here is the general definition of this central concept.

Definition 2 The *spectrum* Σ of a commutative Banach algebra $\mathfrak{A}$ consists of all u in the dual space $\mathfrak{A}^*$ of $\mathfrak{A}$ such that $u(e) = 1$ and $u(xy) = u(x)u(y)$ for all x and y in $\mathfrak{A}$.

In other words, Σ consists of all bounded homomorphisms of $\mathfrak{A}$ onto $\mathbf{C}$.

Consider any u in Σ, and let c be a bound for u. For each x in $\mathfrak{A}$ and n in Z^+, we have

$$|u(x)| = |u(x^n)|^{1/n} \leq (c\|x^n\|)^{1/n} \leq c^{1/n}\|x\| \to \|x\| \qquad \text{as } n \to \infty$$

Thus 1 is a bound for u. Since $|u(e)| = 1 = \|e\|$, it follows that u is in fact normable and that $\|u\| = 1$.

Our study of locally compact abelian groups and selfadjoint commutative operator algebras makes it plausible that Σ is compact. Unfortunately this is *not* true in general. The following proposition is the most we can say.

Proposition 1 *The spectrum Σ of a commutative Banach algebra $\mathfrak{A}$ is the intersection of a decreasing sequence $\{\Sigma_n\}$ of subsets of the unit sphere S^* of $\mathfrak{A}^*$, each of which is either compact or void.*

Proof Since $\|u\| = 1$ for all u in Σ, we have $\Sigma \subset S^*$. Let $\{x_n{}^0\}$ be a dense sequence of elements of $\mathfrak{A}$. Choose a sequence $c_1 > c_2 > \cdots$ of positive constants converging to 0 so that each of the sets

$$\Sigma_n \equiv \{u \in S^*: |u(x_i{}^0 x_j{}^0) - u(x_i{}^0)u(x_j{}^0)| \leq c_n$$
$$\text{for } 1 \leq i, j \leq n, |1 - u(e)| \leq c_n\}$$

is compact or void. We can do this because for each x in $\mathfrak{A}$ the map $u \to u(x)$ is continuous on S^*. Then $\{\Sigma_n\}$ is a decreasing sequence of subsets of S^*. Clearly $\Sigma \subset \bigcap_{n=1}^{\infty} \Sigma_n$. Conversely, if $u \in \bigcap_{n=1}^{\infty} \Sigma_n$, we have $u(x_i{}^0 x_j{}^0) = u(x_i{}^0)u(x_j{}^0)$ for all i and j. Since $\{x_i{}^0\}$ is dense in $\mathfrak{A}$, it follows that u is multiplicative. Therefore $u \in \Sigma$.

We shall see later that each Σ_n is actually compact. Here is an example of a Banach algebra $\mathfrak{A}$ for which Σ is *not* compact. Fix a sequence $r_0 \geq r_1 \geq \cdots$ of integers, so that for each n either $r_n = 0$ or $r_n = 1$, for which we are unable to rule out either the possibility that $r_n = 1$ for all n or the possibility that $r_n = 0$ for some n. Let $\mathfrak{A}$ consist of all sequences $x \equiv \{x_n\}_{n=0}^{\infty}$ of complex numbers for which

$$\|x\| \equiv \sum_{n=0}^{\infty} r_n |x_n|$$

exists. Define the elements x and $y \equiv \{y_n\}_{n=0}^{\infty}$ of $\mathfrak{A}$ to be *equal* if $\|x - y\| = 0$. Then $\mathfrak{A}$ is a Banach space with norm $\| \ \|$. Moreover, if we define the product by

$$xy \equiv \Big\{ \sum_{k=0}^{n} x_k y_{n-k} \Big\}_{n=0}^{\infty}$$

$\mathfrak{A}$ is a Banach algebra. In case $r_n = 1$ for all n, every complex number z with $|z| \leq 1$ defines an element u_z of Σ, as follows:

$$u_z(x) = \sum_{n=0}^{\infty} x_n z^n$$

On the other hand, if there exists an n with $r_n = 0$, then Σ consists of a single element u_0, given by

$$u_0(x) \equiv x_0$$

Since neither of these possibilities is ruled out, Σ is *not* totally bounded, and so *not* compact.

The Hahn-Banach theorem is the fundamental tool for constructing elements of the dual B^* of a Banach space B. The corresponding result in the classical theory of Banach algebras states that every proper closed ideal in a Banach algebra $\mathfrak{A}$ is contained in a maximal ideal. This result is the fundamental tool for constructing maximal ideals in $\mathfrak{A}$, and therefore for constructing elements of the spectrum Σ of $\mathfrak{A}$. Unfortunately this result on the existence of maximal ideals is *not* constructively valid. We must therefore look for a good constructive substitute. Before giving such a substitute, we prove a simple lemma about analytic functions.

Lemma 1 *Let α, β, and R be real numbers, with $0 < \alpha < \beta$ and $R > 0$. Let $\zeta_1, \ldots, \zeta_m$ be an α approximation to the compact subset $\{z : |z| \le R\}$ of $\mathbf{C}$. For each j $(1 \le j \le m)$ let h_j be an analytic function on $U_j \equiv \{z : |z - \zeta_j| < \beta\}$, and let h_{m+1} be an analytic function on $U_{m+1} \equiv \{z : |z| > R\}$, such that $h_{m+1}(z) \to 0$ as $|z| \to \infty$. Then there exists $t > 0$, depending only on R, α, β, $\zeta_1, \ldots, \zeta_m$ and not on the functions $h_1, \ldots, h_{m+1}$, such that $|h_1(\zeta_1)| \le \frac{1}{2}$ whenever $|h_i(z) - h_j(z)| \le t$ for all i and j and all z in $U_i \cap U_j$.*

Proof Choose closed polygonal paths $\gamma_1, \ldots, \gamma_N, \gamma_{N+1}$ having the following properties. First, $O(\gamma_1, \zeta_1) = 1$ and $O(\gamma_i, \zeta_1) = 0$ for $2 \le i \le N$. Second, γ_1 lies in U_1, γ_{N+1} lies in U_{m+1}, and each of the paths γ_i (for $2 \le i \le N$) lies in one of the sets U_j, where j depends on i. Third, each side δ of each of the paths γ_i has the property that $-\delta$ is a side of exactly one of the remaining paths. These conditions imply that

$$h_1(\zeta_1) = (2\pi i)^{-1} \int_{\gamma_1} h_1(z)(z - \zeta_1)^{-1}\, dz$$

and
$$0 = \int_{\gamma_i} h_i(z)(z - \zeta_1)^{-1}\, dz \qquad (2 \le i \le N + 1)$$

Hence

$$(1.1) \qquad |h_1(\zeta_1)| \le (2\pi)^{-1} \left| \sum_{i=1}^{N+1} \int_{\gamma_i} h_i(z)(z - \zeta_1)^{-1}\, dz \right|$$

Let $\delta_1, \ldots, \delta_M$ be linear paths such that each side δ of each of the paths $\gamma_1, \ldots, \gamma_{N+1}$ is equal to exactly one of the paths $\delta_1, \ldots, \delta_M, -\delta_1, \ldots, -\delta_M$. For each k $(1 \le k \le M)$ let i_1 and i_2 $(1 \le i_1, i_2 \le N+1)$ be the unique values of i for which δ_k occurs as a side of γ_{i_1} and $-\delta_k$ occurs as a side of γ_{i_2} (so that i_1 and i_2 are functions of k). Denote the values of j corresponding to i_1 and i_2 by j_1 and j_2, respectively (so that j_1 and j_2 are functions of k). Then (1.1) gives

$$|h_1(\zeta_1)| \le (2\pi)^{-1} \left| \sum_{k=1}^{M} \int_{\delta_k} (h_{j_1}(z) - h_{j_2}(z))(z - \zeta_1)^{-1}\, dz \right|$$

$$< tc^{-1} \sum_{k=1}^{M} l(\delta_k)$$

where c is the minimum distance from ζ_1 to the paths δ_k. Thus to make $|h_1(\zeta_1)| \le \frac{1}{2}$ we need only set

$$t \equiv c \left(2 \sum_{k=1}^{M} l(\delta_k) \right)^{-1}$$

We shall need to consider differentiable functions with values in a normed linear space B. Let K be a compact subset of $\mathbf{C}$, and let the functions f and g from K to B be uniformly continuous. We say that f is *differentiable* on K, with $f' = g$, if for each $\epsilon > 0$ there exists $\delta > 0$ such that

$$\|f(z_2) - f(z_1) - (z_2 - z_1)g(z_1)\| \le \epsilon|z_2 - z_1|$$

whenever $z_1, z_2 \in K$ and $|z_2 - z_1| \le \delta$. We say that f is differentiable on an open set U, with $f' = g$ on U, if $f' = g$ on every compact set $K \subset\subset U$.

The theory of differentiable functions with values in B is much the same as in the scalar case (that is, the case $B = \mathbf{C}$). In particular, we shall need the following facts, whose proofs are left to the reader.

(a) If f is differentiable on an open sphere $S(z_0, r_0)$, then f has a unique expansion

$$f(z) = \sum_{n=0}^{\infty} (z - z_0)^n a_n$$

where the coefficients a_n belong to B, and the series converges uniformly to f on each sphere $S(z_0, r)$ $(r < r_0)$. Moreover, $|a_n| \le (2\pi)^{-1}\|f\|_\gamma r^{-n}$, where γ is the bounding circle of $S(z_0, r)$.

(*b*) If f is differentiable on K (respectively, U), then for each u in B^* the function $u \circ f$ is differentiable on K (respectively, U), with derivative $u \circ f'$.

If, in addition, B is a commutative Banach algebra, the following statements also hold.

(*c*) The product fg of differentiable functions f and g is differentiable, with $(fg)' = f'g + fg'$.

(*d*) If the inverse f^{-1} of a differentiable function f exists and is bounded, then f^{-1} is differentiable and $(f^{-1})' = -f^{-2}f'$.

To develop constructive substitutes for the classical constructions involving ideals, instead of considering the ideal $\{x_1 y_1 + \cdots + x_n y_n : y_1, \ldots, y_n \in \mathfrak{A}\}$ generated by elements $x_1, \ldots, x_n$ of a commutative Banach algebra $\mathfrak{A}$, we consider a *partial ideal*

$$P(x_1, \ldots, x_n; A) \equiv \{x_1 y_1 + \cdots + x_n y_n : y_1, \ldots, y_n \in A\}$$

generated by $x_1, \ldots, x_n$. The set A is required to be a core; a *core* is a totally bounded subset of $\mathfrak{A}$. The set A is called *the core* of $P(x_1, \ldots, x_n; A)$. Each partial ideal P is totally bounded; hence $\rho(e, P)$ exists, for each partial ideal P. The following lemma is the main tool in the constructive theory of commutative Banach algebras. It gives a constructive substitute for the classical result that if I is any proper ideal and x is any element of $\mathfrak{A}$, then there exist a proper ideal J containing I and a complex number z such that $x - ze \in J$.

Lemma 2 *Let C, α, and β be positive constants, with $\alpha < 1$ and $\beta < 1$, and let n be a positive integer. Then there exists a positive constant D, depending only on C, α, β, and n, having the following property. Let $x_1, \ldots, x_n$, and x be any elements of $\mathfrak{A}$, of norms at most C, and A any core, whose norm $\sup\{\|y\| : y \in A\}$ is at most C. Then there exists a core B, of norm at most D, such that if*

$$(1.2) \qquad \rho(e, P(x_1, \ldots, x_n; B)) > \beta$$

then there exists a complex number z, with $|z| \leq \|x\|$, such that

$$(1.3) \qquad \rho(e, P(x_1, \ldots, x_n, x - ze; A)) > \alpha$$

Proof Let C, α, β, and n be given. Let $x_1, \ldots, x_n$, and x be any elements of $\mathfrak{A}$, of norms at most C, and A any core, of norm at most C. To simplify notation, we write $P(U) \equiv P(x_1, \ldots, x_n; U)$ for each core U.

For each z in $\{z: |z| \leq \|x\|\}$, let $\rho(z)$ be the left side of (1.3). Then ρ is a continuous function. Therefore the supremum σ of ρ exists. Either $\sigma > \alpha$ or $\sigma < (1 + \alpha)/2$. In the former case we may take $B \equiv \{0\}$. Assume therefore that $\sigma < (1 + \alpha)/2$. For each z in $\mathbf{C}$ with $|z| \leq \|x\|$ there exist vectors $\mu(z)$ in $P(A)$ and $\lambda(z)$ in A with

$$\|e - \mu(z) - (x - ze)\lambda(z)\| < \frac{1 + \alpha}{2}$$

Write $\delta \equiv C^{-1}(1 - \alpha)/4$. Then for $|z| \leq \|x\|$ and $|z - z'| \leq \delta$ we have

$$\|e - \mu(z) - (x - z'e)\lambda(z)\| < \frac{1 + \alpha}{2} + |z - z'|C = \frac{3 + \alpha}{4}$$

Let $\zeta_1, \ldots, \zeta_m$ be a $\delta/3$ approximation to $\{z: |z| \leq \|x\|\}$. Write $R \equiv \|x\| + \delta/6$. Then $\zeta_1, \ldots, \zeta_m$ is a $\delta/2$ approximation to $\{z: |z| \leq R\}$. Write $U_j \equiv S(\zeta_j, \delta)$ for $1 \leq j \leq m$. For $1 \leq j \leq m$ the function $w_j: U_j \to \mathfrak{A}$ defined by

$$w_j(z) \equiv e - \mu_j - (x - ze)\lambda_j \qquad (z \in U_j)$$

where $\mu_j \equiv \mu(\zeta_j)$ and $\lambda_j \equiv \lambda(\zeta_j)$, is differentiable and uniformly continuous on U_j. Moreover, $\|w_j(z)\| < (3 + \alpha)/4$ for all z in U_j. Write $U_{m+1} \equiv \{z: |z| > R\}$, and let $w_{m+1}: U_{m+1} \to \mathfrak{A}$ be defined by $w_{m+1}(z) \equiv z^{-1}x$. Then w_{m+1} is differentiable and uniformly continuous on U_{m+1}, with $\|w_{m+1}(z)\| \leq |z|^{-1}\|x\| < (\|x\| + \delta/6)^{-1}\|x\|$. Also, $\|w_{m+1}(z)\| \to 0$ as $|z| \to \infty$ We have the representation

$$w_{m+1}(z) = e - \mu_{m+1} - (x - ze)\lambda_{m+1}$$

where $\mu_{m+1} \equiv 0$ and $\lambda_{m+1} \equiv -z^{-1}e$.

For $1 \leq j \leq m + 1$ the function $f_j: U_j \to \mathfrak{A}$ defined by

$$f_j(z) \equiv (e - w_j(z))^{-1}\lambda_j \qquad (z \in U_j)$$

is differentiable and uniformly continuous on U_j. We compute

$$\begin{aligned}
e - f_1(\zeta_1)(x - \zeta_1 e) &= e - (e - w_1(\zeta_1))^{-1}(\mu_1 + (x - \zeta_1 e)\lambda_1) \\
&\qquad\qquad\qquad\qquad + (e - w_1(\zeta_1))^{-1}\mu_1 \\
&= (e - w_1(\zeta_1))^{-1}\mu_1
\end{aligned}$$

Since $\mu_1 \in P(A)$, it follows that

$$(1.4) \qquad\qquad e - f_1(\zeta_1)(x - \zeta_1 e) \in P((e - w_1(\zeta_1))^{-1}A)$$

By Lemma 1, there exists $t > 0$ such that if h_j is an analytic function on U_j $(1 \leq j \leq m + 1)$, with $|h_{m+1}(z)| \to 0$ as $|z| \to \infty$, for which $|h_i - h_j| \leq t$ on $U_i \cap U_j$ $(1 \leq i, j \leq m + 1)$, then $|h_1(\zeta_1)| \leq \frac{1}{2}$.

Consider i and j $(1 \leq i, j \leq m + 1)$ and z in $U_i \cap U_j$. Then

$$
\begin{aligned}
f_i(z) - f_j(z) &= f_i(z)(e - w_j(z))^{-1}(\mu_j + (x - ze)\lambda_j) \\
&\qquad\qquad - f_j(z)(e - w_i(z))^{-1}(\mu_i + (x - ze)\lambda_i) \\
&= f_i(z)(e - w_j(z))^{-1}\mu_j - f_j(z)(e - w_i(z))^{-1}\mu_i
\end{aligned}
$$

Since the functions $f_i(e - w_j)^{-1}$ and $f_j(e - w_i)^{-1}$ are uniformly continuous on $U_i \cap U_j$, it follows that there exists a core B_0 such that

$$(1.5) \qquad \rho(f_i(z) - f_j(z), P(B_0)) \leq \tfrac{1}{2}t\beta(C + R)^{-1}$$

for all z in $U_i \cap U_j$. Moreover, the norm of B_0 is bounded by some constant depending only on C, α, β, and n. We choose B_0 to work for all i and j simultaneously. There is no loss of generality in taking B_0 to be convex and invariant under multiplication by constants z with $|z| \leq 1$. Write

$$\epsilon \equiv \tfrac{1}{4}t\beta(1 - \alpha)C^{-1}(C + R)^{-1}$$

We shall show that the core

$$B \equiv \{y_1 + y_2 \colon y_1 \in (e - w_1(\zeta_1))^{-1}A,\ y_2 \in \epsilon^{-1}(x - \zeta_1 e)B_0\}$$

satisfies the prescribed conditions. Clearly B is a core whose norm is bounded by some constant D depending only on C, α, β, and n. Assume that (1.2) is satisfied. By (1.2) and (1.4),

$$
\begin{aligned}
\beta < \rho(e, P(B)) &\leq \rho(f_1(\zeta_1)(x - \zeta_1 e),\ \epsilon^{-1}(x - \zeta_1 e)P(B_0)) \\
&\leq \|x - \zeta_1 e\|\rho(f_1(\zeta_1),\ \epsilon^{-1}P(B_0))
\end{aligned}
$$

Since $\|x - \zeta_1 e\| \leq C + R$, this gives

$$\rho(f_1(\zeta_1),\ \epsilon^{-1}P(B_0)) > \beta(C + R)^{-1}$$

Since B_0 is convex and invariant under multiplication by scalars z with $|z| \leq 1$, we see that $P(B_0)$ has the same property. By the separation theorem, there exists a complex linear functional u on $\mathfrak{A}$, with $\|u\| = 1$ and

$$u(f_1(\zeta_1)) > \epsilon^{-1}|u(y)| + \beta(C + R)^{-1}$$

for all y in $P(B_0)$. Hence

$$(1.6) \quad \|u\|_{P(B_0)} \leq \epsilon \|f_1(\zeta_1)\| \leq \epsilon \|(e - w_1(\zeta_1))^{-1}\| \|\lambda_1\|$$
$$\leq \epsilon(1 - \|w_1(\zeta_1)\|)^{-1}C \leq \epsilon 2(1 - \alpha)^{-1}C$$
$$= \tfrac{1}{2}t\beta(C + R)^{-1}$$

Write $\qquad h_j \equiv \beta^{-1}(C + R)u \circ f_j \qquad (1 \leq j \leq m + 1)$

so that h_j is an analytic function on U_j. Using (1.5) and (1.6), we find

$$|h_i(z) - h_j(z)| = \beta^{-1}(C + R)|u(f_i(z) - f_j(z))|$$
$$\leq \beta^{-1}(C + R)\{\rho(f_i(z) - f_j(z), P(B_0)) + \|u\|_{P(B_0)}\}$$
$$\leq \beta^{-1}(C + R)\{\tfrac{1}{2}t\beta(C + R)^{-1} + \tfrac{1}{2}t\beta(C + R)^{-1}\} = t$$

By Lemma 1,

$$\tfrac{1}{2} \geq |h_1(\zeta_1)| = \beta^{-1}(C + R)|u(f_1(\zeta_1))| \geq \beta^{-1}(C + R)\beta(C + R)^{-1} = 1$$

Since $\tfrac{1}{2} \geq 1$, we have $\sigma > \alpha$, as was to be proved.

We now give a slightly strengthened version of Lemma 2.

Lemma 3 *Under the hypotheses of Lemma 2, let F be a core of norm at most C. Then the core B of Lemma 2 can be chosen to be independent of $x_1, \ldots, x_n$, and x, as long as these elements are restricted to F. (Thus B depends on A and F.)*

Proof For each $\lambda \equiv \{x_1, \ldots, x_n, x\}$ in the cartesian product F^{n+1} there exists a core B_λ satisfying the requirements of Lemma 2, with β replaced by $\beta/2$ and α by $(1 + \alpha)/2$. Let the finite set $\Gamma \subset F^{n+1}$ be an $\epsilon \equiv \min\{(2D)^{-1}\beta, C^{-1}(1 - \alpha)/4\}$ approximation to F^{n+1}. Write

$$B \equiv \bigcup_{\gamma \in \Gamma} B_\gamma$$

Consider any $\lambda \equiv \{x_1, \ldots, x_n, x\}$ in F^{n+1}, satisfying (1.2). Let $\gamma \equiv \{y_1, \ldots, y_n, y\}$ be any element of Γ with $\rho(\gamma, \lambda) < \epsilon$. Since the norm of B is at most D, we have

$$\rho(e, P(y_1, \ldots, y_n; B)) > \rho(e, P(x_1, \ldots, x_n; B)) - \sum_{i=1}^{n} D\|y_i - x_i\|$$
$$> \beta - D\epsilon \geq \frac{\beta}{2}$$

Hence there exists ζ, with $|\zeta| \leq \|y\|$, such that

$$(1.7) \qquad \rho(e, P(y_1, \ldots, y_n, y - \zeta e; A)) > \frac{1 + \alpha}{2}$$

Since $\|x - y\| \leq \rho(\lambda,\gamma) < \epsilon$, there exists z in $\mathbf{C}$ with $|z| \leq \|x\|$ and $|z - \zeta| < \epsilon$. Since the norm of A is at most C, inequality (1.7) gives

$$\rho(e, P(x_1, \ldots, x_n, x - ze; A))$$
$$> \frac{1 + \alpha}{2} - \sum_{i=1}^{n} C\|x_i - y_i\| - C\|x - ze - (y - \zeta e)\|$$
$$\geq \frac{1 + \alpha}{2} - 2C\epsilon \geq \alpha$$

Lemma 3 can be extended as follows.

Theorem 1 *Let C, α, and β be positive constants, with $\alpha < 1$ and $\beta < 1$. Let n and m be positive integers. Then there exists a positive integer D such that if A and F are any cores, of norms at most C, there exists a core B, of norm at most D, such that if $x_1, \ldots, x_n, x_1{}^0, \ldots, x_m{}^0$ are arbitrary elements of F with*

$$(1.8) \qquad \qquad \rho(e, P(x_1, \ldots, x_n; B)) > \beta$$

then there exist complex numbers $z_1, \ldots, z_m$, with $|z_i| \leq \|x_i{}^0\|$ ($1 \leq i \leq m$), such that

$$\rho(e, P(x_1, \ldots, x_n, x_1{}^0 - z_1 e, \ldots, x_m{}^0 - z_m e; A)) > \alpha$$

Proof The proof is by induction on m. The case $m = 1$ is just Lemma 3. Assume that the result is true for a given m and all values of C, α, β, and n. Fix C, α, β and n. Let D_0 be the value of D corresponding to $2C$, α, β, $n + 1$, and m. Let D_1 be the value of D corresponding to $C + D_0$, β, β, n, and 1. Consider cores A and F, of norms at most C. By the choice of D_0, there exists a core B_0, of norm at most D_0, such that if $x_1, \ldots, x_n, x_1{}^0, \ldots, x_{m+1}^0$ are elements of F and z_1 is a complex number, with $|z_1| \leq \|x_1{}^0\|$, for which

$$(1.9) \qquad \qquad \rho(e, P(x_1, \ldots, x_n, x_1{}^0 - z_1 e; B_0)) > \beta$$

then there exist complex numbers $z_2, \ldots, z_{m+1}$, with $|z_i| \leq \|x_i{}^0\|$ ($2 \leq i \leq m + 1$), such that

$$(1.10) \quad \rho(e, P(x_1, \ldots, x_n, x_1{}^0 - z_1 e, \ldots, x_{m+1}^0 - z_{m+1} e; A)) > \alpha$$

By the choice of D_1 there exists a core B, of norm at most D_1, such that if $x_1, \ldots, x_n, x_1{}^0$ are elements of F with

$$(1.11) \qquad \qquad \rho(e, P(x_1, \ldots, x_n; B)) > \beta$$

then there exists a complex number z_1, with $|z_1| \leq \|x_1^0\|$, satisfying (1.9). Hence if (1.11) is valid, for given elements $x_1, \ldots, x_n, x_1^0, \ldots, x_{m+1}^0$ of F, there exist complex numbers $z_1, \ldots, z_{m+1}$, with $|z_i| \leq \|x_i\|$ $(1 \leq i \leq m+1)$ satisfying (1.10). This completes the induction.

The following result makes use of Theorem 1 to get a good hold on the sets Σ_n.

Theorem 2 *Let $x_1, \ldots, x_n$ be elements of $\mathfrak{A}$, let β be a real number with $0 < \beta < 1$, and let m be a positive integer. Then there exists a core B such that if $z_1, \ldots, z_n$ are any complex numbers with*

$$(1.12) \qquad \rho(e, P(x_1 - z_1 e, \ldots, x_n - z_n e; B)) > \beta$$

there exists u in Σ_m with $|u(x_j) - z_j| < m^{-1}$ $(1 \leq j \leq n)$.

Proof Let $\{x_k^0\}$ be the dense sequence in $\mathfrak{A}$ that was used to construct the sets Σ_k (in Proposition 1). Write

$$A \equiv \{ze: |z| \leq 1\} \cup \bigcup_{i=1}^{m} \{zx_i^0: |z| \leq 1\}$$

Let α be a constant, with $0 < \alpha < 1$, to be fixed later. Let F be the set consisting of all elements $x_i - ze$ $(1 \leq i \leq n, |z| \leq M \equiv \max \{1, \beta^{-1}\|x_1\|, \ldots, \beta^{-1}\|x_n\|\})$, and the elements $x_1^0, \ldots, x_m^0$. Clearly F is totally bounded. Let C be the norm of F. Let B be the union of $\{ze: |z| \leq 1\}$ and the core given by Theorem 1. Consider complex numbers $z_1, \ldots, z_n$ such that (1.12) holds. If $|z_i| > M$ for some i, then $z_i^{-1}e \in B$, and therefore $z_i^{-1}x_i - e \in P(x_1 - z_1 e, \ldots, x_n - z_n e; B)$, contradicting (1.12). Hence the $|z_i|$ are all bounded by M. By the choice of B, there exist complex numbers $z_1^0, \ldots, z_m^0$, with $|z_i^0| \leq \|x_i^0\|$ $(1 \leq i \leq m)$, such that $\rho(e, P) > \alpha$, where

$$P \equiv P(x_1 - z_1 e, \ldots, x_n - z_n e, x_1^0 - z_1^0 e, \ldots, x_m^0 - z_m^0 e; A)$$

By the separation theorem, there exists u in S^* such that $u(e) > \alpha + |u(y)|$ for all y in P. If α is arbitrarily near to 1, then $u(e)$ is arbitrarily near to 1, and u is arbitrarily small on P. Since the points $x_i - z_i e$ $(1 \leq i \leq n)$ and $x_i^0 - z_i^0 e$ $(1 \leq i \leq m)$ all belong to P, it follows that $u(x_i)$ is arbitrarily near to z_i and $u(x_i^0)$ to z_i^0. Since the points $(x_i^0 - z_i^0 e)x_k^0$ $(1 \leq i, k \leq m)$ all belong to P, it follows that $u(x_i^0 x_k^0)$ is arbitrarily near to $z_i^0 u(x_k^0)$, which in turn is arbitrarily near to $u(x_i^0)u(x_k^0)$. Thus the required conditions are all satisfied if α is sufficiently near to 1.

Corollary *Each of the sets Σ_m is compact.*

Proof Since Σ_m is either compact or void, we must show that Σ_m contains at least one point. To this end, apply Theorem 2 with

$$x_1 = x_2 = \cdots = x_m = 0$$

Then (1.12) is satisfied by $z_1 = z_2 = \cdots = z_n = 0$. Hence there exists u in Σ_m.

Our next result is the fundamental tool for solving linear equations in a Banach algebra.

Theorem 3 *Let $x_1, \ldots, x_n$ be elements of $\mathfrak{A}$, let ϵ be a positive constant, and let m be a positive integer such that*

$$(1.13) \qquad |u(x_1)| + \cdots + |u(x_n)| \geq \epsilon \qquad (u \in \Sigma_m)$$

Then there exist $y_1, \ldots, y_n$ in $\mathfrak{A}$ with

$$(1.14) \qquad y_1 x_1 + \cdots + y_n x_n = e$$

Proof By Theorem 2, there exists a core B such that if $\rho(e, P(x_1, \ldots, x_n; B)) > \frac{1}{2}$, there exists u in Σ_m with $|u(x_i)| < n^{-1}\epsilon$ $(1 \leq i \leq n)$, contradicting (1.13). Hence $\rho(e, P(x_1, \ldots, x_n; B)) \leq \frac{1}{2}$. It follows that (1.14) holds for some choice of $y_1, \ldots, y_n$.

Applications of Theorem 3 will be given in the problems.

In the classical theory the *spectral norm*

$$\|x\|_\Sigma = \text{l.u.b.} \ \{|u(x)| : u \in \Sigma\}$$

exists, for each x in $\mathfrak{A}$, and is given by

$$(1.15) \qquad \|x\|_\Sigma = \lim_{n \to \infty} \|x^n\|^{1/n}$$

Because Σ need not be constructively compact, there is no reason to believe that $\|x\|_\Sigma$ is constructively defined, and indeed the pathological Banach algebra constructed above is an instance in which it is not. Nevertheless, there exists a good constructive substitute for (1.15).

Proposition 2 *For each x in $\mathfrak{A}$ the sequences $\{\|x^n\|^{1/n}\}$ and $\{\|x\|_{\Sigma_n}\}$ are equiconvergent, in the sense that for each element a_m of either sequence* *and each $\epsilon > 0$ there exists N in Z^+ such that $b_n \leq a_m + \epsilon$, where b_n is any element of the other sequence with $n \geq N$.*

Proof   Let $\|x^m\|^{1/m}$ be any element of the first sequence and ϵ any positive constant. Choose N in Z^+ so large that (i) there are elements of the finite sequence $x_1^0, \ldots, x_N^0$ arbitrarily close to each of the elements $x, x^2, \ldots, x^m$ and (ii) $u(x_1^0 x_i^0)$ is arbitrarily close to $u(x_1^0)u(x_i^0)$ for $1 \leq i \leq m$ and all u in Σ_N. Then $u(x^m)$ is arbitrarily close to $u(x)^m$, for all u in Σ_n and all $n \geq N$. Hence, $|u(x)|$ is less than $|u(x^m)|^{1/m} + \epsilon \leq \|x^m\|^{1/m} + \epsilon$. In other words, $\|x\|_{\Sigma_n}$ is less than $\|x^m\|^{1/m} + \epsilon$.

Consider, conversely, an element $\|x\|_{\Sigma_m}$ of the second sequence and an $\epsilon > 0$. By Theorem 2 there exists a core B such that for each z in $\mathbf{C}$, with $\rho(e, P(x - ze; B)) > \frac{1}{4}$, there exists u in Σ_m with $|u(x) - z| \leq \epsilon/2$, so that $|z| \leq \|x\|_{\Sigma_m} + \epsilon/2$. Thus if $|z| > \|x\|_{\Sigma_m} + \epsilon/2$, then $\rho(e, P(x - ze; B)) < \frac{1}{2}$, and thus $\|e - (x - ze)y\| < \frac{1}{2}$ for some y in B. Hence $((x - ze)y)^{-1}$ exists, and has norm at most 2. Therefore

$$(x - ze)^{-1} = y((x - ze)y)^{-1}$$

exists, and has norm at most $2\|y\| \leq 2D$, where D is the norm of B. Thus $(x - ze)^{-1}$ is analytic for $|z| > \|x\|_{\Sigma_m} + \epsilon/2$, and is bounded by $2D$. Equivalently, $(e - \zeta x)^{-1}$ is analytic for $|\zeta| < (\|x\|_{\Sigma_m} + \epsilon/2)^{-1}$, and is bounded by $D_0 \equiv 2D(\|x\|_{\Sigma_m} + \epsilon/2)$. For all sufficiently small ζ we have the expansion

$$(e - \zeta x)^{-1} = \sum_{n=0}^{\infty} \zeta^n x^n$$

By (a) (page 339), we see that

$$\|x^n\| \leq D_0 \left(\|x\|_{\Sigma_m} + \frac{\epsilon}{2} \right)^n$$

for all n in Z^+. Hence $\|x^n\|^{1/n} \leq \|x\|_{\Sigma_m} + \epsilon$ for all sufficiently large n.

PROBLEMS

1.   A subset S of a Banach algebra $\mathfrak{A}$ *generates* $\mathfrak{A}$ if $\mathfrak{A}$ is the smallest closed subalgebra of $\mathfrak{A}$ containing $S \cup \{e\}$. Let $\mathfrak{A}$ be a Banach algebra with a single generator x_0. Show that the map $\sigma \to \sigma(x_0)$ is a metric equivalence of Σ with $\{\sigma(x_0) : \sigma \in \Sigma\} \subset \mathbf{C}$.

2.   Call the spectrum Σ of a commutative Banach algebra $\mathfrak{A}$ *firm* (a) if Σ is compact, and (b) if $\rho(\Sigma_n, \Sigma) \to 0$ as $n \to \infty$. Show that if $\mathfrak{A}$ has a single generator and Σ is compact, then Σ is firm.

3. Let $\mathfrak{A}$ be a commutative Banach algebra whose spectrum Σ is firm. Let $x_1, \ldots, x_n$ be elements of $\mathfrak{A}$ and ϵ a positive constant such that (1.13) holds for all u in Σ. Show that there exist $y_1, \ldots, y_n$ in $\mathfrak{A}$ satisfying (1.14).

4. Let $\mathfrak{A}$ be the Banach algebra of all continuous functions on $\{z: |z| \leq 1\}$ that are analytic on $\{z: |z| < 1\}$. Show that Σ is firm.

5. Let G be a discrete locally compact abelian group. Let f be any element of $L_1(G)$ such that $|\hat{f}(\alpha)| \geq c > 0$ for all α in G^*. Show that there exists g in $L_1(G)$ such that $(g * f)(e) = 1$ and $(g * f)(x) = 0$ for all $x \neq e$.

NOTES

Certain classical Banach algebras are *not* Banach algebras in our sense, because the norm is *not* computable. Again, L_∞ is an example. It would be interesting to extend the theory to cover such algebras.

The fact that we work with partial ideals rather than ideals causes ugly complications. Lemma 1 would not be necessary if we could work with ideals. These complications seem somehow extraneous. It may be that the theory has not found its proper form. Another possibility is that the complications are not extraneous, but that they could be avoided by means of a metamathematical result, to the effect that under appropriate conditions there exists a routine way for constructivizing a given mathematical theory. In certain circumstances we might be content with the knowledge that the constructivization exists. Chapter 11 contains the only material in this book in which it is at all plausible that such a metatheorem could be successfully applied.

METRIZABILITY AND SEPARABILITY

How much have we lost by restricting our attention almost exclusively to metric spaces and by our liberal use of separability hypotheses? There are really two questions, depending on whether we take the constructive or the classical point of view. From the constructive point of view, we ask whether there is significant mathematics that cannot be done, or perhaps cannot be done in a good way, within the framework we have developed. From the classical point of view, the problem is to see how much of the nonmetric or nonseparable theory can be recovered from our results, and how much cannot.

In this appendix, we take the constructive point of view. The situation is easily summarized: Nonmetric spaces and nonseparable metric spaces play no significant role in those parts of analysis with which this book is concerned. To illustrate this point, consider the concept of a uniform space, as developed in Probs. 17 to 21 of Chap. 4. A uniform space at first sight appears to be a natural and fruitful concept for constructive mathematics, a promising substitute for the concept of a topological space. In fact, this is not the case. For instance, just to construct a compact uniform space X, such that the assumption that X is metrizable leads to a contradiction, seems to be a hard problem. Here is an attempt that fails. Let I be a compact proper interval and S a set such that the assumption that S is countable leads to a contradiction. Let X consist of all functions from S into I. For all s in S and all f and g in X write $\rho_s(f,g) \equiv |f(s) - g(s)|$. There is no known choice of S which makes X totally bounded (in the sense of Prob. 18 of Chap. 4) relative to the uniform structure defined by the pseudometrics ρ_s, because the known sets S all have the property that there exist points s and t in S for which we are unable either to construct f in X with $f(s) \neq f(t)$ or to prove $f(s) = f(t)$ for all f in X. This means we are unable to show that S is totally bounded in the pseudometric $\rho_s + \rho_t$, and thus we are unable to show that S is totally bounded relative to the given uniform structure.

Of course, important constructively defined uniform spaces that are not necessarily metrizable exist: every locally convex space has a natural uniform structure. At first glance, the concept of a locally convex space would appear to be important for constructive mathematics, since examples exist in profusion. However, in most cases of interest it seems to be unnecessary to make use of any deep facts from the general theory of locally convex spaces. For example, the dual B^* of a separable Banach space B is most conveniently studied in terms of the double norm, rather than the locally convex structure. As another example, consider the theory of distributions. Let $\mathfrak{D}$ denote all infinitely differentiable functions $\varphi \colon \mathbf{R} \to \mathbf{R}$ with compact support. For each sequence $N \equiv \{N_k\}$ of positive integers and each φ in $\mathfrak{D}$, write

$$\|\varphi\|_N \equiv \sum_{k=1}^{\infty} N_k \sup \left\{ \sum_{n=0}^{N_k} |\varphi^{(n)}(x)| \colon |x| \geq k - 1 \right\}$$

where, of course, the sum is actually finite. These seminorms $\| \ \|_N$ generate a locally convex structure on $\mathfrak{D}$. With this structure, $\mathfrak{D}$ is called the space of test functions. Classically $\mathfrak{D}$ is complete (in the sense of Prob. 19 of Chap. 4), but constructively the completion of $\mathfrak{D}$ consists of all infinitely differentiable functions $\varphi \colon \mathbf{R} \to \mathbf{R}$ such that the norm $\|\varphi\|_N$ exists for each N. This difficulty could be overcome by redefining $\mathfrak{D}$, although such a procedure seems slightly artificial. The same difficulty arises, in more acute form, with the dual space $\mathfrak{D}^*$ (the space of distributions). Here again the constructive completion of $\mathfrak{D}^*$ cannot be identified with $\mathfrak{D}^*$, and in this case it seems definitely artificial to enlarge $\mathfrak{D}^*$ to make it complete. Although detailed studies are needed, tentatively we conclude that the constructive theory of distributions should rely heavily on the concept of sequential convergence, defined ad hoc both for sequences in $\mathfrak{D}$ and for sequences in $\mathfrak{D}^*$, rather than on general theorems about locally convex spaces.

A final remark: From the constructive point of view, there is no loss of generality in taking all metric spaces to be separable. Not only are the spaces that naturally arise separable spaces, but the author knows of no constructively defined nonvoid metric space that can be proved to be nonseparable.

ASPECTS OF CONSTRUCTIVE TRUTH

The role of contradiction in constructive mathematics deserves further comment. As remarked in Chap. 1, there are two possible points of view. Suppose that in the course of proving a certain proposition P we have constructed an integer n, which we know must have one of finitely many values $1, 2, \ldots, N$. Which of these N values n actually assumes will of course depend on the particular numerical data for the proposition P. For given data, n is finitely computable. We may prove P, using an argument by cases: there are N cases to consider (call them cases $C_1, C_2, \ldots, C_N$, depending on the value of n), and we are at liberty to give a separate argument in each case. Suppose that one of the cases C_m leads to a contradiction—that is, the assumption $n = m$ allows us to deduce the equality $0 = 1$. Then either we may rule this case out on the ground that it cannot occur for any permissible choice of the numerical data, or we may be more meticulous and prove that in this case too the proposition P is valid.

Let us examine the first possibility. To rule out the case C_m because it leads to a contradiction seems to require a belief in the consistency of constructive mathematics, and so raises the question of the constructivist attitude to questions of consistency. For the constructivist, consistency is not a hobgoblin. It has no independent value; it is merely a consequence of correct thought. The consistency of some particular formal system, even if it were constructively proved, would not be a very interesting result for the constructive mathematician.

Since consistency is a consequence of correct thought, an inconsistency should be regarded as a consequence of incorrect thought. The constructive mathematician, being human, will not be surprised if one of his results leads to a contradiction. In this sense, he does not believe in the consistency of constructive mathematics. An inconsistency will cause him to examine his thinking to find out where he went wrong. On the other hand, a constructive proof which leads to a contradiction *is wrong*, perhaps by definition. In this sense, he does believe in the consistency of constructive mathematics, and therefore has no hesitation in ruling out a case C_m that leads to a contradiction.

The practice of ruling out a case C_m that does not occur involves no loss in computational meaning, because the character of the proof for the cases that do occur is not affected. In spite of this, certain skeptics seem to think that this practice compromises the constructivist position. These skeptics are entitled to prove every result in all cases, if they wish, including those cases that lead to a contradiction. We have not chosen this course, because to us it seems silly to insist on finishing a proof that already contains a mistake. Nevertheless, it would be easy to rewrite this book along such lines. In some instances, definitions would have to be modified.

Some mathematicians would still be dissatisfied, if the book were rewritten as described, because to them it also compromises the constructivist position to prove a theorem as a consequence of a contradiction. The book could be rewritten again, with these mathematicians in mind. Each occurrence of a contradiction would be cleverly disguised; all cases involving a contradiction would be made to appear affirmative. (The book as it stands undoubtedly contains instances of arguments in which certain cases that are not treated as contradictory do in fact lead to contradictions.)

Contrary to the opinion expressed above, Hilbert thought that a constructive proof of the consistency of a sufficiently powerful formal system F would be of great value, even to the constructive mathematician, because it would guarantee the constructive validity of every theorem T proved in F whose statement (standing by itself) makes constructive sense. For instance, a constructive proof of the consistency of such a formal system F together with a proof in F of the theorem T_1 that every even integer $n \geq 4$ is the sum of two primes could automatically be transformed into a constructive proof of T_1. On the other hand, a constructive proof of the consistency of F, together with a proof in F of the theorem T_2 that there exist 100 consecutive digits equal to 7 in the decimal expansion of π, would not necessarily give rise to a constructive proof of T_2, because the statement T_2 (standing by itself) does not make constructive sense. (In order for T_2 to make constructive sense, an upper bound for the location of the 100 digits in question would have to be given. In other words, computing one digit after another in the decimal expansion of π until we either find 100 consecutive sevens or get tired and quit is not a finitely performable computation.) Hilbert's implied belief that there are a significant number of interesting theorems whose statements (standing alone) are constructive but whose proofs are not constructive (or cannot easily be made constructive) has not been justified. In fact we do not know of even one such theorem.

It is worthwhile to probe the dictum that all mathematics should have computational meaning. It would be a mistake to attempt to formalize the notion of computational meaning. Rather let us ask how the computational content of a given mathematical result is to be realized. From one point of view, a constructively valid proof already constitutes a realization of the computational content of the result being proved. In other words, a proof *is* a computation. Thus there are many ways to compute. The most primitive way is to count digits one by one. A more sophisticated method is to use decimal notation. Using exponential notation, we can perform more sophisticated computations yet. Every constructive proof can be regarded as the development of a (possibly) new notation, and the verification that certain computations performed with the aid of that notation produce certain results. Of course, all computations can theoretically be performed by the process of counting digits one by one, or by decimal arithmetic if we prefer. In practice, there are human limitations; for example, it would be folly to try to compute the national debt by counting pennies one by one. The national debt nevertheless is very real.

Mathematics that has computational meaning may require the exercise of judgment before that meaning is discharged. Whether a given rule generates a constructively defined sequence of integers; whether a given hypothesis is satisfied; whether a given phrase in the English language defines a measure space—these matters all require judgment. If the judgment is not unanimous, either the mathematics is bad or the judgment of at least one of the judges is bad. This is the difference between man and the presently available computers: We exercise judgment, and the computers do not. Because the computer is lacking in judgment, the theorems of constructive mathematics do not in general represent computer programs. They represent person programs, which in some instances can be transformed into computer programs and in other instances cannot.

To what extent can the results of constructive analysis be realized as computer programs, rather than person programs? More realistically, since the necessity for judgment can never be eliminated completely, we seek to realize each constructive result T as a computer program requiring minimal preparation and supervision by the operator of the computer. Let D represent the data of the result T, and let E represent the quantities whose existence is asserted by T. Let R represent the totality of all relations asserted by T to hold among the quantities D and E. One way we might realize T as a computer program is as follows. First, we represent the data D as a sequence of integers, called the *input sequence*. In other words, we define a function that assigns an input sequence I_D to each permissible choice of data D, distinct sequences being assigned to distinct sets of data. Next, we define a function that assigns to each sequence J of integers, called an *output sequence*, values E_J for the quantities whose existence is asserted by T. Finally, we use the proof of T to program the computer so that whenever an input sequence I_D corresponding to given data D is fed in term by term, the computer will produce term by term an output sequence J such that D and E_J together satisfy all of the relations R.

This vague description is intended merely as an indication of some of the steps that might be involved in realizing a constructive result T as a computer program. It is not clear how the number of terms of the input sequence required to generate a given term of the output sequence would be specified. Neither is it clear how E_J is to be recovered from the output sequence J; presumably any particular integer involved in the structure of E_J could be recovered from a certain finite portion of J by a routine computation. Hopefully the relations R required to hold between D and E could be transformed into a sequence of finitely computable numerical relations, each involving only finitely many terms of the input and output sequences, so that a second computer, whose input sequence would consist of both the input sequence and the output sequence of the first computer, could be programmed to check the resulting relations one by one. One should also consider the possibility of having a computer check that the input sequence I actually represents a set D of permissible data; for instance, the computer might be programmed to indicate whether the portion of the input sequence I already fed in corresponds to some set D of permissible data. In practice, of course, the input and output sequences of a computer are sequences of binary digits, rather than sequences of integers. Therefore, when we ask such questions as "How many terms of the input sequence are required to generate a given term of the output sequence?" we should probably be talking about binary digits. This would introduce additional complications.

It is clear that many of the results in this book could be programmed for a computer, by some such procedure as that indicated above. In particular, it is likely that most of the results of Chaps. 2, 4, 5, 9, 10, and 11 could be presented as computer programs. As an example, a complete separable metric space X can be described by a sequence of real numbers, and therefore by a sequence of integers, simply by listing the distances between each pair of elements of a given countable dense set.

The results of Chap. 3, on the other hand, resist computerization. A typical result of set theory has too much scope for its numerical content to be described by a single computer program.

At first glance, measure theory also appears to resist computerization, but a closer inspection indicates that most of the material of Chaps. 6, 7, and 8 could be programmed for a computer. More accurately, Chaps. 6, 7, and 8 could be rewritten so that most of the resulting material could be programmed for a computer. One possibility for the rewrite is the following. In place of the definition given in the text, define a measure space to be a complete separable metric space M satisfying some appropriate axioms. [The correspondence with the original definition is that M consists of the integrable sets $\mathfrak{M}$, with equality taken to mean equality almost everywhere and with metric ρ defined by $\rho(A,B) \equiv \mu(A - B) + \mu(B - A)$.] One of the axioms should assert that M has a partial ordering $\leq$ with least element ϕ_0. The *measure* of an element A of M is $\mu(A) \equiv \rho(A,\phi_0)$. In place of the definition given in the text, define a simple function to be a formal linear combination of elements of M, with real coefficients. Define equality and inequality (almost everywhere) of simple functions in the obvious way. Define a measurable function to be a sequence of simple functions that is Cauchy almost uniformly, and define equality and inequality (almost everywhere) of measurable functions. Define the integral of a simple function, and use it to put a metric on the set of simple functions. Let the space of integrable functions be the completion of the space of simple functions. With these definitions as a start, it should not be hard to put the results of Chaps. 7 and 8 in a form suitable for computerization. As to Chap. 6, it might be rewritten along the following lines. Define a measure μ on a locally compact space X as before. Theorems 3 and 5 remain unchanged, as does Theorem 6 in essence. Define weak integrability of a compact set K as before, and show that if K_1 and K_2 are weakly integrable compact sets relative to a positive measure μ, with associated sequences $\{f_n{}^1\}$ and $\{f_n{}^2\}$, respectively, the limit

$$\mu(K_1 \cap K_2) \equiv \lim_{n \to \infty} \int\!\!\int f_n{}^1 f_n{}^2 \, d\mu$$

exists. Write

$$\rho(K_1, K_2) \equiv \mu(K_1) + \mu(K_2) - \mu(K_1 \cap K_2)$$

Show that this defines a pseudometric on the weakly integrable compact sets. The completion M of the corresponding metric space will be the measure space defined by μ.

As written, this book is person-oriented rather than computer-oriented. It would be of great interest to have a computer-oriented version. Without such a version, it is hard to predict with any confidence what form computer-oriented abstract analysis will eventually assume. A thoughtful computer-oriented presentation should uncover many interesting phenomena.

Certain classical, nonconstructive results can profitably be considered from a computer-oriented point of view. To give the idea, we first look at an elementary example. Let T be the classical result that for every (uniformly) continuous function $f \colon [0,1] \to \mathbf{R}$ there exists x_0 in $[0,1]$ with $f(x_0) \geq f(x)$ for all x in $[0,1]$. The possible computer inputs corresponding to this theorem are sequences I of integers describing (a) the values of f at the rational points of $[0,1]$ and (b) the values of the modulus of continuity of f at the positive rational numbers. The possible computer outputs are sequences J of integers describing a real number x_0 belonging to $[0,1]$. If the theorem T were constructive, we would expect to be able to program a computer to produce term by term the output sequence J_I corresponding to an arbitrary input sequence I, the input sequence I being fed in term by term. This is impossible; even by classical reasoning it is necessary in general to know the entire input sequence I before the production of the corresponding output sequence J_I can begin. To go deeper, we must introduce the relation sequence R, which describes the relation of the input sequence I to the corresponding output sequence J_I. It is a sequence of finitely computable integer-valued functions. Any given term of R is a function of a certain finite set of terms of the input sequence and a certain finite set of terms of the output sequence. The function f described by a given input sequence I assumes its maximum value at the point x_0 described by a given output sequence J if and only if all terms of the corresponding relation sequence $R(I,J)$ vanish. In fact, for each positive integer k there exists a positive integer N_k such that if the first N_k terms of the relation sequence $R(I,J)$ vanish, then $f(x_0) \geq f(x) - k^{-1}$ for all x in [0,1]. An examination of the proof of T leads us to a computer program P each of whose input sequences I, in addition to the previous information, describes a positive integer k. The output sequences are not changed. The program P associates to each input sequence I an output sequence J such that the first N_k terms of $R(I,J)$ vanish. In other words, P is a computer program for the constructive result that for each positive integer k and each continuous function $f: [0,1] \to \mathbf{R}$ there exists x_0 in [0,1] with $f(x_0) \geq f(x) - k^{-1}$ for all x in [0,1]. This result is a special case of the Corollary to Proposition 8 of Chap. 4, and the proof of that corollary was obtained by essentially the method we have just discussed—an inspection of the classical proof to extract as much constructive information as possible.

Many classical results can be constructivized in much the same way as the above theorem. A good example is Brouwer's fixed-point theorem (which we have not considered in this book). A less obvious example is the theorem T that every ideal in a commutative Banach algebra $\mathfrak{A}$ with unit is contained in a maximal ideal, a constructive version of which was presented in Chap. 11. For the present discussion, we take $\mathfrak{A}$ to be separable. The possible computer inputs for this theorem are sequences I of integers describing (a) a seminorm $\| \quad \|_I$ on the vector space S of all finitely nonzero sequences of complex numbers, and (b) a multiplication operation on the completion $\mathfrak{A}_I$ of S with respect to $\| \quad \|_I$ that makes $\mathfrak{A}_I$ into a commutative Banach algebra, with unit equal to a fixed element e of S, such that the subset S_0 of S consisting of all $s \equiv \{s_n\}$ with $s_n = 0$ for all odd n generates an ideal E_I of $\mathfrak{A}_I$. The possible computer outputs are sequences J of integers describing a linear functional φ_J from S to $\mathbf{C}$, with $\varphi_J(e) = 1$, that vanishes on S_0. As before, we have no computer program that transforms an input sequence I into an output sequence J so that φ_J extends to a bounded multiplicative linear functional on $\mathfrak{A}_I$. There exists a relation sequence R, with the property that the linear functional φ_J described by a given output sequence J extends to a bounded multiplicative linear functional on the Banach algebra $\mathfrak{A}_I$ described by a given input sequence I if and only if all terms of the sequence $R(I,J)$ vanish. In fact, for given elements $x_1, \ldots, x_n$ of S_0, a given element x of S, a given finite subset A of S, and a given constant α $(0 < \alpha < 1)$, there exists an integer $N = N(x_1, \ldots, x_n, x, A, \alpha)$ such that

$$\rho_I(e, P(x_1, \ldots, x_n; x - \varphi_J(x)e; A)) > \alpha$$

whenever the first N terms of the relation sequence $R(I,J)$ vanish, where I is any input sequence, J is any output sequence, and ρ_I refers to the distance in the Banach algebra $\mathfrak{A}_I$. An examination of the proof of T leads us to a computer program P, each of whose input sequences I, in addition to the information already described, determines a positive integer n and an $(n + 3)$-tuple $(x_1, \ldots, x_n, x, A, \alpha)$ of the type just described. The output sequences are not changed. The program P associates to each input sequence I an output sequence J such that the first $N(x_1, \ldots, x_n, x, A, \alpha)$ terms of $R(I,J)$ vanish. Thus P is a computer program corresponding to the constructive result that if E is an ideal in a separable commutative Banach algebra $\mathfrak{A}$ with unit, if $x_1, \ldots, x_n$ are in E, if x is in $\mathfrak{A}$, if A is a finite subset of $\mathfrak{A}$, and if α is a constant $(0 < \alpha < 1)$, then there exists a complex number z such that

$$\rho(e, P(x_1, \ldots, x_n, x - ze; A)) > \alpha$$

This result is a weakened version of Lemma 2 of Chap. 11. It would be possible to extract from the proof of T a computer program corresponding to Lemma 2 itself. In fact this is essentially how Lemma 2 was proved—by extracting as much constructive information as possible from the proof of the classical theorem T.

Such classical results as Brouwer's fixed-point theorem and the existence of a point at which a continuous real-valued function on a compact set attains its maximum are easy to constructivize; in fact, given the proper preparation, the constructivization of these results is almost routine. In these instances the proofs of the constructive versions are not significantly more involved than their classical prototypes. On the other hand, Lemma 2 of Chap. 11 and its final version, Theorem 1, have proofs which are considerably less perspicuous than their classical prototypes, in spite of the fact that a mathematician with the proper preparation will intuitively appreciate that the proof of the classical theorem T can be transformed into a constructive proof of the constructive Theorem 1 of Chap. 11. Perhaps we have not yet put the relation of the classical theorem T to its constructive version in the proper perspective. It is possible that there exists a more or less routine procedure which for (a) a classical theorem T of a certain type, when presented together with its proof in an appropriate form, and (b) a constructive version T' of T, related to T in a certain way, guarantees we can translate the proof of T into a constructively valid proof of T'. Such a procedure, of general applicability, would be extremely important. This is what we had in mind in bringing up the possibility of a metamathematical result in the notes to Chap. 11. The investigation required to achieve such a result will probably be painstaking rather than deep.

Various mathematicians have devised methods for classifying (classical) real numbers according to their logical complexity—we mention only Russell's theory of types, Weyl's book [16], certain ideas of Poincaré [13], and the work of various modern logicians. Echoes of such type-theoretic classifications appear in constructive analysis. The real numbers as defined in Chap. 2 are the first stage of a hierarchy of number systems. The construction of the numbers of the second stage is indicated in Probs. 6 and 7 of Chap. 4, where they are called fickle real numbers. By using fickle real numbers, we could probably put some of the results of Chap. 8 in a form bearing a closer superficial resemblance to their classical counterparts. Thus we could hope to recover convergence in the ergodic theorem and the martingale theorem, but the limit would be a function with values in the set of fickle real numbers, and convergence would have to be suitably defined. Similarly, we might prove Theorem 1 of Chap. 8 without the extra hypothesis about the existence of the least upper bounds $\sigma(f_n)$, but the measures μ_1 and μ_2 would have fickle values. Occurrences, either explicit or implicit, of numbers belonging to stages higher than the second in the hierarchy seem to be very rare. Problems 7 and 8 of Chap. 8 suggest that a constructive version of the classical result about the decomposition of a measure into absolutely continuous and singular parts might be formulated in terms of numbers of stage three. It is not clear that further study of the number hierarchy would be worth the effort. To present the ergodic theorem, for instance, in the framework of fickle real numbers, would undoubtedly involve a loss of meaning compared to its presentation as an upcrossing inequality (Theorem 6 of Chap. 8).

REFERENCES

No attempt has been made to provide individual references for well-known classical results. General references are [1] (for Chaps. 6 and 7), [4] (for Chaps. 4, 5, and 9), [5] (for Chap. 8), [8] (for Chaps. 6 and 7), [12] (for Chaps. 10 and 11), and [15] (for Chaps. 3, 4, and 11).

Good presentations of the constructive viewpoint are hard to find. For the philosophy of intuitionism, the reader is referred to [9], [17], and Chap. 4 of [6], and to Brouwer's papers, for which a complete bibliography is given in [6]. Brouwer's theory of the continuum is summarized in the expository article [2]. A formal treatment of his theory is given in [11]. Additional points of view, with some constructive content but less closely related than intuitionism to the philosophy of this book, can be found in [13], [14], [16], and Chap. 3 of [6]. For opposed philosophies, see [7] and [10].

1. Berberian, S.: "Measure and Integration," The Macmillan Company, New York, 1965.

2. Brouwer, L. E. J.: Points and Spaces, *Can. J. Math.*, vol. 6 (1954), pp. 1–17.

3. Chacon, R. V.: Convergence of Operator Averages, pp. 89–120 of the "Proceedings of a Symposium in Ergodic Theory," Academic Press Inc., New York, 1963.

4. Dieudonne, J.: "Foundations of Modern Analysis," Academic Press Inc., New York, 1960.

5. Doob, J. L.: "Stochastic Processes," John Wiley & Sons, Inc., New York, 1953.

6. Fraenkel, A., and Y. Bar-Hillel: "Foundations of Set Theory," North Holland Publishing Company, Amsterdam, 1958.

7. Godel, K.: What Is Cantor's Continuum Problem? *Am. Math. Monthly,* vol. 54 (1947), pp. 515–525.

8. Halmos, P. R.: "Measure Theory," D. Van Nostrand Company, Inc., New York, 1950.

9. Heyting, A.: "Intuitionism. An Introduction," North Holland Publishing Company, Amsterdam, 1956.

10. Hilbert, D.: Neubegründung der Mathematik. Erste Mitteilung, *Abhandl. Math. Seminar Univ.,* vol. 1 (1922), pp. 157–177.

11. Kleene, S. C., and R. Vesley: "The Foundations of Intuitionistic Mathematics," North Holland Publishing Company, Amsterdam, 1965.

12. Loomis, L.: "Abstract Harmonic Analysis," D. Van Nostrand Company, Inc., New York, 1953.

13. Poincaré, H.: "Dernières Pensées," Paris, 1913. Translated as "Mathematics and Science: Last Essays," Dover Publications, Inc., New York, 1963.

14. Shanin, N. A.: On the Constructive Interpretation of Mathematical Judgements (Russian), *Tr. Mat. Inst. Akad. Nauk. SSSR,* vol. 52 (1958), pp. 226–311. Translated in *Am. Math. Soc. Transl.,* series 2, vol. 23 (1963).

15. Simmons, G.: "Introduction to Topology and Modern Analysis," McGraw-Hill Book Company, New York, 1963.

16. Weyl, H.: "Das Kontinuum," Leipzig, 1918. Reprinted by Chelsea Publishing Company, New York, 1932.

17. Weyl, H.: Uber die neue Grundlagenkrise der Mathematik, *Math. Z.,* vol. 10 (1921), pp. 39–79.

SYMBOLS

This is not intended to be a complete list of the symbols that occur on this book. Most of the entries represent a nonstandard notation that is used at least once in isolation from its definition.

363

INDEX

Printed in Dunstable, United Kingdom